# ADVANCED POWER RECTIFIER CONCEPTS

B. Jayant Baliga

# ADVANCED POWER RECTIFIER CONCEPTS

Springer

B. Jayant Baliga
Power Semiconductor Research Center
North Carolina State University
Raleigh, NC 27695-7924, USA
bjbaliga@unity.ncsu.edu

ISBN: 978-0-387-75588-5 e-ISBN: 978-0-387-75589-2
DOI: 10.1007/978-0-387-75589-2

Library of Congress Control Number: 2008943855



Printed on acid-free paper

springer.com

# Dedication

*The author would like to dedicate this book to his wife, Pratima, for her unwavering support throughout his career devoted to the enhancement of the performance and understanding of power semiconductor devices.*

# Preface

Today the semiconductor business exceeds \$200 billion with about 10 percent of the revenue derived from power semiconductor devices and smart power integrated circuits. Power semiconductor devices are recognized as a key component of all power electronic systems. It is estimated that at least 50 percent of the electricity used in the world is controlled by power devices. With the wide spread use of electronics in the consumer, industrial, medical, and transportation sectors, power devices have a major impact on the economy because they determine the cost and efficiency of systems. After the initial replacement of vacuum tubes by solid state devices in the 1950s, semiconductor power devices have taken a dominant role with silicon serving as the base material. These developments have been referred to as the *Second Electronic Revolution*.

Bipolar power devices, such as bipolar transistors and thyristors, were first developed in the 1950s. Due to the many advantages of semiconductor devices when compared with vacuum tubes, there was a constant demand for increasing the power ratings of these devices. Their power ratings and switching frequency increased with advancements in the understanding of the operating physics, the availability of larger diameter, high resistivity silicon wafers, and the introduction of more advanced lithography capability. During the next 20 years, the technology for the bipolar devices reached a high degree of maturity. By the 1970s, bipolar power transistors with current handling capability of hundreds of amperes and voltage blocking capability of over 500 volts became available. More remarkably, technology was developed capable of manufacturing an individual power thyristor from an entire 4-inch diameter silicon wafer with voltage ratings over 5000 volts.

In the 1970s, the power MOSFET product was first introduced by International Rectifier Corporation. Although initially hailed as a replacement for all bipolar power devices due to its high input impedance and fast switching speed, the power MOSFET has successfully cornered the market for low voltage (< 100 V) and high switching speed (> 100 kHz) applications but failed to make serious inroads in the high voltage arena. This is because the on-state resistance of power MOSFETs increases very rapidly with increase in the breakdown voltage. The resulting high conduction loss, even when using larger more expensive die, degrades the overall system efficiency.

The large on-state voltage drop for high voltage silicon power MOSFETs and the large drive current needed for silicon power bipolar transistors encouraged the development of the insulated gate bipolar transistor (IGBT)[1]. First commercialized in the early 1980s, the IGBT has become the dominant device used in all medium and high power electronic systems in the consumer, industrial, transportation, and military systems, and even found applications in the medical sector.

In conjunction with the development of improved power switches, there has been a need to improve the performance of power rectifiers. The ability to operate power systems at higher frequencies was limited by the poor switching performance of power rectifiers in the 1980s[2]. The advanced rectifier concepts discussed in this monograph evolved during this time to enable significant improvements in their switching characteristics. These advanced rectifier concepts targeted both low voltage applications where silicon unipolar devices can be utilized and high voltage applications where silicon bipolar devices are required. The advanced concepts proposed for silicon bipolar rectifiers can be effectively utilized for devices with reverse blocking voltages up to 5000 volts. For recently proposed microgrids, power rectifiers with even larger reverse blocking voltages of up to 15-20 kV are needed for the development of high frequency solid-state-transformers. This application can be served with the advanced concepts described in this book with silicon carbide as the base semiconductor material.

Due to these developments, it is anticipated that there will be an increasing need for technologists trained in the discipline of designing and manufacturing power semiconductor devices. This monograph complements my recently published textbook which dealt with only the basic power rectifier structures due to space limitations[3]. For the convenience of readers, some portions of the chapters on 'Schottky Rectifiers' and 'P-i-N Rectifiers' from the textbook have been reproduced in this monograph. As in the case of the textbook, analytical expressions that describe the behavior of the advanced power rectifier concepts have been rigorously derived using the fundamental semiconductor Poisson's, continuity, and conduction equations in this monograph. The electrical characteristics of all the power rectifiers discussed in this book can be computed using these analytical solutions as shown by typical examples provided in each section. In order to corroborate the validity of these analytical formulations, I have included the results of two-dimensional numerical simulations in each section of the book. The simulation results are also used to further elucidate the physics and point out two-dimensional effects whenever relevant. Due to increasing interest in the utilization of wide band-gap semiconductors for power devices, the book includes the analysis of silicon carbide structures.

In the first chapter, a broad introduction to potential applications for power devices is provided. The electrical characteristics for ideal power rectifiers are then defined and compared with those for typical devices. The second chapter provides a detailed analysis of the Schottky rectifier structure which is borrowed from the textbook. On-state current flow via thermionic emission is described followed by the impact of image force barrier lowering on the reverse leakage current. These phenomena influence the selection of the barrier height to optimize the power losses as described in the chapter. The influence of the tunneling current component is also included in this chapter due to its importance for silicon carbide Schottky rectifiers.

The subsequent chapters are devoted to various advanced power rectifier structures. The unipolar device structures are first covered in the chapters on the 'Junction Barrier controlled Schottky (JBS) Rectifier', the 'Trench Schottky Barrier controlled Schottky (TSBS) Rectifier', and the 'Trench MOS Barrier controlled Schottky (TMBS) Rectifier'. The JBS rectifier concept is attractive for reducing the reverse leakage current in Schottky power rectifiers while retaining a low on-state voltage drop. This concept is also suitable for integration of the Schottky rectifier with the power MOSFET structure[4]. The TSBS concept is particularly suitable for reducing the leakage current in silicon carbide Schottky rectifiers. The TMBS concept provides yet another alternative to reducing the leakage current in silicon Schottky rectifiers. This concept is not suitable for application to silicon carbide structures due to the high electric field generated in the oxide.

The above concepts are applicable to the development of unipolar devices. When the reverse blocking voltage becomes large (more than 200 volts for silicon devices and 5000 volts for silicon carbide devices), it is advantageous to utilize bipolar current flow in power rectifiers to reduce the on-state voltage drop. Chapter 6, which is based upon portions borrowed from the textbook, describes the physics of operation of high voltage P-i-N rectifiers. The theory for both low-level and high-level injection conditions during on-state current flow is described here. The impact of this on the reverse recovery phenomenon during turn-off is then analyzed. The influence of end region recombination is included in the analysis.

For the development of high voltage silicon power rectifiers with reduced reverse recovery charge, the concept of merging the P-i-N and Schottky diodes was proposed in the 1980s[5]. Although first met with skepticism as having the worst attributes of both the P-i-N and Schottky rectifiers, the concept has now been embraced by the semiconductor industry as having the best characteristics of both devices with products available in the marketplace for motor control applications. Chapter 7 provides a detailed analysis of the MPS concept with analytical formulations developed for the on-state carrier distribution, the on-state voltage drop, and the reverse recovery characteristics. In this monograph, it is also demonstrated that the MPS concept can be extended to silicon carbide power rectifiers with judicious choice of the Schottky contact width and barrier height.

An alternate approach to improving the reverse recovery characteristics in power rectifiers is by utilizing the SSD structure[6]. In this concept, the highly doped

$P^+$ region of the P-i-N rectifier structure is confined to a portion of the cell structure with a shallower lightly doped P- region used for the rest of the anode region. The low injection efficiency of the P- region suppresses the injection of minority carriers (holes) resulting in a reduced hole concentration near the anode. When the doping concentration in the P- region is made small, the characteristics of the SSD structure approach those of the MPS rectifier structure. As the doping concentration of this region is increased, the characteristics of the SSD structure approach those of the P-i-N rectifier structure[7]. The operation of silicon and silicon carbide rectifiers with the SSD structure is described in Chapter 8.

Several other power rectifier structures have also been proposed in the literature. It has been demonstrated that the stored charge in the P-i-N rectifier structure can be reduced by decreasing the doping concentration in the $P^+$ anode region[8]. However, this produces a large increase in the on-state voltage drop especially at surge current levels. A power rectifier called SPEED has been proposed that is similar to the SSD structure with a deep lightly doped P region[9]. The reverse recovery characteristic of this structure is not as good as that for the SSD and MPS rectifier structures. For the reasons cited in this paragraph, these alternate power rectifier structures have not been included in this monograph.

I am hopeful that this monograph will be useful for researchers in academia and to product designers in the industry. It can also be used for the teaching of courses on solid state devices as a supplement to my textbook[3].

Prof. B. Jayant Baliga
December 2008

**References**

[1] B.J. Baliga, "How the Super-Transistor Works", Scientific American Magazine, Special Issue on 'The Solid-State-Century', pp. 34-41, January 22, 1988.

[2] B.J. Baliga, "Power Semiconductor Devices for Variable-Frequency Drives", Proceedings of the IEEE, Vol. 82, pp. 1112-1122, 1994.

[3] B.J. Baliga, "Fundamentals of Power Semiconductor Devices", Springer Scientific, New York, 2008.

[4] B.J. Baliga and D.A. Girdhar, "Paradigm Shift in Planar Power MOSFET Technology", Power Electronics Technology Magazine, pp. 24-32, November 2003.

[5] B.J. Baliga, "Analysis of a High Voltage Merged P-i-N/Schottky (MPS) Rectifier", IEEE Electron Device Letters, Vol. EDL-8, pp. 407-409, 1987.

[6] Y. Shimizu, et al, "High-Speed Low-Loss P-N Diode having a Channel Structure", IEEE Transactions on Electron Devices, Vol. ED-31, pp. 1314-1319, 1984.

[7] M. Mehrotra and B.J. Baliga, "Comparison of High Voltage Power Rectifier Structures", IEEE International Symposium on Power Semiconductor Devices and ICs, Abstract 7.11, pp. 199-204, 1993.

[8] M. Naito, H. Matsuzaki, and T. Ogawa, "High Current Characteristics of Asymmetrical p-i-n Diodes having Low Forward Voltage Drops", IEEE Transactions on Electron Devices, Vol. ED-23, pp. 945-949, 1976.

[9] H. Sclangenotto, et al, "Improved Recovery of Fast Power Diodes with Self-Adjusting P-emitter Efficiency", IEEE Electron Device Letters, Vol. 10, pp. 322-324, 1989.

# Contents

# Chapter 1

# Introduction

Solid state devices began to displace vacuum tubes in the 1950s for various power control applications. Power devices are required for applications that operate over a broad spectrum of power levels and frequencies. In Fig. 1.1, the applications for power devices are shown as a function of operating frequency. High power systems, such as HVDC power distribution and locomotive drives, requiring the control of megawatts of power operate at relatively low frequencies. As the operating frequency increases, the power ratings decrease for the devices with typical microwave devices handling about 100 watts. Today, all of these applications are served by silicon devices.

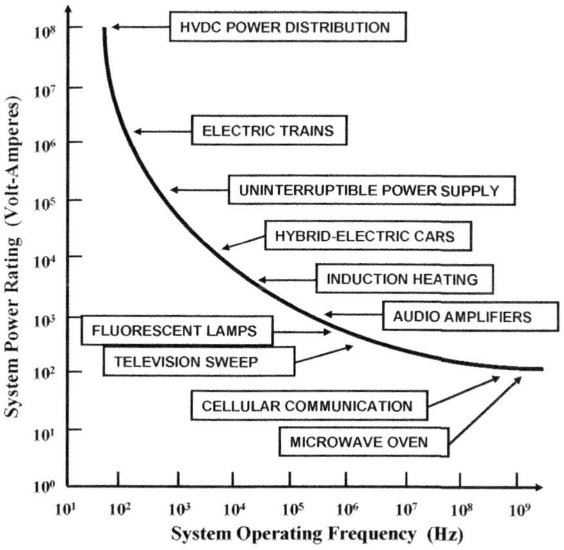


Fig. 1.1 Applications for Power Devices.

B.J. Baliga, *Advanced Power Rectifier Concepts*, DOI: 10.1007/978-0-387-75589-2_1,

When the operating voltage of the power circuit is relatively small (< 100 volts), silicon unipolar devices offer the best performance[1]. Bipolar silicon devices have better characteristics for applications where the circuit operating voltage is relatively large (> 100 volts). The development of power devices from wide band gap semiconductors, such as silicon carbide, allows extending the operating voltage of unipolar structures to at least 5000 volts[2]. In modern power circuits, it is common-place to utilize power transistors as switches to regulate the power flow to the load while using power rectifiers to control the direction of current flow. With the advent of high performance power switches (power MOSFETs and IGBTs), the performance of power rectifiers often limits the operation of power circuits. This book focuses on advanced concepts for power rectifiers that allow reducing power losses in circuits enabling an increase in system efficiency.

## 1.1 Ideal Power Switching Waveforms

An ideal power device must be capable of controlling the flow of power to loads with zero power dissipation. The loads encountered in systems may be inductive in nature (such as motors and solenoids), resistive in nature (such as heaters and lamp filaments), or capacitive in nature (such as transducers and LCD displays). Most often, the power delivered to a load is controlled by turning-on a power switch on a periodic basis to generate pulses of current that can be regulated by a control circuit. The ideal waveforms for the power delivered through a power switch are shown in Fig. 1.2. During each switching cycle, the switch remains on for a time

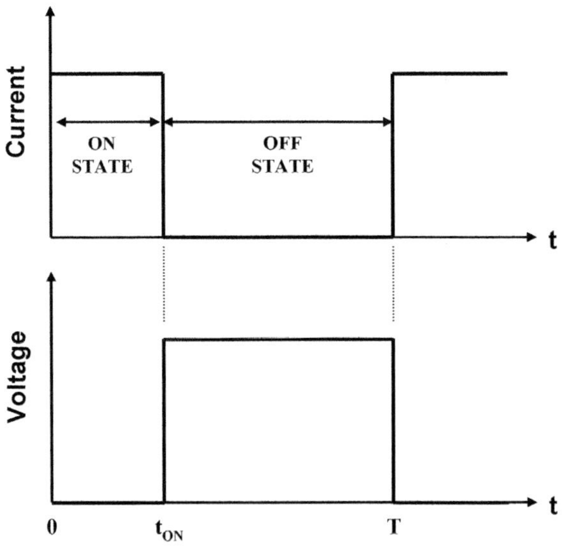


Fig. 1.2 Ideal Switching Waveforms for Power Delivery.

upto $t_{ON}$ and maintains an off-state for the remainder of the period T. This produces pulses of current that flow through the circuit including the power rectifiers. For an ideal power rectifier, the voltage drop during the on-state is zero resulting in no power dissipation. Similarly, during the off-state, the (leakage) current in the ideal power rectifier is zero resulting in no power dissipation. In addition, it is assumed that the ideal power rectifier makes the transition between the on-state and off-state instantaneously resulting in no power loss as well.

Typical power rectifiers exhibit a finite voltage drop in the on-state and leakage current flow in the off-state. In addition, bipolar power rectifiers exhibit a large reverse recovery current when switching from the on-state to the off-state. This produces considerable power dissipation not only in the rectifier but also the power transistor used to control the current flow. After the invention and development of the insulated gate bipolar transistor in the 1980s[3], it became apparent that the performance of power rectifiers is limiting the performance of high voltage power circuits for motor control[4]. This led to the innovations in power rectifiers that are discussed in this book.

## 1.2 Ideal and Typical Power Rectifier Characteristics

Silicon power rectifiers have served the industry for well over five decades but cannot be considered to have ideal device characteristics. An ideal power rectifier should exhibit the current-voltage (*i-v*) characteristics shown in Fig. 1.3. In the forward conduction mode, the first quadrant of operation the figure, it should be able to carry any amount of current with zero on-state voltage drop. In the reverse blocking mode, the third quadrant of operation in the figure, it should be able to hold off any value of voltage with zero leakage current. Further, the ideal rectifier should be able to switch between the on-state and the off-state with zero switching time.

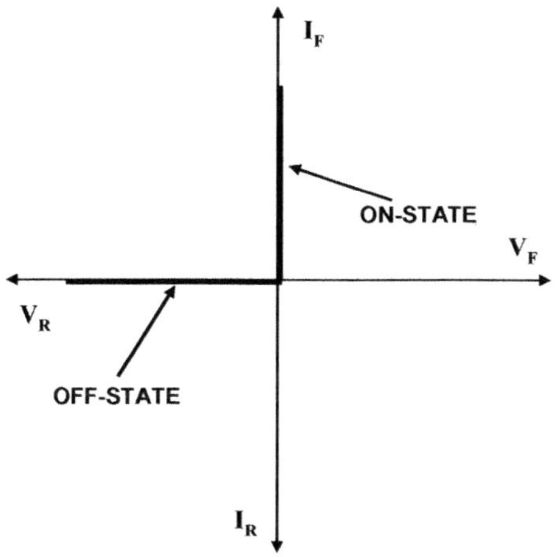


Fig. 1.3 Characteristics of an Ideal Power Rectifier.

Actual power rectifiers exhibit the *i-v* characteristics illustrated in Fig. 1.4. They have a finite voltage drop ($V_{ON}$) when carrying current on the on-state leading to 'conduction' power loss. They also have a finite leakage current ($I_{OFF}$) when blocking voltage in the off-state creating off-state power loss. In addition, bipolar power rectifiers exhibit a large reverse current flow for a short duration to remove the stored charge created within the structure by the on-state current flow. The doping concentration and thickness of the drift region of the device must be carefully chosen with a design target for the breakdown voltage (BV). Moreover, the power dissipation in power rectifiers increases when their voltage rating is increased due to an increase in the on-state voltage drop.

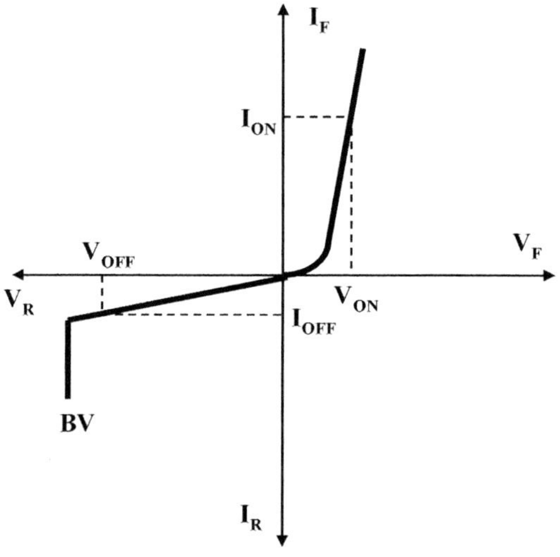


Fig. 1.4 Characteristics of a Typical Power Rectifier.

## 1.3 Unipolar Power Rectifiers

Bipolar power devices operate with the injection of minority carriers during on-state current flow. These carriers must be removed when the switching the device from the on-state to the off-state introducing significant power losses that degrade the power management efficiency. It is therefore preferable to utilize unipolar current conduction in a power device. The commonly used unipolar power diode structure is the Schottky rectifier that contains a metal-semiconductor barrier to produce current rectification. The power Schottky rectifier structure also contains a drift region, as show in Fig. 1.5, which is designed to support the reverse blocking voltage. The resistance of the drift region increases rapidly with increasing blocking voltage capability as discussed later in this chapter. Silicon Schottky rectifiers are commercially available with blocking voltages of up to 100 volts. Beyond this value, the on-state voltage drop of silicon Schottky rectifiers becomes too large for practical applications. Silicon P-i-N rectifiers are favored for designs with larger breakdown voltages due to their low on-state voltage drop despite slower switching properties.

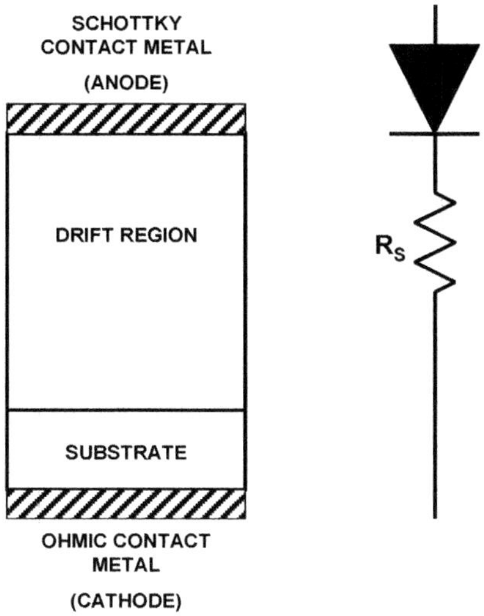


Fig. 1.5 The Power Schottky Rectifier Structure and its Equivalent Circuit.

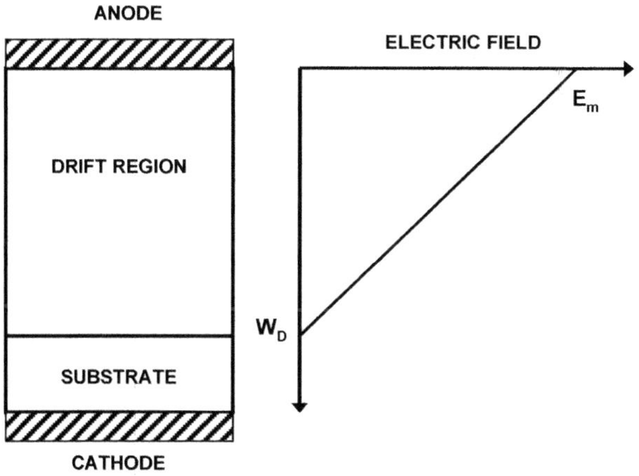


Fig. 1.6 The Ideal Drift Region and its Electric Field Distribution.

The properties (doping concentration and thickness) of the *ideal drift region* can be analyzed by assuming an abrupt junction profile with high doping concentration on one side and a low uniform doping concentration on the other side, while neglecting any junction curvature effects by assuming a parallel-plane configuration. The resistance of the ideal drift region can then be related to the basic properties of the semiconductor material[5]. The solution of Poisson's equation leads to a triangular electric field distribution, as shown in Fig. 1.6, within a uniformly doped drift region with the slope of the field profile being determined by the doping concentration. The maximum voltage that can be supported by the drift region is determined by the maximum electric field ($E_m$) reaching the critical electric field ($E_c$) for breakdown for the semiconductor material. The critical electric field for breakdown and the doping concentration then determine the maximum depletion width ($W_D$).

The specific resistance (resistance per unit area) of the ideal drift region is given by:

$$R_{on.sp} = \left( \frac{W_D}{q\mu_n N_D} \right) \quad \textbf{[1.1]}$$

Since this resistance was initially considered to be the lowest value achievable with silicon devices, it has historically been referred to as the *ideal specific on-resistance of the drift region*. The depletion width under breakdown conditions is given by:

$$W_D = \frac{2BV}{E_C} \quad \textbf{[1.2]}$$

where BV is the desired breakdown voltage. The doping concentration in the drift region required to obtain this breakdown voltage is given by:

$$N_D = \frac{\varepsilon_S E_C^2}{2qBV} \quad \textbf{[1.3]}$$

Combining these relationships, the specific resistance of the ideal drift region is obtained:

$$R_{on-ideal} = \frac{4BV^2}{\varepsilon_S \mu_n E_C^3} \quad \textbf{[1.4]}$$

The denominator of this equation ($\varepsilon_S.\mu_n.E_C^3$) is commonly referred to as *Baliga's Figure of Merit for Power Devices*. It is an indicator of the impact of the semiconductor material properties on the resistance of the drift region. The dependence of the drift region resistance on the mobility (assumed to be for electrons here because in general they have higher mobility values than for holes) of the carriers favors semiconductors such as Gallium Arsenide. However, the much stronger (cubic) dependence of the on-resistance on the critical electric field for breakdown favors wide band gap semiconductors such as silicon carbide[2].

One of the fundamental problems encountered with silicon Schottky rectifiers is the large reverse leakage current flow when the device is designed with small on-state voltage drop. This is associated with the Schottky barrier lowering and pre-breakdown avalanche multiplication phenomena which produce an increase in the leakage current by on order of magnitude when the reverse voltage increases from zero to typical operating levels. This increase in the leakage current can be suppressed by shielding the Schottky contact from the high electric field generated within the semiconductor.

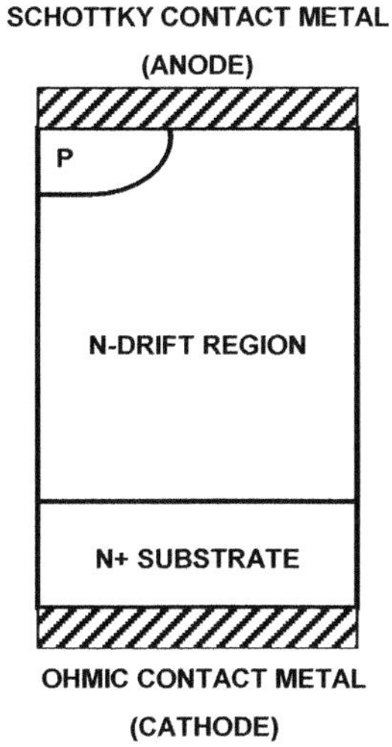


Fig. 1.7 The Junction Barrier controlled Schottky (JBS) Rectifier Structure.

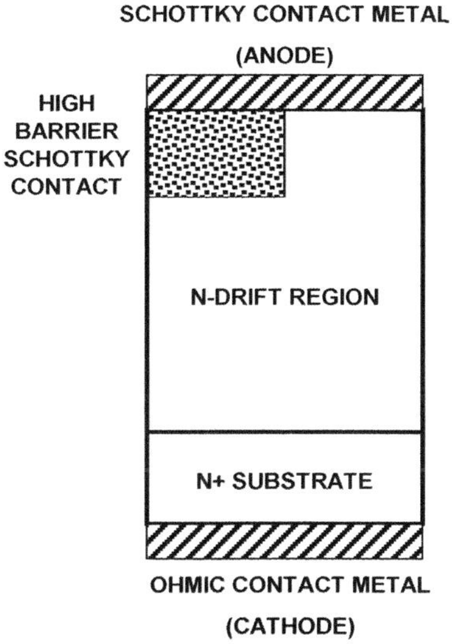


Fig. 1.8 The Trench Schottky Barrier controlled Schottky (TSBS) Rectifier Structure.

One approach to reducing the leakage current in power Schottky rectifiers is by incorporation of a P-N junction as illustrated in Fig. 1.7. This concept, called the Junction Barrier controlled Schottky (JBS) rectifier[6,7], has been effectively utilized for improving the performance of low voltage silicon devices and high voltage silicon carbide devices. In the case of the silicon carbide structures, the shielding of the Schottky contact reduces the leakage current by many orders of magnitude by mitigating the tunneling phenomenon at the Schottky contact[2].

An alternate method for shielding the anode Schottky contact from the high electric fields in the semiconductor is by using a second Schottky contact with larger barrier height. It is preferable to locate the Schottky contact with the larger barrier height within a trench, as shown in Fig. 1.8, to enhance the shielding effect. This structure is consequently called the Trench Schottky Barrier controlled Schottky (TSBS) rectifier. Although first proposed[8] for the improvement of silicon devices, this idea has found greater interest for silicon carbide structures due to the larger barrier heights available in wide band gap semiconductors.

In principle, it is also possible to shield the anode Schottky contact by incorporating an MOS region as illustrated in Fig. 1.9. It is preferable to form the MOS structure within a trench to enhance the shielding phenomenon[9]. This approach is feasible for silicon devices with a mature MOS technology. In the case of silicon carbide, this approach is not advisable because very high electric field is generated in the oxide that can lead to its destructive failure. In the case of silicon devices, this idea has been combined with the charge-coupling phenomenon to allow a significant reduction in the resistance of the drift region. Such structures will be discussed in a subsequent monograph on the topic of charge coupled power devices.

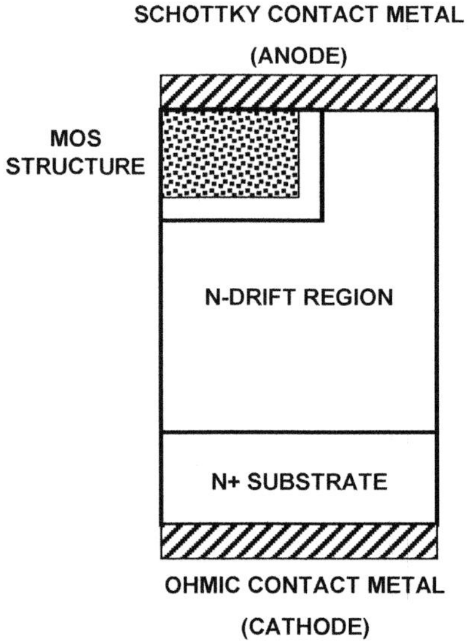


Fig. 1.9 The MOS Barrier controlled Schottky (MBS) Rectifier Structure.

## 1.4 Bipolar Power Rectifiers

The resistance of the drift region becomes very large for silicon unipolar Schottky rectifiers when the blocking voltage is increased beyond 100 volts. In contrast, very high voltage bipolar devices are feasible in silicon because of conductivity modulation of the drift region by the carriers injected during on-state current flow. The P-i-N rectifier structure that has been widely utilized by the industry for the last 50 decades is shown in Fig. 1.10. In the on-state, the P-N junction becomes forward biased allowing the injection of minority carriers whose concentration far exceeds the doping concentration of the drift region. An equal concentration of majority carriers is also prevalent in the drift region under these conditions to satisfy charge neutrality. Typical silicon P-i-N rectifiers have on-state voltage drops in the range of 1 to 2 volts even when designed to support large voltage in the reverse blocking mode. Commercial devices with blocking voltage capability of up to 10 kV have been developed.

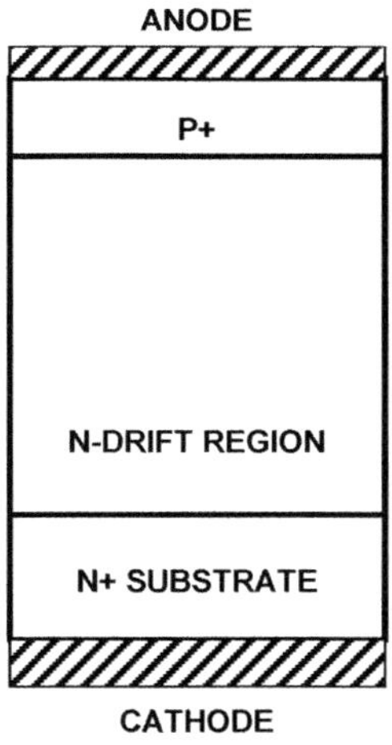


Fig. 1.10 The P-i-N Rectifier Structure.

The main drawback of the silicon P-i-N rectifier structure is its slow switching speed with a large reverse recovery current. This transient produces large power dissipation in the rectifier and the power switch controlling the transient. One approach to reducing the reverse recovery current is by reducing the lifetime using deep level impurities[1]. Another approach is to combine a Schottky contact and the P-N junction as illustrated in Fig. 1.11. Although this structure is identical in appearance to the JBS rectifier structure, it operates in the bipolar mode in the case of high voltage devices. This structure should not be considered as a simple parallel combination of a Schottky rectifier and a P-i-N rectifier because this would result in the worst performance attributed of both structures. In contrast, by merging the physics of current flow in the Schottky rectifier and the P-i-N rectifier, it is possible to realize the best attributes of both structures[10].

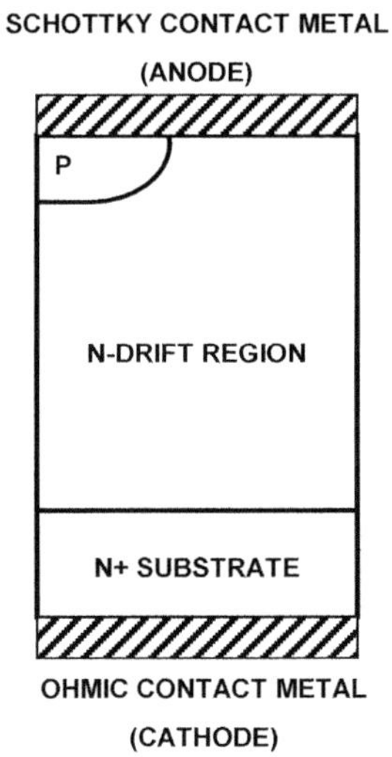


Fig. 1.11 The Merged P-i-N Schottky (MPS) Rectifier Structure.

## 1.5 Typical Power Rectifier Applications

Power rectifiers are used to control the direction of current flow in circuits. Most often, they are used in conjunction with power transistors, such as MOSFETs and IGBTs. Two typical examples for the application of high performance power rectifiers are provided in this section to emphasize the characteristics of importance from an application standpoint. The first example is in the 'Buck' converter used to reduce the voltage from one DC level to a smaller value. This circuit is popular for distributing power within a computer. The second example is in variable speed motor drives. This application is popular for operating induction motors with variable loads leading to large gains in efficiency.

### 1.5.1 DC-DC Buck Converter

The buck converter can be regarded as a DC-to-DC transformer because of its ability to reduce a DC input voltage to a smaller DC output voltage. As mentioned above, one popular application of the Buck-converter is to provide power to various loads inside a computer from the back-plane DC power supply. The back-plane power supply has a typical voltage range of 17-20 volts depending on the type of computer. Loads, such as disk drives require a DC voltage of 5-12 volts. In contrast, integrated circuits in the computer, such as the microprocessor and graphics chips, require a lower DC voltage in the range of 1-2 volts.

The commonly used Buck-converter circuit used for the DC-to-DC voltage conversion in computers is shown in Fig. 1.12. Due to the relatively low operating voltage in this circuit, a power MOSFET is typically used as the switch. When the transistor is turned on by the control circuit, current flows from the DC input source through the inductor to the load connected at the output terminals. When the transistor is switched off by the control circuit, the load current circulates through the rectifier and the inductor. The regulation of the DC output voltage can be achieved by adjusting the on-time of the transistor[11]. The switching waveforms for the transistor are similar to those shown in Fig. 1.2. The waveforms for the rectifier are a complement of those for the transistor.

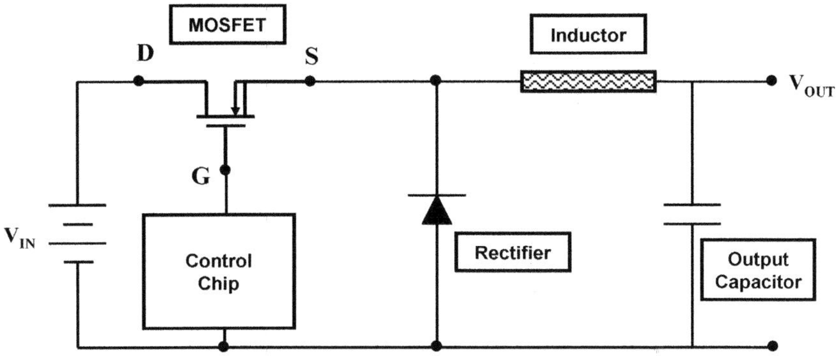


Fig. 1.12 The Buck DC-DC Converter Circuit.

There is a trend towards increasing the operating frequency of the DC-DC converter in an effort to reduce the size of the inductor. The power MOSFETs used as switches have inherent high frequency operating capability because of unipolar current flow. It is advantageous to also use unipolar current flow in the power rectifier to enable operation at high frequencies with low switching losses. It is also desirable to reduce the on-state voltage drop for the rectifier to reduce conduction losses when current is flowing through this path. These features are inherent in the silicon power Schottky rectifier structure. However, care must be taken to retain a sufficiently low leakage current at the maximum operating temperature to prevent thermal runaway.

### 1.5.2 Variable-Frequency Motor Drive

A significant increase in the efficiency for running motors can be achieved by used variable-frequency motor drives in place of constant speed drives with dampers to regulate the output. The most commonly used topology converts the constant frequency input AC power to a DC bus voltage and then use an inverter stage to produce the variable frequency output power[12]. The circuit diagram for a three-phase motor drive system is shown in Fig. 1.13. Six IGBTs are used with six fly-back rectifiers in the inverter stage to deliver the variable frequency power to the motor windings. A pulse-width-modulation (PWM) scheme is used to generate the variable frequency AC voltage waveform that is fed to the motor windings[13].

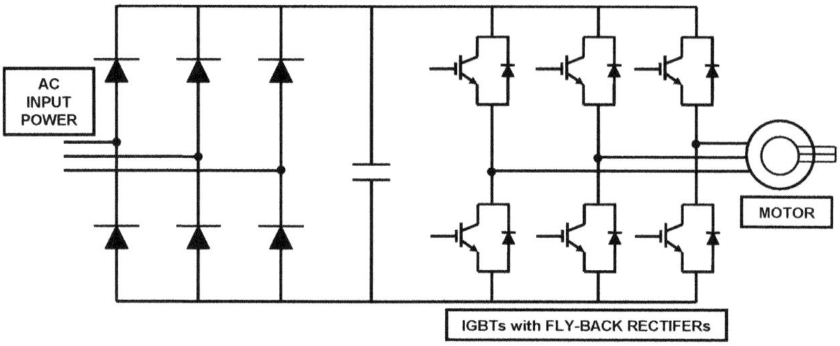


Fig. 1.13 Variable Frequency Motor Drive Circuit.

During each cycle of the PWM period, the current in the motor winding can be considered to remain approximately constant. This allows linearlization of the waveforms for the current and voltage experienced by the IGBTs and the rectifiers. Typical waveforms for the transistor and the fly-back diode are illustrated in Fig. 1.14. The large reverse recovery current typically observed in silicon P-i-N rectifiers during the time interval from $t_1$ to $t_3$ produces high power dissipation not only in the diodes but also in the transistors[1,4]. This power loss can be eliminated by replacing the silicon P-i-N rectifiers with silicon carbide Schottky rectifiers.

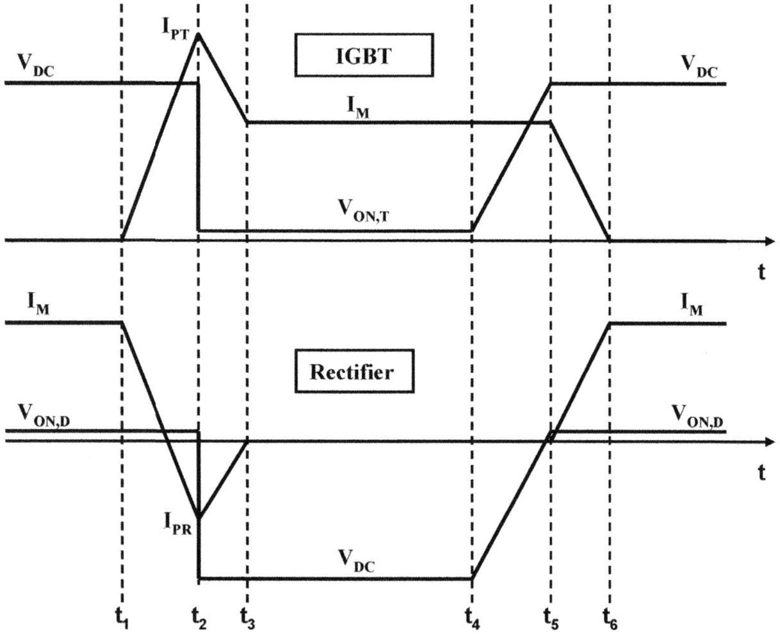


Fig. 1.14 Linearized Waveforms for the PWM Motor Drive Circuit.

## 1.6 Summary

The desired characteristics for power semiconductor rectifiers have been reviewed in this chapter. The characteristics of typical devices have been compared with those for the ideal case. Two high volume applications for the rectifiers have been briefly described. Various unipolar and bipolar power device structures that are suitable for these applications have been introduced here. These structures are discussed in detail in subsequent chapters of this book.

## References

[1] B.J. Baliga, "Fundamentals of Power Semiconductor Devices", Springer Science, New York, 2008.
[2] B.J. Baliga, "Silicon Carbide Power Devices", World Scientific Press, 2006.
[3] B.J. Baliga, "Evolution of MOS-Bipolar Power Semiconductor Technology", Proceedings of the IEEE, pp. 409-418, 1988.
[4] B.J. Baliga, "Power Semiconductor Devices for Variable-Frequency Drives", Proceedings of the IEEE, pp. 1112-1122, 1994.
[5] B.J. Baliga, "Semiconductors for High Voltage Vertical Channel Field Effect Transistors", J. Applied Physics, Vol. 53, pp. 1759-1764, 1982.
[6] B.J. Baliga, "Pinch Rectifier", U.S. Patent 4,641,174, Issued February 3, 1987.
[7] B.J. Baliga, "The Pinch Rectifier: A Low Forward Drop, High Speed Power Diode", IEEE Electron Device Letters, Vol. EDL-5, pp. 194-196, 1984.
[8] L. Tu and B.J. Baliga, "Schottky Barrier Rectifier including Schottky Barrier Regions of differing Barrier Heights", U.S. Patent 5,262,668, November 6, 1993.
[9] B.J. Baliga, "New Concepts in Power Rectifiers", pp. 471-481 in 'Physics of Semiconductor Devices', World Scientific Press, 1985.
[10] B.J. Baliga, "Analysis of a High-Voltage Merged P-i-N/Schottky (MPS) Rectifier", IEEE Electron Device Letters, Vol. EDL-8, pp. 407-409, 1987.
[11] N. Mohan, T.M. Undeland, and W.P. Robbins, "Power Electronics", pp. 164-172, John Wiley and Sons, New York, 1995.
[12] J.D. van Wyk, "Power Electronic Converters for Drives", pp. 81-137, in 'Power Electronics and Variable Frequency Drives', Edited by B.K. Bose, IEEE Press, 1997.
[13] J. Holtz, "Pulse Width Modulation for Electronic Power Conversion", pp. 138-208, in 'Power Electronics and Variable Frequency Drives', Edited by B.K. Bose, IEEE Press, 1997.

# Chapter 2

# Schottky Rectifiers

A Schottky rectifier is formed by making an electrically non-linear contact between a metal and the semiconductor drift region. The Schottky rectifier is an attractive unipolar device for power electronics applications due to its relatively low on-state voltage drop and its fast switching behavior. It has been widely used in power supply circuits with low operating voltages due to the availability of excellent devices based upon silicon technology. In the case of silicon, the maximum breakdown voltage of Schottky rectifiers has been limited by the increase in the resistance of the drift region. Commercially available devices are generally rated at breakdown voltages of less than 100 volts.

Many applications described in chapter 1 require fast switching rectifiers with low on-state voltage drop that can also support over 500 volts. The much lower resistance of the drift region for silicon carbide enables development of such Schottky rectifiers with very high breakdown voltages[1]. These devices not only offer fast switching speed but also eliminate the large reverse recovery current observed in high voltage silicon P-i-N rectifiers. This reduces switching losses not only in the rectifier but also in the IGBTs used within the power circuits[2].

In this chapter, the basic structure of the power Schottky rectifier is first introduced to define its constituent elements. The current transport mechanisms that are pertinent to power devices are elucidated for both the forward and reverse mode of operation. In the first quadrant of operation, the thermionic emission process is dominant for power Schottky rectifiers. In the third quadrant of operation, the influence of Schottky barrier lowering has a strong impact on the leakage current for silicon devices. In the case of silicon carbide devices, the influence of tunneling current must also be taken into account when performing the analysis of the reverse leakage current.

B.J. Baliga, *Advanced Power Rectifier Concepts*, DOI: 10.1007/978-0-387-75589-2_2,

The information in this chapter is intended to provide the context and perspective for the discussion of advanced device concepts in subsequent chapters. A more detailed discussion of the fundamental concepts relevant to power Schottky rectifiers is provided in a recently published textbook[3]. This includes optimization of the Schottky barrier height depending upon the duty cycle and the maximum junction temperature.

## 2.1 Power Schottky Rectifier Structure

The basic one-dimensional structure of the metal-semiconductor or Schottky rectifier structure is shown in Fig. 2.1 together with electric field profile under reverse bias operation. The applied voltage is supported by the drift region with a triangular electric field distribution if the drift region doping is uniform. The maximum electric field occurs at the metal contact. The device undergoes breakdown when this field becomes equal to the critical electric field for the semiconductor.

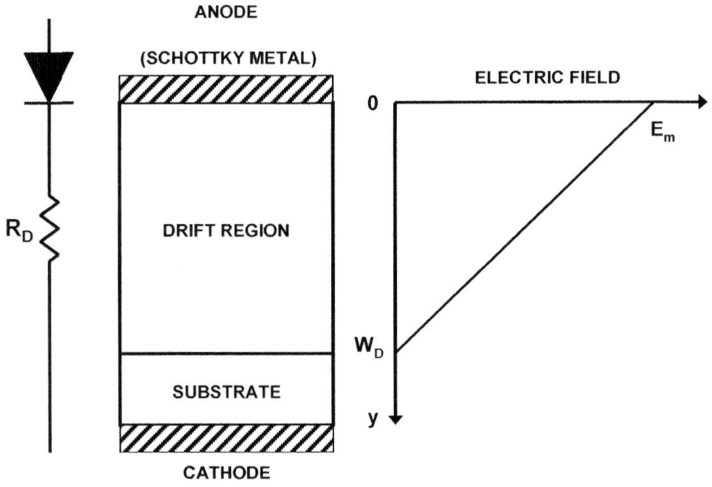


Fig. 2.1 Electric Field Distribution in a Schottky Rectifier.

When a negative bias is applied to the cathode, current flow occurs in the Schottky rectifier by the transport of electrons over the metal-semiconductor contact and through the drift region as well as the substrate. The on-state voltage drop is determined by the voltage drop across the metal-semiconductor interface and the ohmic voltage drop in the resistance of the drift region, the substrate and its ohmic contact.

At typical on-state operating current density levels, the current transport is dominated by majority carriers. Consequently, there is insignificant minority carrier stored charge within the drift region in the power Schottky rectifier. This enables switching the Schottky rectifier from the on-state to the reverse blocking off-state in a rapid manner by establishing a depletion region within the drift region. The fast switching capability of the Schottky rectifier enables operation at high frequencies with low power losses making this device popular for high frequency switch mode power supply applications. With the advent of commercially available high voltage Schottky rectifiers based upon silicon carbide, they are expected to be utilized in motor control applications as well.

A useful relationship for obtaining the Schottky barrier height is:

$$\Phi_{BN} = \Phi_M - \chi_S \quad [2.1]$$

where $\phi_M$ is the metal work function and $\chi_S$ is the electron affinity of the semiconductor. The potential difference between the Fermi level in the semiconductor ($E_{FS}$) and the Fermi level in the metal ($E_{FM}$) is called the *contact potential* ($V_C$) which is given by:

$$qV_C = (E_{FS} - E_{FM}) = \Phi_M - \Phi_S = \Phi_M - (\chi_S + E_C - E_{FS}) \quad [2.2]$$

where $\phi_S$ is the semiconductor work function and $E_C$ is the conduction band edge. The built-in potential ($V_{bi}$) at the Schottky contact (equal to the contact potential) creates a depletion region within the semiconductor at zero bias given by:

$$W_0 = \sqrt{\frac{2\varepsilon_S V_{bi}}{qN_D}} \quad [2.3]$$

## 2.2 Forward Conduction

Current flow across the metal-semiconductor junction can be produced by the application of a negative bias to the N-type semiconductor region. Current flow across the interface then occurs mainly due to majority carriers – electrons for the case of an N-type semiconductor. The current flow via the thermionic emission process is the dominant current transport mechanism in silicon and silicon carbide Schottky power rectifiers. In the case of high mobility semiconductors, such as silicon, gallium arsenide and silicon carbide, the thermionic emission theory can be used to describe the current flow across the Schottky barrier interface[4]:

$$J = AT^2 e^{-(q\Phi_{BN}/kT)}\left[e^{(qV/kT)} - 1\right] \quad [2.4]$$

where A is the effective Richardson's constant, T is the absolute temperature, k is Boltzmann's constant, and V is the applied bias. An effective Richardson's constant of 110, 140, and 146 $A/cm^2$-$^oK^2$ can be used for n-type silicon[6], gallium arsenide[6], and 4H silicon carbide[3], respectively. This expression, based upon the superimposition of the current flux from the metal and the semiconductor[5] which balance out at zero bias, holds true for both positive and negative voltages applied to the metal contact.

When a forward bias is applied (positive values for V in Eq. [2.4]), the first term in the square brackets of the equation becomes dominant allowing calculation of the forward current density:

$$J_F = AT^2 e^{-(q\Phi_{BN}/kT)} e^{(qV_{FS}/kT)} \qquad [2.5]$$

where $V_{FS}$ is the forward voltage drop across the Schottky contact. In the case of power Schottky rectifiers, a thick lightly doped drift region must be placed below the Schottky contact as illustrated in Fig. 2.1 to allow supporting the reverse blocking voltage. A resistive voltage drop ($V_R$) occurs across this drift region which increases the on-state voltage drop of the power Schottky rectifier beyond $V_{FS}$. In case of current transport by the thermionic emission process, there is no modulation of the resistance of the drift region because minority carrier injection is neglected. Due to the small thickness (typically less than 50 microns) of the drift region for power Schottky diodes, it is grown on top of a heavily doped $N^+$ substrate as a handle during processing and packaging of the devices. The resistance contributed by the substrate ($R_{SUB}$) must be included in the analysis because it can be comparable to that of the drift region especially for silicon carbide devices. In addition, the resistance of the ohmic contact ($R_{CONT}$) to the cathode may make a substantial contribution to the on-state voltage drop.

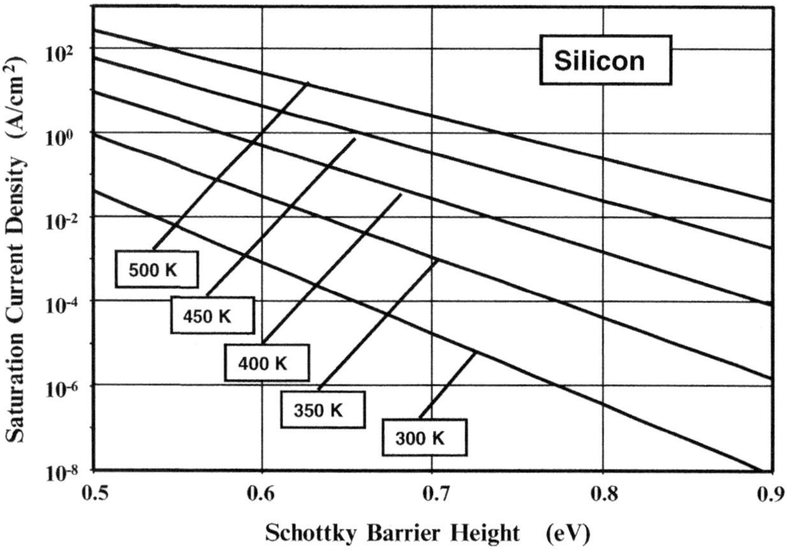


Fig. 2.2 Saturation Current Density for Silicon Schottky Barrier Rectifiers.

The on-state voltage drop ($V_F$) for the power Schottky rectifier, after including the resistive voltage drop, is given by:

$$V_F = V_{FS} + V_R = \frac{kT}{q} \ln\left(\frac{J_F}{J_S}\right) + R_{S,SP} J_F \qquad [2.6]$$

where $J_F$ is the forward (on-state) current density, $J_S$ is the saturation current density, and $R_{S,SP}$ is the total series specific resistance. In this expression, the saturation current is given by:

$$J_S = AT^2 e^{-(q\Phi_{BN}/kT)} \qquad [2.7]$$

and the total series specific resistance is given by:

$$R_{S,SP} = R_{D,SP} + R_{SUB} + R_{CONT} \qquad [2.8]$$

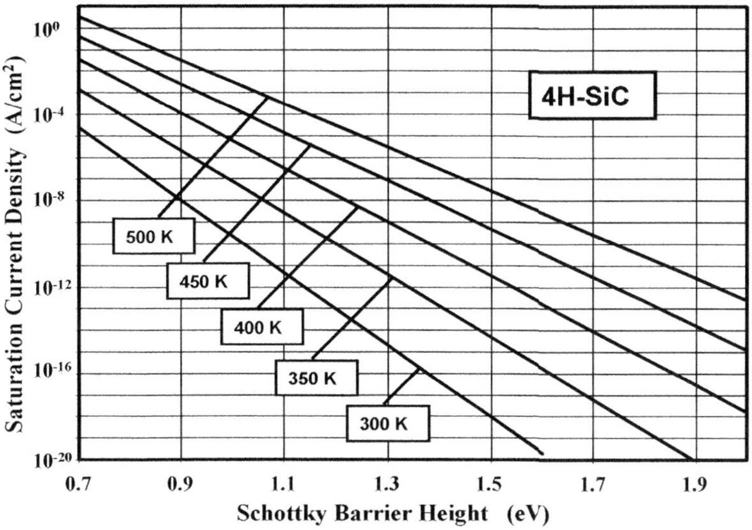


Fig. 2.3 Saturation Current Density for 4H-SiC Schottky Barrier Rectifiers.

The saturation current is a strong function of the Schottky barrier height and the temperature as shown in Fig. 2.2 for silicon devices. The barrier heights chosen for this plot are in the range for typical metal contacts with silicon. The saturation current density increases with increasing temperature and reduction of the barrier height. This has an influence not only on the on-state voltage drop but an even greater impact on the reverse leakage current as discussed in the next section. A corresponding plot is provided in Fig. 2.3 for 4H-SiC devices. A larger range of Schottky barrier heights has been selected for this graph because this is typical for the wide band gap semiconductor.

As discussed in chapter 1, the specific on-resistance of the drift region is given by:

$$R_{on-ideal} = \frac{4BV^2}{\varepsilon_S \mu_n E_C^3} \qquad [2.9]$$

The specific on-resistance of the drift region for 4H-SiC is approximately 2000 times smaller than for silicon devices for the same breakdown voltage. Their values are given by:

$$R_{D,SP} = R_{on-ideal}(Si) = 5.93x10^{-9}\,BV^{2.5} \qquad \textbf{[2.10]}$$

and

$$R_{D,SP} = R_{on-ideal}(4H-SiC) = 2.97x10^{-12}\,BV^{2.5} \qquad \textbf{[2.11]}$$

In addition, it is important to include the resistance associated with the thick, highly doped $N^+$ substrate because this is comparable to that for the drift region in some instances. The specific resistance of the $N^+$ substrate can be determined by taking the product of its resistivity and thickness. For silicon, $N^+$ substrates with resistivity of 1 mΩ-cm are available. If the thickness of the substrate is 200 microns, the specific resistance contributed by the $N^+$ substrate is 2 x $10^{-5}$ Ω-$cm^2$. For silicon carbide, the available resistivity of the $N^+$ substrates is substantially larger. For the available substrates with a typical resistivity of 0.02 Ω-cm and thickness of 200 microns, the substrate contribution is 4 x $10^{-4}$ Ω-$cm^2$. The specific resistance of the ohmic contact to the $N^+$ substrate can be reduced to less than 1 x $10^{-6}$ Ω-$cm^2$ with adequate attention to increasing the doping concentration at the contact and by using ohmic contact metals with low barrier heights.

The calculated forward conduction characteristics for silicon Schottky rectifiers are shown in Fig. 2.4 for various breakdown voltages. For this figure, a Schottky barrier height of 0.7 eV was chosen because this is a typical value used in actual power devices. It can be seen that the series resistance of the drift region

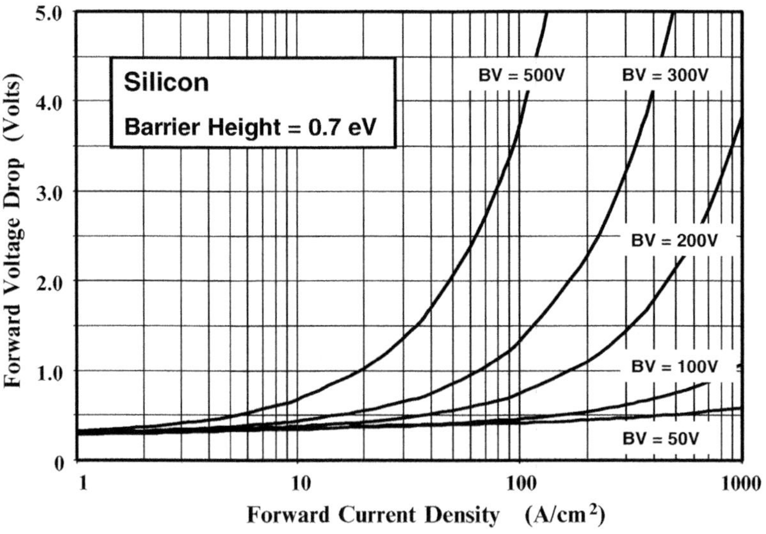


Fig. 2.4 Forward Characteristics of Silicon Schottky Rectifiers.

does not adversely impact the on-state voltage drop for the device with a breakdown voltage of 50 volts at a nominal on-state current density of 100 $A/cm^2$. However, this resistance becomes significant when the breakdown voltage exceeds 100 volts, limiting the application of silicon Schottky rectifiers to systems, such as switch-mode power supply circuits, operating at voltages below 100 V.

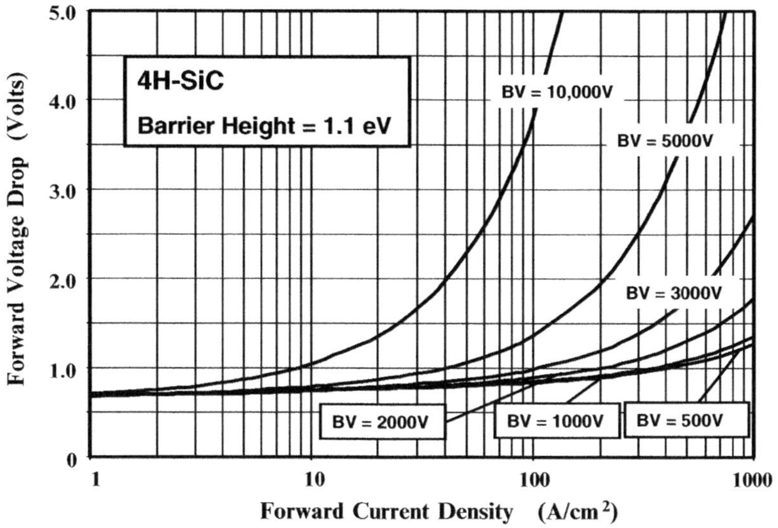

Fig. 2.5 Forward Characteristics of 4H-SiC Schottky Rectifiers.

The significantly smaller resistance of the drift region enables scaling of the breakdown voltage of silicon carbide Schottky rectifiers to much larger voltages typical of medium and high power electronic systems, such as those used for motor control. The forward characterstics of high voltage 4H-SiC Schottky rectifiers are shown in Fig. 2.5 for the case of a Schottky barrier height of 1.1 eV. The $N^+$ substrate resistance used for these calculations was $4 \times 10^{-4}$ $\Omega$-$cm^2$. It can be seen that the drift region resistance does not produce a significant increase in on-state voltage drop until the breakdown voltage exceeds 3000 volts. From these results, it can be concluded that silicon carbide Schottky rectifiers are excellent companion diodes for medium and high power electronic systems that utilize Insulated Gate Bipolar Transistors (IGBTs). Their fast switching speed and absence of reverse recovery current can reduce power losses and improve the efficiency in motor control applications[2].

The choice of the Schottky barrier height has a strong impact on the on-state voltage drop. For typical power Schottky rectifiers, the on-state voltage drop increases in proportion to the magnitude of the Schottky barrier height. It is therefore attractive to use a low Schottky barrier height for power rectifiers in order to reduce the on-state voltage drop. For silicon devices with low blocking voltages, the forward voltage drop for the Schottky diode decreases with increasing temperature due to a reduction in the voltage across the Schottky contact. For 4H-SiC Schottky rectifiers with high blocking voltages, the on-state voltage drop increases with increasing temperature due to an increase in the resistive voltage drop across the drift region. A more detailed discussion of the impact of the Schottky barrier height on the on-state voltage drop can be found in the textbook[3]. The design of a power Schottky rectifier for any application requires selection of the Schottky barrier height to minimize the on-state power losses while avoiding excessive leakage current in the reverse blocking mode. The reverse blocking characteristics of Schottky rectifiers are discussed in the next section.

## 2.3 Reverse Blocking

When a reverse bias is applied to the Schottky rectifier, the voltage is supported across the drift region with the maximum electric field located at the metal-semiconductor contact as shown in Fig. 2.1. Since no voltage can be supported within the metal, the reverse blocking capability of the Schottky rectifier is governed by the physics for the abrupt P-N junction[3]. If a parallel-plane breakdown voltage is assumed, the drift region doping and width for a silicon device are given by:

$$N_D = 2x10^{18}\left(BV_{PP}\right)^{-4/3} \quad \textbf{[2.12]}$$

and

$$W_D = 2.58x10^{-6}\left(BV_{PP}\right)^{7/6} \quad \textbf{[2.13]}$$

In the case of actual power Schottky rectifiers, the breakdown voltage is constrained by breakdown at the edges. Edge terminations have been developed to raise the breakdown voltage of Schottky rectifiers close to the parallel-plane value[1,3].

Due to the relatively small barrier height utilized in silicon Schottky rectifiers, the thermionic emission component is dominant. The leakage current for the Schottky rectifier can be obtained by using Eq. [2.4] and substituting a negative bias of magnitude $V_R$. The leakage current is then determined by the saturation current:

$$J_L = -AT^2 e^{-(q\Phi_{BN}/kT)} = -J_S \quad \textbf{[2.14]}$$

which is a strong function of the Schottky barrier height and the temperature. In order to reduce the leakage current and minimize power dissipation in the blocking state, a large Schottky barrier height is required. Further, a very rapid increase in leakage current occurs with increasing temperature, as shown in Fig. 2.6. If the power dissipation due to the leakage current becomes dominant, the resulting increase in the device temperature produces a positive feedback mechanism which can lead to unstable operation of the Schottky rectifier due to thermal runaway.

This destructive failure mechanism for power Schottky rectifiers must be avoided by sufficiently increasing the Schottky barrier height even though this increases the on-state voltage drop. A larger Schottky barrier height is warranted for power Schottky rectifiers that must operate at higher ambient temperatures. The leakage current in power Schottky rectifiers is actually much greater than the saturation current due to the Schottky barrier lowering and tunneling phenomena.

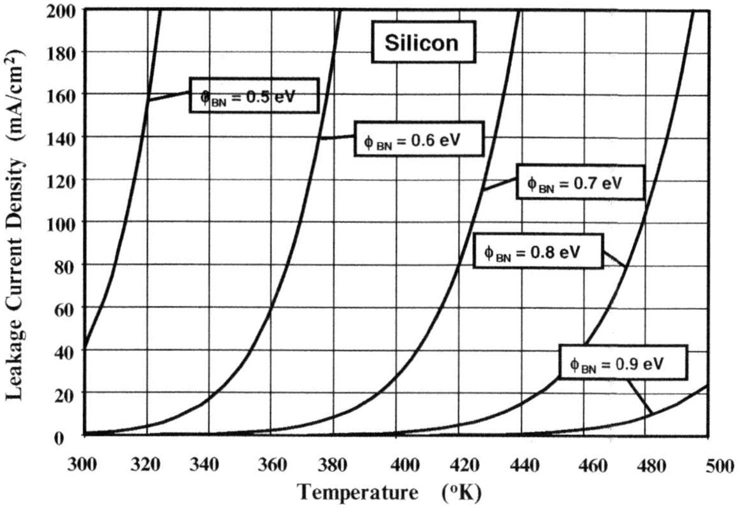


Fig. 2.6 Temperature Dependence of the Leakage Current for Silicon Schottky Rectifiers.

### 2.3.1 Schottky Barrier Lowering

Based upon the above analysis, the leakage current of the Schottky rectifier should be independent of the magnitude of the applied reverse bias voltage. However, actual power Schottky rectifiers exhibit a significant increase in the leakage current with increasing reverse bias voltage. This increase in the leakage current is far greater than the space charge generation current within the expanding depletion region with increasing reverse bias voltage.

Under reverse blocking operation, it has been found that there is a reduction of the Schottky barrier height due the image force lowering phenomenon[6]. The barrier lowering is found to be determined by the maximum electric field ($E_M$) at the metal-semiconductor interface[3]:

$$\Delta\phi_{BN} = \sqrt{\frac{qE_M}{4\pi\varepsilon_S}} \qquad [2.15]$$

For a one-dimensional structure, the maximum electric field is related to the applied reverse bias voltage ($V_R$) by:

$$E_M = \sqrt{\frac{2qN_D}{\varepsilon_S}(V_R + V_{bi})} \qquad [2.16]$$

As an example, the reductions of the barrier height for silicon and 4H-SiC Schottky rectifiers are shown in Fig. 2.7 for the case of a drift region doping concentration of 1 x $10^{16}$ $cm^{-3}$. The reduction of the Schottky barrier height is 0.065 eV at the maximum reverse bias voltage for the silicon structure. Since the low specific on-resistance of the drift region in silicon carbide devices is associated with the much larger electric field in the material before the on-set of impact ionization, the Schottky barrier lowering in silicon carbide rectifiers can be expected to be significantly larger than in silicon devices. For the case of a drift region doping level of 1 x $10^{16}$ $cm^{-3}$, the barrier lowering is found to be three times larger in silicon carbide at the corresponding breakdown voltage as shown in Fig. 2.7. This can lead to a much larger increase in the leakage current with increasing reverse bias voltage for silicon carbide devices. In preparing Fig. 2.7, the reverse voltage was normalized to the breakdown voltage because of the different breakdown voltages for the silicon (50 volts) and silicon carbide devices (3000 volts).

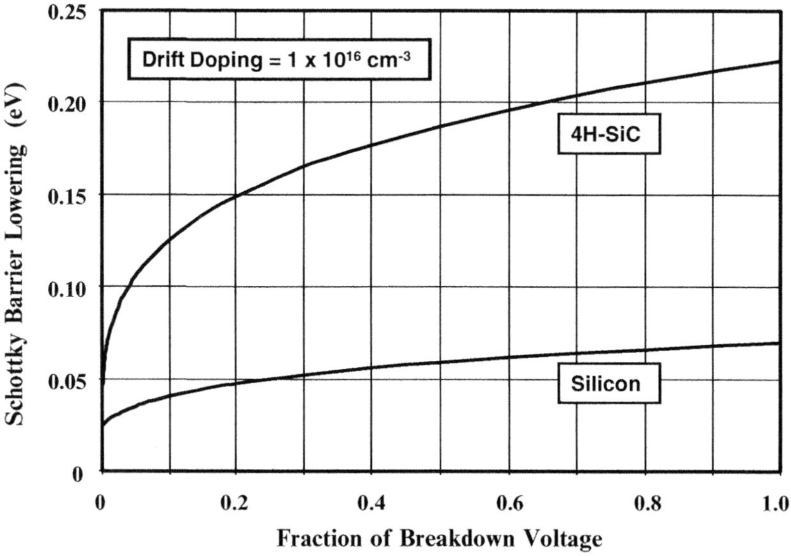


Fig. 2.7 Schottky Barrier Lowering for Silicon and 4H-SiC Schottky Rectifiers.

The leakage current for the Schottky rectifier including the effect of Schottky barrier lowering is given by:

$$J_L = -AT^2 e^{-q(\phi_{BN} - \Delta\phi_{BN})/kT} \qquad [2.17]$$

The leakage currents calculated with and without the Schottky barrier lowering effect are compared for the case of a silicon device with a breakdown voltage of 50 volts in Fig. 2.8. In making these plots, the leakage current due to space-charge-generation was neglected because it is much smaller than the leakage current across the metal-semiconductor contact. It can be seen that the leakage current is enhanced by a factor of five times due to the barrier lowering phenomenon as the reverse voltage increase and approaches the breakdown voltage. The actual reverse leakage current for silicon Schottky rectifiers has been found to increase by an even greater degree than predicted by the Schottky barrier lowering phenomenon.

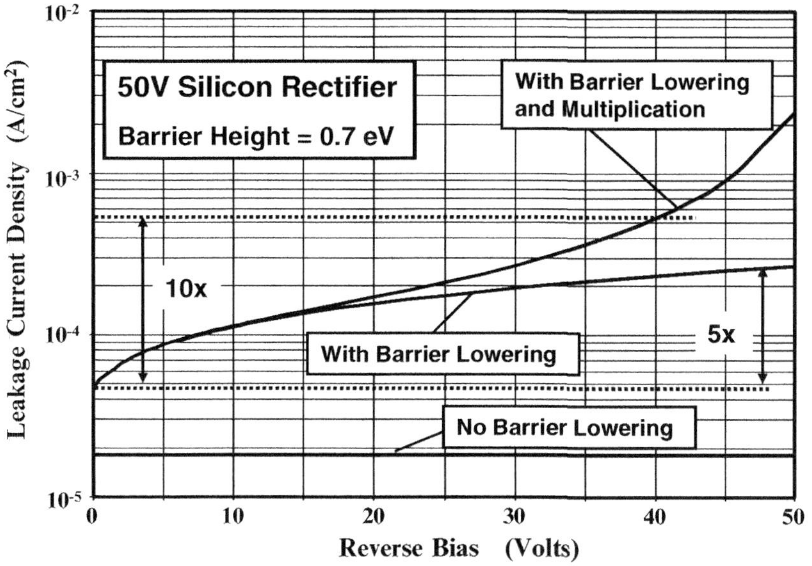


Fig. 2.8 Leakage Current Density for a 50V Silicon Schottky Rectifier.

### 2.3.2 Pre-Breakdown Avalanche Multiplication

The large increase in leakage current in actual silicon power Schottky rectifiers can be accounted for by including the effect of pre-breakdown avalanche multiplication of the large number of free carriers being transported through the Schottky rectifier structure at the high electric fields associated with reverse bias voltages close to the breakdown voltage[7]. The total number of electrons that reach the edge of the depletion region are larger that those crossing the metal-semiconductor contact by a factor $M_n$. The multiplication coefficient ($M_n$) is determined from the maximum electric field ($E_M$) at the metal-semiconductor contact:

$$M_n = \left\{1 - 1.52\left[1 - \exp\left(-7.22x10^{-25} E_m^{4.93} W_D\right)\right]\right\}^{-1} \quad [2.18]$$

where $W_D$ is the depletion layer width. The leakage current density for a silicon Schottky rectifier with drift region doping concentration of $1 \times 10^{16}$ $cm^{-3}$ is shown in Fig. 2.8 after including the influence of the pre-breakdown multiplication coefficient. The effect of including the multiplication coefficient is apparent at high voltages when the electric field approaches the critical electric field for breakdown. The leakage currents obtained, after including the effects of Schottky barrier lowering and pre-breakdown multiplication, are consistent with the characteristics of commercially available silicon devices, which exhibit an order of magnitude increase in leakage current from low reverse bias voltages to the rated voltage (about 80 percent of the breakdown voltage).

### 2.3.3 Silicon Carbide Schottky Rectifiers

The enhanced Schottky barrier lowering in silicon carbide devices leads to a more rapid increase in leakage current with increasing reverse bias as shown in Fig. 2.9. The leakage current is predicted by this model to increase by about three orders of magnitude when the reverse voltage approaches the breakdown voltage. However, the observed increase in leakage current with applied reverse bias voltage for high voltage silicon carbide Schottky rectifiers is much greater than can be accounted for with the Schottky barrier lowering model[8,9,10] despite the much larger barrier lowering effect. The experimentally observed increase in leakage current is about 6 orders of magnitude with increase in reverse bias voltage.

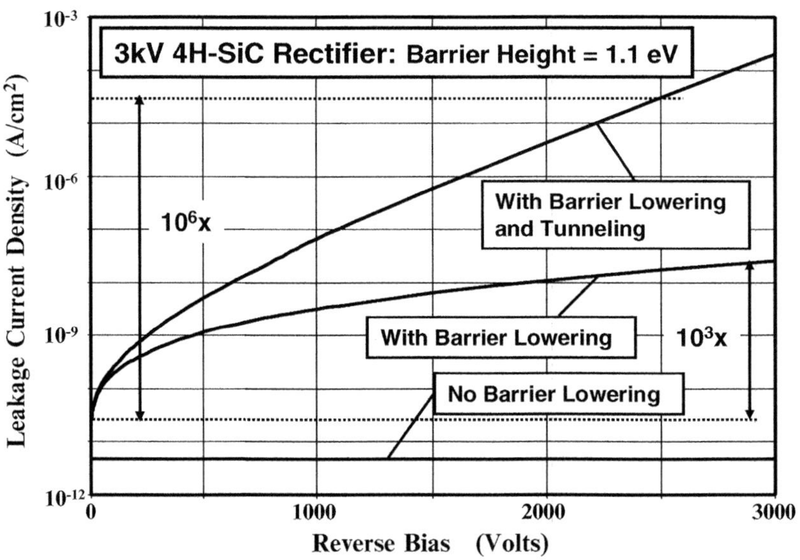


Fig. 2.9 Leakage Current Density for a 3kV 4H-SiC Schottky Rectifier.

In order to explain the more rapid increase in leakage current observed in silicon carbide Schottky rectifiers, it is necessary to include the field emission (or tunneling) component of the leakage current[11]. The thermionic field emission model for the tunneling current leads to a barrier lowering effect proportional to the square of the electric field at the metal-semiconductor interface. When combined with the thermionic emission model, the leakage current density can be written as:

$$J_S = AT^2 \exp\left(-\frac{q\phi_{BN}}{kT}\right).\exp\left(\frac{q\Delta\phi_{BN}}{kT}\right).\exp\left(C_T E_M^2\right) \quad \textbf{[2.19]}$$

where $C_T$ is a tunneling coefficient. A tunneling coefficient of 8 x $10^{-13}$ $cm^2/V^2$ was found to yield an increase in leakage current by six orders of magnitude as shown in Fig. 2.9 consistent with the experimental observations. Thus, the inclusion of the tunneling model enhances the leakage current by another three orders of magnitude beyond that due to the Schottky barrier lowering phenomenon.

As discussed above, the leakage current in silicon carbide Schottky rectifiers increases much more rapidly with reverse voltage than in silicon devices. Fortunately, larger barrier heights can be utilized in silicon carbide devices when compared with silicon devices to reduce the absolute magnitude of the leakage current density because an on-state voltage drop of 1 to 1.5 volts is acceptable for such high voltage structures. This enables maintaining an acceptable level of power dissipation in the reverse blocking mode. For example, in the case of the 3kV 4H-SiC Schottky diode discussed above, the reverse power dissipation at room temperature is less than 1 $W/cm^2$ compared with an on-state power dissipation of 100 $W/cm^2$. The expected increase in leakage current with temperature must of course be taken into account in order to ensure than the reverse power dissipation remains below the on-state power dissipation for stable operation. The leakage current can be suppressed by shielding the Schottky contact[1] using the JBS rectifier structure[12] originally proposed for silicon devices.

## 2.4 Summary

The physics of operation of the power Schottky rectifier structure has been described in this chapter to provide background information for the discussion of the advanced concepts in this book. For power devices with relatively high breakdown voltages, the dominant current conduction mechanism is by the thermionic emission process. For power Schottky rectifiers, it is necessary to include the impact of the series resistance of the drift region on the on-state voltage drop. This series resistance limits the performance of silicon rectifiers to a breakdown voltage of less than 100 volts. In addition, the Schottky barrier lowering and pre-breakdown avalanche multiplication must be taken into consideration when analyzing the leakage current for silicon devices because they can enhance the leakage current by an order of magnitude for high reverse bias voltages. In the case of silicon carbide

Schottky rectifiers, it is possible to extend the breakdown voltage to at least 3000 volts due to the much smaller resistance in the drift region. However, the reverse leakage current in silicon carbide devices is significantly enhanced by the tunneling current at high reverse bias voltages. The textbook[3] should be consulted for a more detailed treatment of the basic power Schottky rectifier.

The leakage current in silicon and silicon carbide Schottky rectifiers can be suppressed by shielding the Schottky barrier to reduce the electric field at the metal-semiconductor interface. Various approaches to achieve this by using either P-N junctions, MOS-regions, or a second metal with larger barrier height are discussed in subsequent chapters of this book.

## References

[1] B.J. Baliga, "Silicon Carbide Power Devices", World Scientific Publishing Company, Singapore, 2005.

[2] B.J. Baliga, "Power Semiconductor Devices for Variable Frequency Drives", Proceedings of the IEEE, Vol. 82, pp. 1112-1122, 1994.

[3] B.J. Baliga, "Fundamentals of Power Semiconductor Devices", Springer Scientific, New York, 2008.

[4] S.M. Sze, "Physics of Semiconductor Devices", pp. 254-258, 2nd Edition, John Wiley and Sons, 1981.

[5] H.A. Bethe, "Theory of the Boundary Layer of Crystal Rectifiers", MIT Radiation Laboratory Report, Vol. 43, p. 12, 1942.

[6] E.H. Rhoderick and R.H. Williams, "Metal-Semiconductor Contacts", Second Edition, pp. 35-38, Oxford Science Publications, 1988.

[7] S.L. Tu and B.J. Baliga, "On the Reverse Blocking Characteristics of Schottky Power Diodes", IEEE Transactions on Electron Devices, Vol. 39, pp. 2813-2814, 1992.

[8] M. Bhatnagar, P.K. McLarty, and B.J. Baliga, "Silicon Carbide High Voltage (400V) Schottky Barrier Diodes", IEEE Electron Device Letters, Vol. 13, pp. 501-503, 1992.

[9] F. Dahlquist, et al, "A 2.8kV, Forward Drop JBS Diode with Low Leakage", Silicon Carbide and Related Materials – 1999, Material Science Forum, Vol. 338-342, pp. 1179-1182, 2000.

[10] Y. Sugawara, K. Asano, and R. Saito, "3.6kV 4H-SiC JBS Diodes with Low Ron", Silicon Carbide and Related Materials – 1999, Material Science Forum, Vol. 338-342, pp. 1183-1186, 2000.

[11] T. Hatakeyama and T. Shinohe, "Reverse Characteristics of a 4H-SiC Schottky Barrier Diode", Silicon Carbide and Related Materials – 2001, Material Science Forum, Vol. 389-393, pp. 1169-1172, 2002.

[12] B.J. Baliga, "The Pinch-Rectifier: A Low Forward Drop, High Speed Power Diode", IEEE Electron Device Letters, Vol. EDL-5, pp. 194-196, 1984.

# Chapter 3

# Junction Barrier Controlled Schottky Rectifiers

In the case of Schottky rectifiers, it is necessary to trade-off the on-state (or conduction) power loss against the reverse blocking power loss by optimizing the Schottky barrier height[1]. As the Schottky barrier height is reduced, the on-state voltage drop decreases producing smaller conduction power loss. At the same time, the smaller barrier height produces an increase in the leakage current leading to larger reverse blocking power loss. The power loss can be minimized by reducing the Schottky barrier height at the expense of a reduced maximum operating temperature. This optimization process is exacerbated by the rapid increase in the leakage current with increasing reverse bias voltage due to the Schottky barrier lowering and pre-breakdown multiplication phenomena.

The first method proposed to ameliorate the barrier lowering effect in vertical silicon Schottky rectifiers utilized shielding by incorporation of a P-N junction[2,3]. Since the basic concept was to create a potential barrier to shield the Schottky contact against high electric fields generated in the semiconductor by using closely spaced $P^+$ regions around the contact, this structure was named the *'Junction-Barrier controlled Schottky (JBS) rectifier'*. In the JBS rectifier, the on-state current is designed to flow in the un-depleted gaps between the $P^+$ regions when the diode is forward biased to preserve unipolar operation. In addition to reducing the leakage current, the presence of the $P^+$ regions improves the ruggedness of the diodes. Detailed optimization of the silicon JBS rectifier characteristics has been achieved with sub-micron dimensions for forming the $P^+$ regions[4]. Minimization of the area consumed by the $P^+$ regions reduces the spreading resistance of the drift region below the Schottky contact resulting in a lower on-state voltage drop.

In the previous chapter, it was also demonstrated that the increase in leakage current with reverse bias voltage is much stronger for the silicon carbide

B.J. Baliga, *Advanced Power Rectifier Concepts*, DOI: 10.1007/978-0-387-75589-2_3, 

Schottky rectifiers. This is due to a larger Schottky barrier lowering effect associated with the larger electric field in silicon carbide drift regions and the onset of thermionic field emission (or tunneling) current. It is therefore obvious that the methods proposed to suppress the electric field at the Schottky contact in silicon diodes will have even greater utility in silicon carbide rectifiers.

This chapter discusses the application of the junction barrier controlled shielding concept to reduce the electric field at the metal-semiconductor contact in silicon and 4H-SiC Schottky rectifiers. In addition to analytical models, the physics of operation of these structures is analyzed by using two-dimensional numerical simulations. It is demonstrated that the JBS concept improves the performance of silicon devices and can be extended to silicon carbide devices[5] to achieve even greater improvements in performance.

## 3.1 Junction Barrier Schottky (JBS) Rectifier Structure

The Junction Barrier controlled Schottky (JBS) rectifier structure is illustrated in Fig. 3.1. It consists of a $P^+$ region placed around the Schottky contact to generate a potential barrier under the metal-semiconductor contact in the reverse blocking mode to shield the Schottky contact. The magnitude of the potential barrier depends upon the separation between the P-N junctions and the junction depth. A smaller separation and larger junction depth favors an increase in the magnitude of the potential barrier leading to a greater reduction of the electric field at the Schottky contact. A reduction of the electric field at the Schottky contact produces a smaller barrier lowering and field emission effect, which is beneficial for reducing the leakage current at high reverse bias voltages.

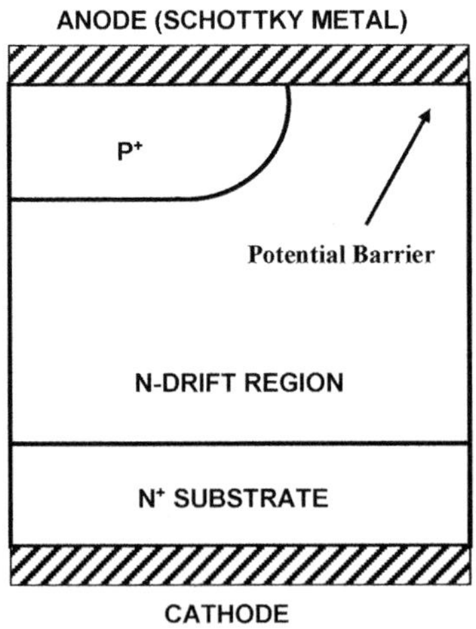


Fig. 3.1 Junction Barrier controlled Schottky (JBS) Rectifier Structure.

The space between the $P^+$ regions is also chosen so that there is an undepleted region below the Schottky contact during on-state operation to enable unipolar conduction through the structure. In the JBS rectifier, the voltage drop across the diode is insufficient to forward bias the P-N junction. Silicon devices with low breakdown voltages operate with on-state voltage drops (around 0.45 volts) well below the 0.7 volts needed for inducing strong injection across the P-N junction. The margin is even larger for silicon carbide due to its much larger band gap. Typical silicon carbide Schottky rectifiers have on-state voltage drops of less than 1.5 volts due to the low specific resistance of the drift region. This is well below the 3 volts required to induce injection from the P-N junction. Consequently, the JBS concept is well suited for development of silicon carbide structures with very high breakdown voltages.

The $P^+$ region is usually formed by ion-implantation of P-type dopants using a mask with appropriate spacing to leave room for the Schottky contact. For processing convenience, the same metal layer is used to make an ohmic contact to the highly doped $P^+$ region as well as to make the Schottky barrier contact to the N- drift region. The space between the $P^+$ regions must be optimized to obtain the best compromise between reducing the on-state voltage drop and reducing the leakage current.

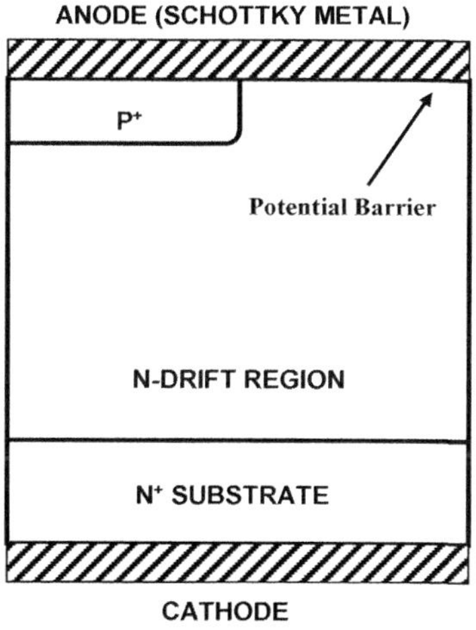


Fig. 3.2 Silicon Carbide Junction Barrier controlled Schottky (JBS) Rectifier Structure.

In silicon devices, the P-N junction is formed with an annealing cycle that creates a planar junction with its extension in the lateral direction as shown in Fig. 3.1. The additional area consumed by the lateral diffusion must be accounted for during the analytical modeling of the JBS rectifier structure. In addition, the cylindrical shape of the junction allows encroachment of the cathode potential towards the Schottky contact producing an enhancement of the electric field at the contact with increasing reverse bias voltage. In the case of silicon carbide devices, there is no significant diffusion during the annealing of the ion implants. A rectangular shape for the P-N junction is appropriate for these structures as illustrated in Fig. 3.2.

It is worth pointing out that the reverse blocking voltage is supported in the JBS rectifier structure by the depletion region formed below the $P^+$/N junction. This junction is usually used to simultaneously form the edge termination for silicon diodes. For these low voltage silicon structures, it is usually sufficient to utilize a cylindrical edge termination with a field plate to enhance the breakdown voltage. The breakdown voltage is therefore reduced to about 80 percent of the ideal parallel-plane value. The doping concentration for the drift region must be computed after accounting for this reduction in the breakdown voltage:

$$N_D = \left( \frac{5.34x10^{13}}{BV_{PP}} \right)^{4/3} \qquad \textbf{[3.1]}$$

where $BV_{PP}$ is the breakdown voltage for the parallel-plane case after accounting for the edge termination.

An error that is often made during design of these devices is to make the thickness of the drift region below the P-N junction equal to the depletion width for the ideal parallel-plane junction with the above doping concentration. In actuality, the maximum depletion width is limited to that associated with the breakdown voltage (BV) of the structure, as given by:

$$t = W_D(BV) = \sqrt{\frac{2\varepsilon_S BV}{qN_D}} \qquad \textbf{[3.2]}$$

The thickness of the drift region required below the P-N junction is therefore less than the depletion width for the ideal parallel-plane junction with the above doping concentration. The resistance of the drift region below the Schottky contact is consequently enhanced above that for the ideal parallel-plane case due to current transport between the junctions and the lower doping concentration of the drift region.

## 3.2 Forward Conduction Models

Analysis of the on-state voltage drop of the JBS rectifier requires taking into consideration the current constriction at the Schottky contact due to the presence of the $P^+$ region and the enhanced resistance of the drift region[6] due to current spreading from the Schottky contact to the $N^+$ substrate. Several models can be developed for the spreading resistance, which are appropriate to different JBS rectifier designs. In addition, the different shape for the P-N junction in silicon carbide warrants a unique model as well. These models are developed below. In all the models, it is assumed that the voltage drop in the on-state for the JBS rectifier is well below the voltage required to induce current flow across the P-N junction.

### 3.2.1 Silicon JBS Rectifier: Forward Conduction Model A

The current flow pattern in the forward conduction mode for Model A for a silicon JBS rectifier is illustrated in Fig. 3.3 with the shaded area. The model takes into account the presence of a depletion layer at the P-N junction which increases the current density at the Schottky contact. The current through the Schottky contact flows only within the un-depleted portion (with dimension 'd') of the drift region at the top surface. Consequently, the current density at the Schottky contact ($J_{FS}$) is related to the cell (or cathode) current density ($J_{FC}$) by:

$$J_{FS} = \left(\frac{p}{d}\right) J_{FC} \quad [3.3]$$

where p is the cell pitch. The dimension 'd' is determined by the cell pitch (p), the size of the ion-implant window (2s), the junction depth of the $P^+$ region, and the on-state depletion width ($W_{D,ON}$):

$$d = p - s - x_J - W_{D,ON} \quad [3.4]$$

In deriving this equation, it has been assumed that the lateral diffusion is equal to the junction depth. Depending up on the lithography used for device fabrication to minimize the size (dimension 's') of the $P^+$ region, as well as the junction depth

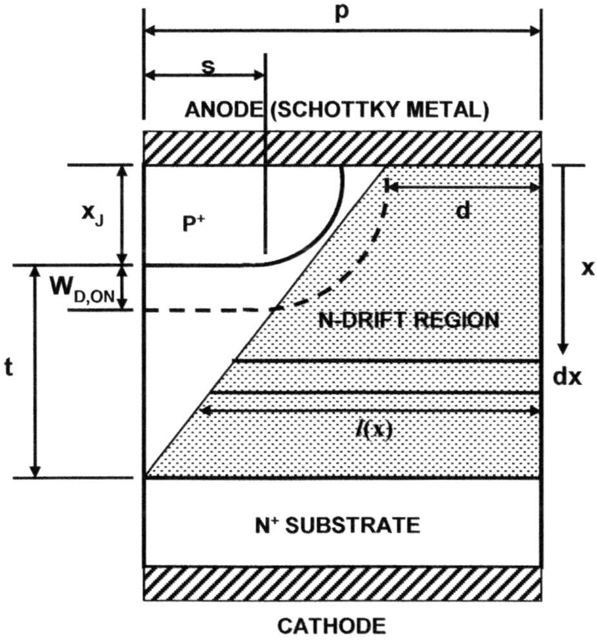


Fig. 3.3 Current Flow Pattern in the Silicon JBS Rectifier Structure during Operation in the On-State: Model A.

($x_J$) created during the diffusion process, the current density at the Schottky contact can be enhanced by a factor of two or more. This must be taken into account when computing the voltage drop across the Schottky contact given by:

$$V_{FS} = \phi_B + \frac{kT}{q} \ln\left(\frac{J_{FS}}{AT^2}\right) \tag{3.5}$$

After crossing the Schottky contact, the current flows through the un-depleted portion of the drift region. The resistance of the drift region is larger than the specific on-resistance discussed in the previous chapter due to current spreading from the Schottky contact to the $N^+$ substrate as illustrated for Model A in the figure. In this model, it is assumed that the current spreads from the Schottky contact region (width d) to the entire cell pitch (p) at the bottom of the drift region. In order to obtain this spreading resistance, consider a sliver of the conduction region with a thickness dx located at a depth of x from the surface. The width $l$(x) of this segment is given by:

$$l(x) = d + \frac{(p-d).x}{(x_J + t)} \tag{3.6}$$

The resistance of the segment is given by:

$$dR_{drift} = \frac{\rho_D .dx}{l(x).Z} = \frac{\rho_D .(x_J + t).dx}{[d.(x_J + t) + (p-d).x]Z} \tag{3.7}$$

where Z is the length of the cell in the direction orthogonal to the cross-section shown in the figure. The resistance of the drift region can be obtained by integration of this resistance between the surface (x = 0) and the $N^+$ substrate (x = $x_J$ + t):

$$R_{drift} = \frac{\rho_D}{Z}\left(\frac{x_J + t}{p-d}\right).\ln\left(\frac{p}{d}\right) \tag{3.8}$$

The specific resistance for the drift region can be calculated by multiplying the cell resistance by the cell-area (p.Z):

$$R_{sp,drift} = \rho_D .p.\left(\frac{x_J + t}{p-d}\right).\ln\left(\frac{p}{d}\right) \tag{3.9}$$

In addition, it is important to include the resistance associated with the thick, highly doped $N^+$ substrate because this is comparable to that for the drift region in some instances. The specific resistance of the $N^+$ substrate can be determined by taking the product of its resistivity and thickness. For silicon, $N^+$ substrates with resistivity of 1 mΩ-cm are available. The specific resistance contributed by the $N^+$ substrate with a typical thickness of 200 microns is 2 x $10^{-5}$ Ω-cm$^2$.

The on-state voltage drop for the JBS rectifier at a forward cell current density $J_{FC}$, including the substrate contribution, is then given by:

$$V_F = \phi_B + \frac{kT}{q}\ln\left(\frac{J_{FS}}{AT^2}\right) + \left(R_{sp,drift} + R_{sp,subs}\right).J_{FC} \quad [3.10]$$

When computing the on-state voltage drop using this equation, it is satisfactory to make the approximation that the depletion layer width can be computed by subtracting an on-state voltage drop of about 0.45 volts from the built-in potential of the P-N junction. In addition, it is important to recognize that the doping profile at the junction is linearly graded leading to half the depletion occurring on the P-side of the junction. Consequently:

$$W_{D,ON} = 0.5.\sqrt{\frac{2\varepsilon_S\left(V_{bi}-0.45\right)}{qN_D}} \quad [3.11]$$

where $V_{bi}$ is the built-in potential of the abrupt $P^+$/N junction. In the case of silicon JBS rectifiers with breakdown voltage in the range of 30 to 60 volts, the depletion width ($W_{D,ON}$) is about 0.1 to 0.15 microns.

### 3.2.2 Silicon JBS Rectifier: Forward Conduction Model B

The current flow pattern in Model B for the forward conduction mode of a silicon JBS rectifier is illustrated in Fig. 3.4 with the shaded area. The current density at the Schottky contact ($J_{FS}$) is enhanced as already discussed for Model A due to the presence of the $P^+$ diffusion and the depletion layer at the P-N junction. This increases the voltage drop across the Schottky contact as defined by Eq. [3.3] – Eq. [3.5]. After crossing the Schottky contact, the current flows through the un-depleted portion of the drift region between the junctions. In Model B, it is assumed that the current flows through a region with a uniform width 'd' until it reaches the bottom of the depletion region and then spreads to the entire cell pitch (p) at the bottom of the drift region.

The net resistance to current flow can be calculated by adding the resistance of the two segments. The resistance of the first segment of uniform width 'd' is given by:

$$R_{D1} = \frac{\rho_D.\left(x_J + W_{D,ON}\right)}{d.Z} \quad [3.12]$$

The spreading resistance of the second segment can be derived by using the same approach used for Model A:

$$R_{D2} = \frac{\rho_D}{Z}\left(\frac{t - W_{D,ON}}{p-d}\right).\ln\left(\frac{p}{d}\right) \quad [3.13]$$

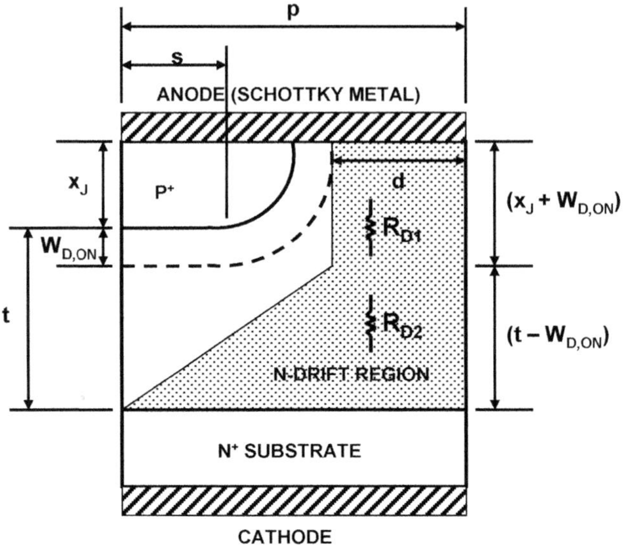


Fig. 3.4 Current Flow pattern in the Silicon JBS Rectifier Structure during Operation in the On-State: Model B.

The net spreading resistance of the drift region is then given by:

$$R_{drift} = \frac{\rho_D.(x_J + W_{D,ON})}{d.Z} + \frac{\rho_D}{Z}\left(\frac{t - W_{D,ON}}{p-d}\right).\ln\left(\frac{p}{d}\right) \quad [3.14]$$

The specific resistance for the drift region can be calculated by multiplying the cell resistance by the cell-area (p.Z):

$$R_{sp,drift} = \frac{\rho_D.p.(x_J + W_{D,ON})}{d} + \rho_D.p.\left(\frac{t - W_{D,ON}}{p-d}\right).\ln\left(\frac{p}{d}\right) \quad [3.15]$$

The on-state voltage drop for the JBS rectifier at a forward cell current density $J_{FC}$, including the substrate contribution, is then given by:

$$V_F = \phi_B + \frac{kT}{q}\ln\left(\frac{J_{FS}}{AT^2}\right) + (R_{sp,drift} + R_{sp,subs}).J_{FC} \quad [3.16]$$

When computing the on-state voltage drop using this equation, it is satisfactory to make the approximation that the depletion layer width can be computed by subtracting an on-state voltage drop of about 0.45 volts from the built-in potential of the P-N junction. In addition, it is important to recognize that the doping profile at the junction is linearly graded leading to half the depletion occurring on the P-side of the junction. Consequently:

$$W_{D,ON} = 0.5.\sqrt{\frac{2\varepsilon_S (V_{bi} - 0.45)}{qN_D}} \quad [3.17]$$

where $V_{bi}$ is the built-in potential of the abrupt $P^+/N$ junction. In the case of silicon JBS rectifiers with breakdown voltage in the range of 30 to 60 volts, the depletion width ($W_{D,ON}$) is about 0.1 to 0.15 microns.

### 3.2.3 Silicon JBS Rectifier: Forward Conduction Model C

When the ion-implant window (2s) for the $P^+$ region is reduced with improved lithographic design rules, the current path in the drift region can overlap before reaching the $N^+$ substrate. The current flow pattern for this case (Model C) is illustrated in Fig. 3.5 with the shaded area. The current density at the Schottky contact ($J_{FS}$) is enhanced in Model C as already discussed for Model B due to the presence of the $P^+$ region and the depletion layer at the P-N junction. This increases the voltage drop across the Schottky contact as defined by Eq. [3.3] – Eq. [3.5]. After flowing across the Schottky contact, the current flows through the un-depleted portion of the drift region between the junctions. In Model C, it is assumed that the current flows through a region with a uniform width 'd' until it reaches the bottom of the depletion region and then spreads to the entire cell pitch (p) at a 45 degree spreading angle. The current paths overlap at a distance ($s + x_J + W_{D,ON}$) from the bottom of the depletion region. The current then flows through a uniform cross-sectional area.

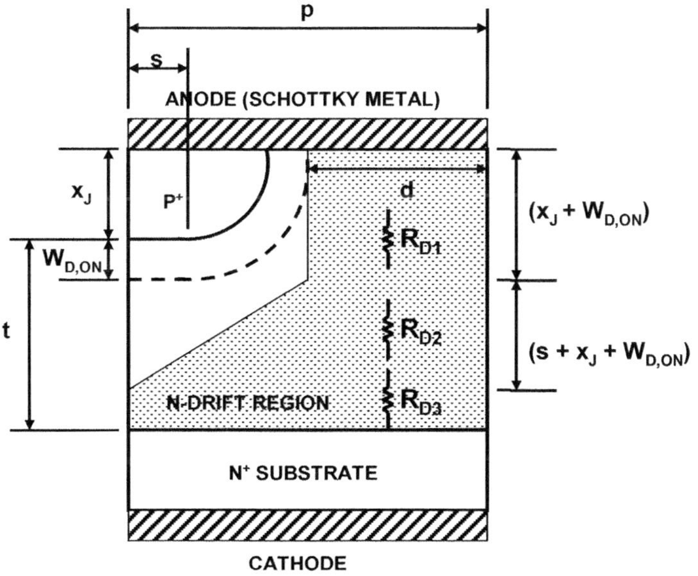


Fig. 3.5 Current Flow pattern in the Silicon JBS Rectifier Structure during Operation in the On-State: Model C.

The net resistance to current flow can be calculated by adding the resistance of the three segments. The resistance of the first segment of uniform width 'd' is given by:

$$R_{D1} = \frac{\rho_D.(x_J + W_{D,ON})}{d.Z} \quad \textbf{[3.18]}$$

The resistance of the second segment can be derived by using the same approach used for Model A. However, the width of the segment for the current flow path in this case is given by:

$$l(x) = d + x \quad \textbf{[3.19]}$$

because of the 45 degree spreading angle. Using this expression, the resistance for the second segment is given by:

$$R_{D2} = \frac{\rho_D}{Z}\ln\left(\frac{p}{d}\right) \quad \textbf{[3.20]}$$

The resistance of the third segment with a uniform cross-section of width p is given by:

$$R_{D3} = \frac{\rho_D.(t - s - x_J - 2W_{D,ON})}{p.Z} \quad \textbf{[3.21]}$$

The specific resistance for the drift region can be calculated by multiplying the cell resistance ($R_{D1} + R_{D2} + R_{D3}$) with the cell-area (p.Z):

$$R_{sp,drift} = \frac{\rho_D.p.(x_J + W_{D,ON})}{d} + \rho_D.p.\ln\left(\frac{p}{d}\right) + \rho_D.(t - s - x_J - 2W_{D,ON}) \quad \textbf{[3.22]}$$

The on-state voltage drop for the JBS rectifier at a forward cell current density $J_{FC}$, including the substrate contribution, is then given by:

$$V_F = \phi_B + \frac{kT}{q}\ln\left(\frac{J_{FS}}{AT^2}\right) + (R_{sp,drift} + R_{sp,subs}).J_{FC} \quad \textbf{[3.23]}$$

When computing the on-state voltage drop using this equation, it is satisfactory to make the approximation that the depletion layer width can be computed by subtracting an on-state voltage drop of about 0.45 volts from the built-in potential of the P-N junction. In addition, it is important to recognize that the doping profile at the junction is linearly graded leading to half the depletion occurring on the P-side of the junction. Consequently:

$$W_{D,ON} = 0.5.\sqrt{\frac{2\varepsilon_S(V_{bi} - 0.45)}{qN_D}} \quad \textbf{[3.24]}$$

where $V_{bi}$ is the built-in potential of the abrupt $P^+/N$ junction. In the case of silicon JBS rectifiers with breakdown voltage in the range of 30 to 60 volts, the depletion width ($W_{D,ON}$) is about 0.1 to 0.15 microns.

### 3.2.4 Silicon JBS Rectifier: Example

In order to understand the differences between the above models for forward current flow in the JBS rectifier structure, it is instructive to consider a specific example of a device with breakdown voltage of 50 volts. If the edge termination limits the breakdown to 80 percent of the ideal value, the parallel-plane breakdown voltage is 62.5 volts. This voltage can be supported by a depletion region width of 2.85 microns in the drift region with doping concentration of 8 x $10^{15}$ $cm^{-3}$. The depletion region width in the N-drift region calculated for this case under the assumption of an on-state voltage drop of 0.45 volts is 0.13 microns. If the JBS structure has a cell pitch (p) of 1.25 microns, and a $P^+$ region with depth of 0.5 microns is created using an ion-implant window (2s) of 0.5 microns, the dimension 'd' is found to be 0.37 microns. Consequently, the current density at the Schottky contact region where the current is transported is enhanced by a factor 3 times when compared with the cathode (or average cell) current density.

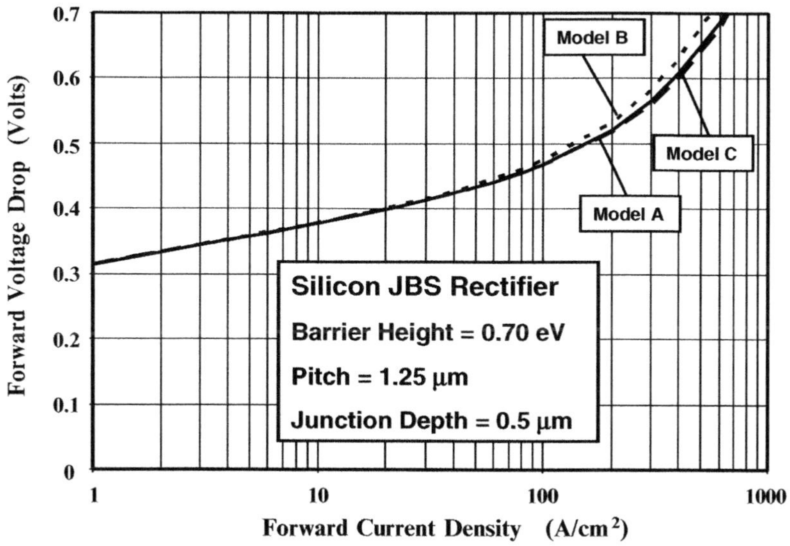


Fig. 3.6 Forward Characteristics of a 50V Silicon JBS Rectifier Structure.

The on-state *i-v* characteristics obtained by using the three models for the above JBS rectifier structure are shown in Fig. 3.6. A Schottky barrier height of 0.70 eV was used for this analysis. Although Model B is slightly pessimistic when compared with the other two models, their predictions are very close to each other for determination of the on-state voltage drop. The on-state voltage drop for this

JBS rectifier structure is found to be 0.467 volts at an on-state current density of 100 A/cm$^2$.

The impact of changing the cell pitch (p) on the on-state characteristics can be predicted by using any one of the three models. The results obtained by using model C are shown in Fig. 3.7 for the case of a Schottky barrier height of 0.7 eV. The *i-v* characteristic of a planar Schottky rectifier with the same drift region parameters is included for comparison. The impact on the on-state voltage drop is small as long as the pitch is more than 1.25 microns. However, the on-state voltage drop increases by 10 percent when the pitch is reduced to 1.00 microns because of current constriction at the Schottky contact. Since a smaller cell pitch and space 'd' favors a smaller leakage current as discussed below, it is necessary to carefully choose, and precisely control, the cell pitch in the JBS rectifier in order to optimize its characteristics.

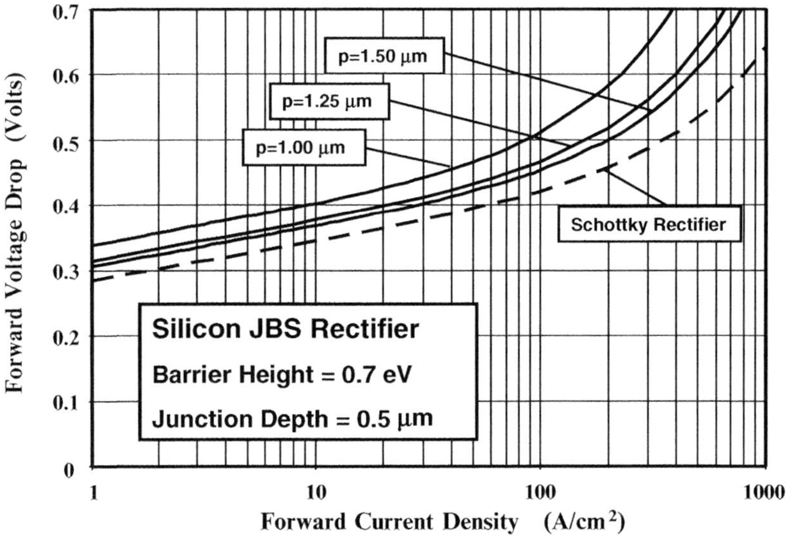


Fig. 3.7 Forward Characteristics of 50V Silicon JBS Rectifier Structures.

The degree of shielding of the Schottky contact by the P-N junction also depends upon the depth of the junction. A deeper junction improves the *aspect ratio* (defined later) of the region between the planar junctions. A larger aspect ratio has been found to provide greater shielding in the case of the vertical junction field effect transistor leading to a larger blocking gain[7]. However, a deeper junction increases the cell pitch and the resistance of the first segment in Model C leading to a larger on-state voltage drop. This is illustrated in Fig. 3.8 where the analytically calculated forward *i-v* characteristics are shown for two junction depths. Increasing the junction depth from 0.5 to 1 micron increases the on-state voltage drop from 0.467 to 0.487 volts.

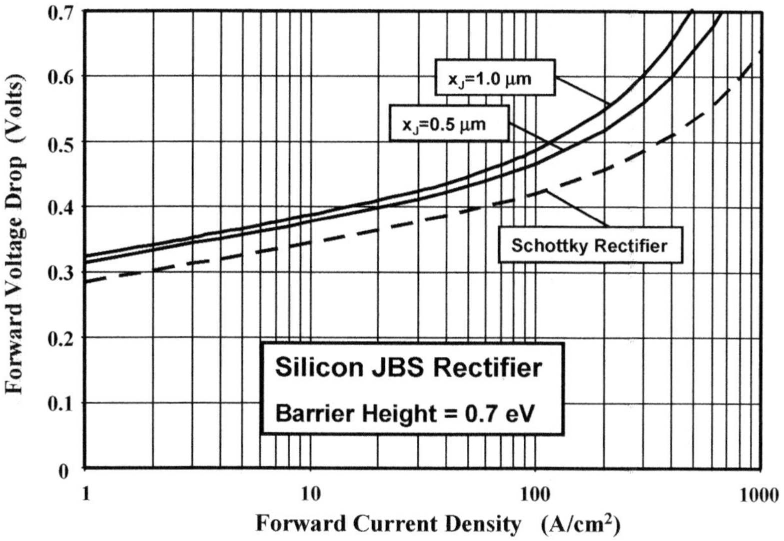


Fig. 3.8 Forward Characteristics of 50V Silicon JBS Rectifier Structures.

### Simulation Example

In order to validate the above model for the on-state characteristics of the silicon JBS rectifier, the results of two-dimensional numerical simulations on a 50 V structure are described here. The structure had a drift region thickness of 3 microns with a doping concentration of 8 x $10^{15}$ $cm^{-3}$. The $P^+$ region had a depth of 0.5 microns with an ion-implant window (dimension 's' in Fig. 3.5) of 0.25 microns. A three-dimensional view of the doping profile for the structure is provided in Fig. 3.1E. The $P^+$ region is located on the upper left-hand side of the figure. The junction extends to 0.75 microns along the top surface, leaving 0.5 microns for the metal-semiconductor contact. However, a portion of this region is depleted by the junction potential.

The on-state *i-v* characteristics, obtained by using a barrier height of 0.7 eV, for the JBS rectifier with a cell pitch 'p' of 1.25 microns are shown in Fig. 3.2E. This plot includes the total current (cathode current) flowing through the structure together with the current flowing through the Schottky contact (dotted line) and the $P^+$ region (dashed line). It can be observed that the current flowing through the $P^+$ region is extremely small unless the on-state voltage drop exceeds 0.7 volts. At an on-state current density of 100 A/$cm^2$, the on-state voltage drop is 0.45 volts indicating that there is no significant injection from the junction and all the current is flowing via the Schottky contact. The current flowing via the $P^+$ region is 6 orders of magnitude less than the current flowing via the Schottky contact. This ensures minimal impact of injection of any minority carriers from the P-N junction on the reverse recovery behavior of the JBS rectifier structure.

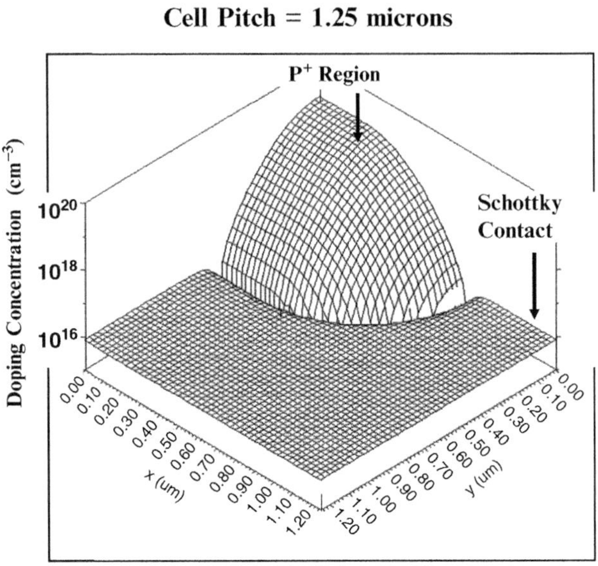


Fig. 3.1E Doping Profile for the 50V Silicon JBS Rectifier.

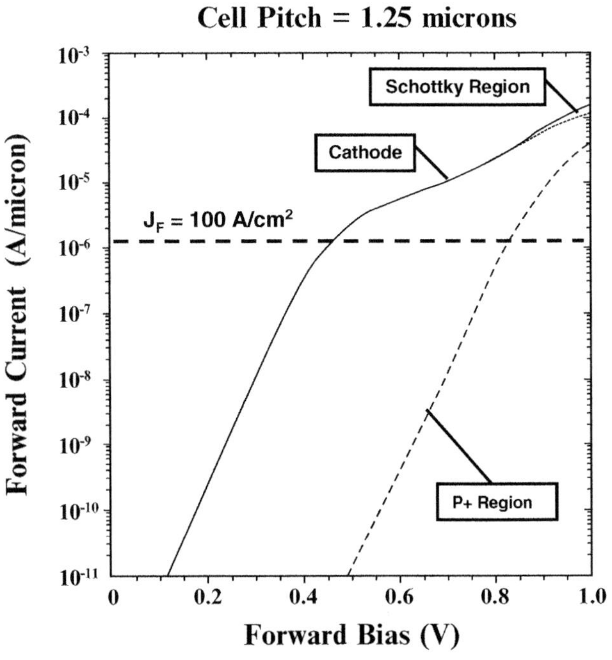


Fig. 3.2E On-State Characteristics for a 50V Silicon JBS Rectifier.

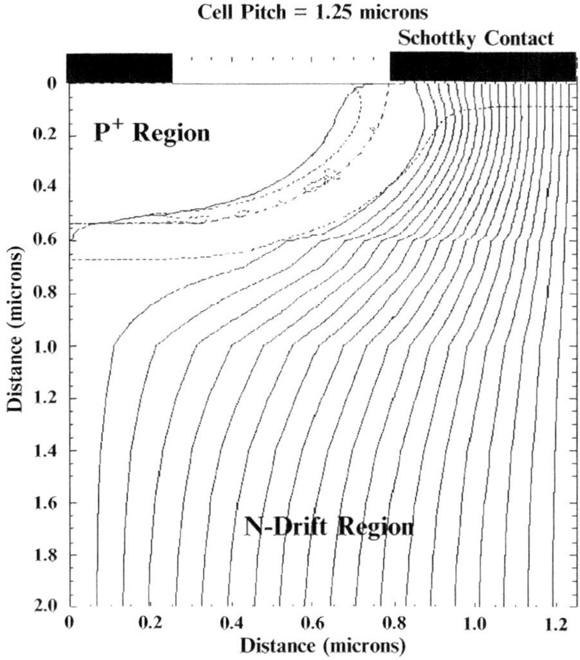


Fig. 3.3E On-State Current Distribution in a 50V Silicon JBS Rectifier.

The current flow-lines for the above JBS rectifier structure, shown in Fig. 3.3E at an on-state voltage drop of 0.5 volts, can be used to confirm that it operates in a unipolar current conduction mode. It can be seen that all the current flow-lines converge to the Schottky contact located on the upper right-hand-side of the structure demonstrating that no current flows via the $P^+$ region. The current flow pattern is consistent with Model C with an approximately uniform cross-section until the bottom of the depletion region followed by a spreading of the current at an angle of about 45 degrees. The current then becomes uniformly distributed below a depth of 1.5 microns from the surface. This justifies the use of a three region model for the series resistance of the drift region in Model C.

The on-state *i-v* characteristics of 50 V JBS rectifiers with cell pitch of 1.00 and 1.25 microns are compared with that of a Schottky rectifier with cell pitch of 1.25 microns in Fig. 3.4E. All the structures had a Schottky contact with barrier height of 0.7 eV. In the JBS rectifiers, the $P^+$ region was implanted through a window (2s) of 0.5 microns and had a depth of 0.5 microns. It can be observed that the on-state voltage drop for the JBS rectifiers is increased by the presence of the $P^+$ region. After accounting for the difference in the areas of the structures, the on-state voltage drop at an on-state current density of 100 A/cm$^2$ was found to be 0.45 V, and 0.49 V for the JBS rectifier structures when compared with 0.41 volts for the Schottky rectifier. As expected, there is an inflection in the curves for the JBS rectifiers at an on-state voltage drop of about 0.8 volts due to the on-set of injection from the P-N junction, which is not observed for the Schottky rectifier. These results are very close to those predicted by the analytical model (see Fig. 3.7) confirming its utility for analysis of the JBS rectifier structure in the on-state.

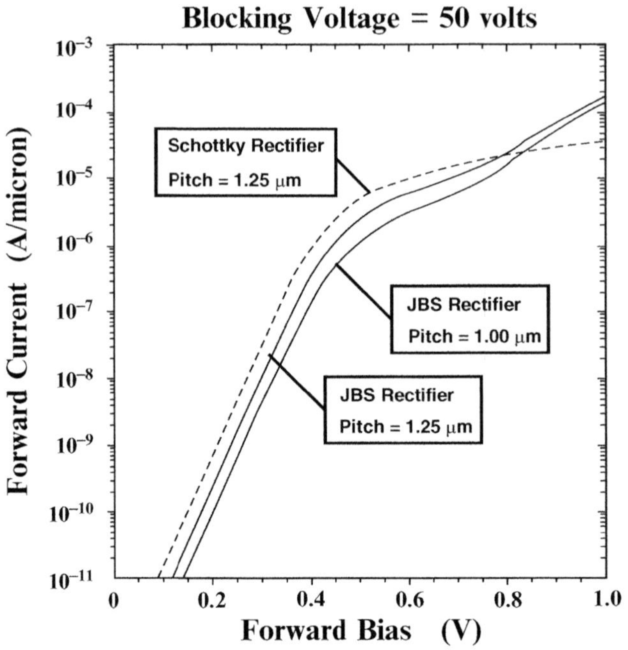


Fig. 3.4E On-State Characteristics for 50V Silicon JBS Rectifiers.

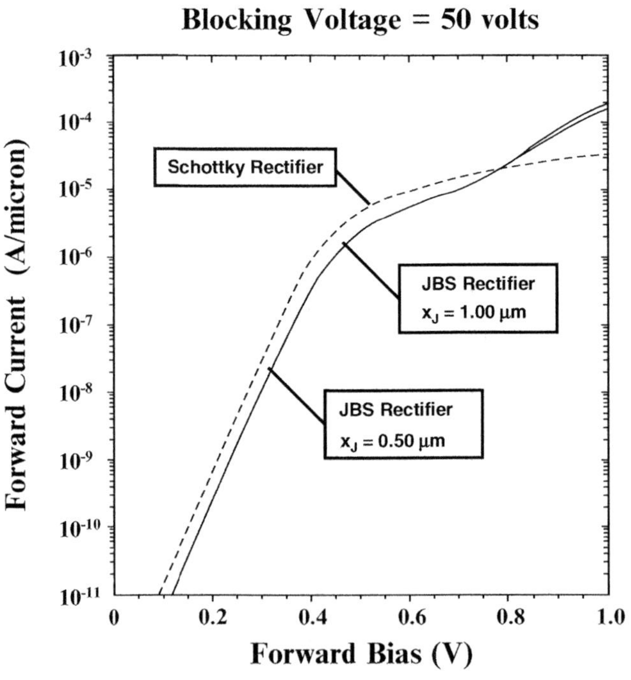


Fig. 3.5E On-State Characteristics for 50V Silicon JBS Rectifiers.

In the JBS rectifier, an increase in the junction depth improves the shielding of the Schottky contact. However, it also increases the resistance of the current path from the Schottky contact to the cathode. The impact of increasing the junction depth from 0.5 to 1.0 micron in the JBS rectifier structure was examined using numerical simulations by simultaneously increasing the cell pitch by 0.5 microns to retain the same Schottky contact size. The on-state *i-v* characteristics for this structure are compared with the structure with 0.5 micron junction depth in Fig. 3.5E. The i-v characteristics of the structures coincide because they have the same Schottky contact size and their cell resistances are almost equal. However, it is important to recognize that the cell area for the two structures is not equal. After accounting for this difference in cell areas, the on-state voltage drop at the same on-state current density of 100 A/cm$^2$ is not equal for the JBS rectifier structures. The on-state voltage drop for the structure with a junction depth of 1.0 microns is found to be larger than that for the structure with junction depth of 0.5 microns by 0.02 volts. This is in excellent agreement with the predictions of the analytical model (see Fig. 3.8).

### 3.2.5 Silicon Carbide JBS Rectifier: Forward Conduction Model

As discussed earlier for the silicon JBS rectifier structure, the three models produce close to the same on-state characteristics. Only a single model for the silicon carbide JBS rectifier is therefore developed in this section based upon the current flow pattern illustrated in Fig. 3.9 with the shaded area. This model, similar to Model C for the silicon JBS rectifier, accounts for the absence of lateral diffusion in silicon carbide junctions. In this model, it is assumed that the current

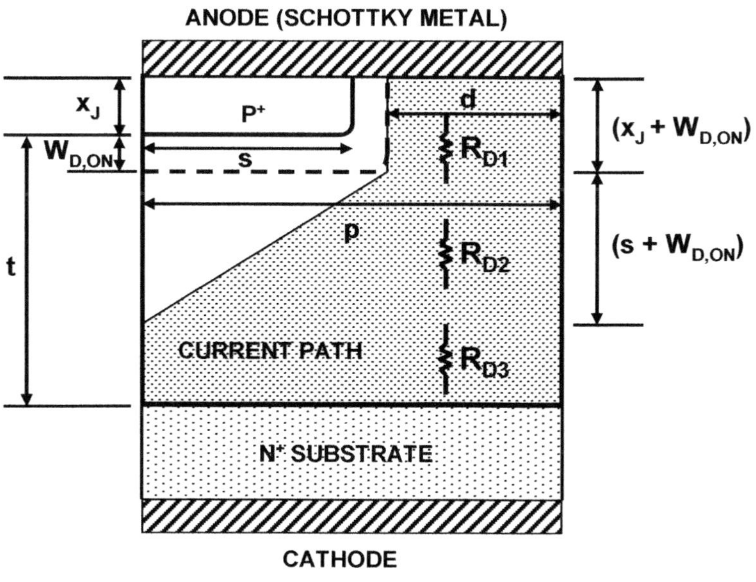


Fig. 3.9 Current Flow Pattern in the Silicon Carbide JBS Rectifier Structure during Operation in the On-State.

through the Schottky contact flows only within the un-depleted portion (with dimension 'd') of the drift region at the top surface. Consequently, the current density at the Schottky contact ($J_{FS}$) is related to the cell (or cathode) current density ($J_{FC}$) by:

$$J_{FS} = \left(\frac{p}{d}\right) J_{FC} \quad [3.25]$$

where p is the cell pitch. The dimension 'd' is determined by the cell pitch (p), the size of the ion-implant window (s), and the depletion width ($W_{D,ON}$):

$$d = p - s - W_{D,ON} \quad [3.26]$$

In deriving this equation, it has been assumed that there is no straggle in the ion-implant. Depending up on the lithography used for device fabrication to minimize the size (dimension 's') of the $P^+$ region, the current density at the Schottky contact can be enhanced by a factor of two or more. This must be taken into account when computing the voltage drop across the Schottky contact:

$$V_{FS} = \phi_B + \frac{kT}{q} \ln\left(\frac{J_{FS}}{AT^2}\right) \quad [3.27]$$

After flowing across the Schottky contact, the current flows through the un-depleted portion of the drift region. In the model, it is assumed that the current flows through a region with a uniform width 'd' until it reaches the bottom of the depletion region and then spreads to the entire cell pitch (p) at a 45 degree spreading angle. The current paths overlap at a distance (s + $W_{D,ON}$) from the bottom of the depletion region. The current then flows through a uniform cross-sectional area.

The net resistance to current flow can be calculated by adding the resistance of the three segments. The resistance of the first segment of uniform width 'd' is given by:

$$R_{D1} = \frac{\rho_D.(x_J + W_{D,ON})}{d.Z} \quad [3.28]$$

The resistance of the second segment can be derived by using the same approach used for Model C for the silicon JBS rectifier:

$$R_{D2} = \frac{\rho_D}{Z} \ln\left(\frac{p}{d}\right) \quad [3.29]$$

The resistance of the third segment with a uniform cross-section of width p is given by:

$$R_{D3} = \frac{\rho_D.(t - s - 2W_{D,ON})}{p.Z} \qquad [3.30]$$

The specific resistance for the drift region can be calculated by multiplying the cell resistance ($R_{D1} + R_{D2} + R_{D3}$) by the cell-area (p.Z):

$$R_{sp,drift} = \frac{\rho_D.p.(x_J + W_{D,ON})}{d} + \rho_D.p.\ln\left(\frac{p}{d}\right) + \rho_D.(t - s - 2W_{D,ON}) \qquad [3.31]$$

In addition, it is important to include the resistance associated with the thick, highly doped $N^+$ substrate because this is substantially larger than for silicon devices. The specific resistance of the $N^+$ substrate can be determined by taking the product of its resistivity and thickness. For 4H-SiC, the lowest available resistivity for $N^+$ substrates is 20 mΩ-cm. If the thickness of the substrate is 200 microns, the specific resistance contributed by the $N^+$ substrate is 4 x $10^{-4}$ Ω-cm$^2$. The on-state voltage drop for the JBS rectifier at a forward cell current density $J_{FC}$, including the substrate contribution, is then given by:

$$V_F = \phi_B + \frac{kT}{q}\ln\left(\frac{J_{FS}}{AT^2}\right) + (R_{sp,drift} + R_{sp,subs}).J_{FC} \qquad [3.32]$$

The depletion layer width for silicon carbide is also substantially larger than for silicon due to the larger built-in potential. However, these structures operate at a relatively larger on-state voltage drop of around 1 volt. Consequently, when computing the on-state voltage drop using the above equation, it is satisfactory to make the approximation that the depletion layer width can be computed by subtracting an on-state voltage drop of 1 volt from a built-in potential of about 3.2 volts for the P-N junction:

$$W_{D,ON} = \sqrt{\frac{2\varepsilon_S(V_{bi} - 1.0)}{qN_D}} \qquad [3.33]$$

where $V_{bi}$ is the built-in potential of the P-N junction. Due to the abrupt junctions formed in silicon carbide by the ion implant process, it is appropriate to assume that the entire depletion occurs on the lightly doped N-side of the junction. For a 4H-SiC JBS rectifier fabricated using a doping concentration of 1 x $10^{16}$ cm$^{-3}$, corresponding to a breakdown voltage of about 3000 V, the zero bias depletion width is about 0.5 microns. Based upon this, it can be concluded that it is even more important to take the depletion width into account for silicon carbide JBS rectifiers.

The forward characteristics of 3kV JBS rectifiers, calculated using the above analytical model with a Schottky barrier height of 0.8 eV, are shown in Fig. 3.10 with the cell pitch (p) as a parameter. The width of the $P^+$ region (2s) was kept at 1.0 microns for this analysis because the small diameters for the silicon carbide wafers preclude the use of smaller geometries. A junction depth of 0.5 microns was used with a drift region thickness of 20 microns below the junction to support the 3000 volts. In comparison with the Schottky rectifier characteristics (shown by the dashed line in the figure), the increase in on-state voltage drop at a forward current density of 100 A/cm$^2$ is small (less than 0.1 volts) as long as the cell pitch is more than 1.1 microns. This cell pitch is sufficient to obtain substantial reduction of the electric field at the metal-semiconductor contact as shown later in the chapter.

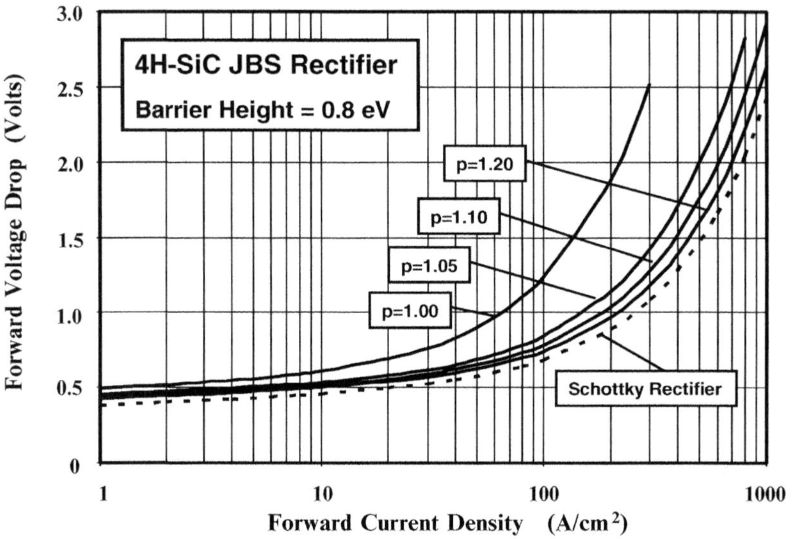


Fig. 3.10 Forward Characteristics of 3kV 4H-SiC JBS Rectifiers.

## Simulation Example

In order to validate the above model for the on-state characteristics of the silicon carbide JBS rectifier, the results of two-dimensional numerical simulations on a 3000 V structure are described here. The structure had a drift region thickness of 20 microns with a doping concentration of 1 x 10$^{16}$ cm$^{-3}$. The P$^+$ region had a depth of 0.5 microns with an ion-implant window (dimension s in Fig. 3.9) of 0.5 microns. No lateral diffusion was assumed to occur for the implanted region.

The forward *i-v* characteristics of a 3000 V JBS 4H-SiC rectifier obtained from the numerical simulations are shown in Fig. 3.6E for the case of a cell pitch of 1.25 microns and contact with work-function of 4.5 eV. This corresponds to a Schottky barrier height of about 0.8 eV based upon an electron affinity of 3.7 eV for 4H-SiC. The on-state voltage drop at a current density of 100 A/cm$^2$ is 0.7 volts which is similar to the results obtained using the analytical model in the previous section providing validation for the model. It is worth pointing out that the diode exhibits the desirable positive temperature coefficient for the on-state voltage drop with no kinks in the characteristics.

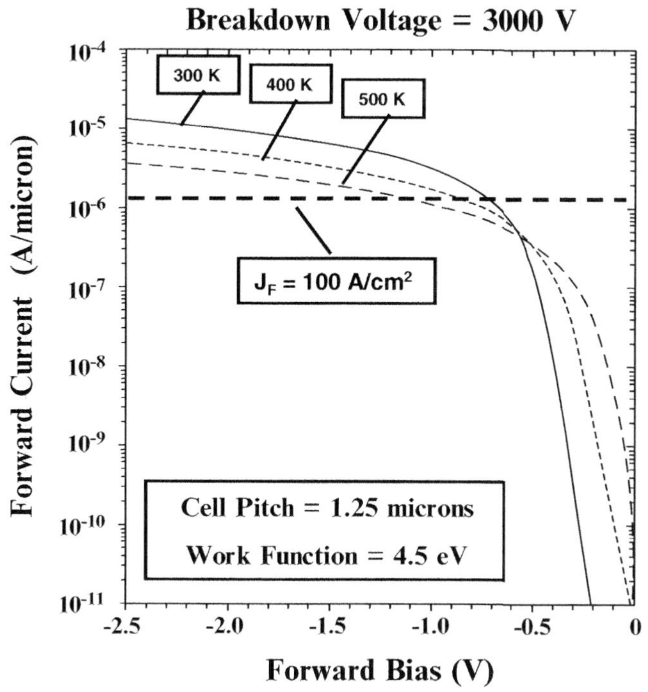


Fig. 3.6E Forward Characteristics of 3 kV 4H-SiC JBS Rectifier.

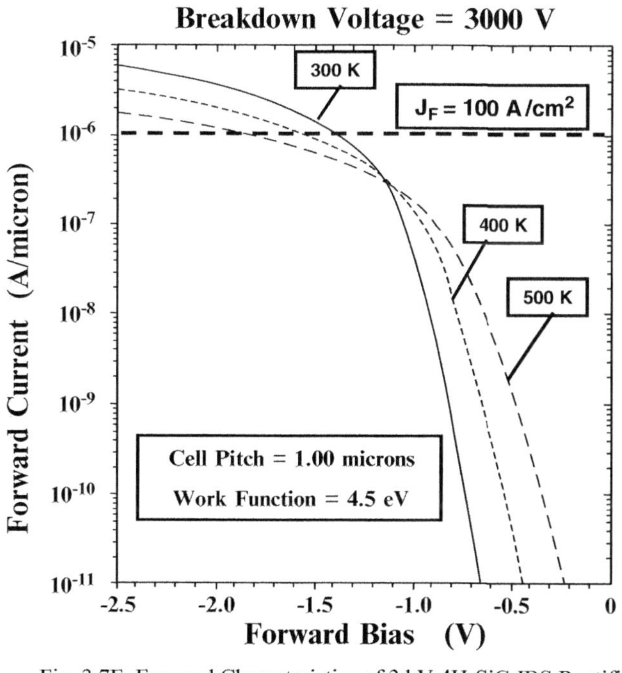


Fig. 3.7E Forward Characteristics of 3 kV 4H-SiC JBS Rectifier.

However, the forward characteristics were found to degrade substantially as shown in Fig. 3.7E, when the cell pitch was reduced to 1.0 microns, as predicted by the analytical model. The on-state voltage drop increases by 0.7 volts when the cell pitch is reduced, which is also consistent with the analytical model. The reason for the increase in the on-state voltage drop is the much greater constriction of the current at the Schottky contact for the smaller cell pitch because the dimension 'd' is reduced to only 0.014 microns for the pitch of 1.00 microns. It will be shown later in this chapter that a cell pitch of 1.2 microns is adequate to suppress the electric field at the Schottky contact.

## 3.3 JBS Rectifier Structure: Reverse Leakage Model

The reverse leakage current in the JBS rectifier is reduced when compared with the Schottky rectifier due to the smaller electric field at the metal-semiconductor interface. In addition, the area of the Schottky contact is a fraction of the total cell area resulting in a smaller reverse current contribution. In the case of the silicon JBS rectifier, the reduced electric field at the Schottky contact suppresses the barrier lowering effect. In the case of the silicon carbide JBS rectifier, the reduced electric field not only decreases the barrier lowering but also mitigates the influence of thermionic field emission. The impact of the reduction of the electric field at the Schottky contact in the JBS rectifier structure on the leakage current is analyzed in this section. The presence of the P-N junction also diverts the avalanche multiplication current from the Schottky contact reducing the pre-breakdown multiplication effect.

### 3.3.1 Silicon JBS Rectifier: Reverse Leakage Model

In the case of the JBS rectifier, the leakage current model must take into account the smaller Schottky contact area within the cell and the influence of the smaller electric field generated at the Schottky contact due to the shielding from the P-N junction. The leakage current for the silicon JBS rectifier is therefore given by:

$$J_L = \left(\frac{p - s - x_J}{p}\right) AT^2 \exp\left(-\frac{q\phi_b}{kT}\right).\exp\left(\frac{q\beta\Delta\phi_{bJBS}}{kT}\right) \quad [3.34]$$

where β is a constant to account for the smaller barrier lowering closer to the P-N junction as discussed below. In the previous chapter on Schottky rectifiers, it was demonstrated that the high electric field at the Schottky contact produces a reduction of the effective barrier height due to the image force lowering phenomenon. In contrast with the Schottky rectifier, the barrier lowering for the JBS rectifier is determined by the reduced electric field $E_{JBS}$ at the contact:

$$\Delta\phi_{bJBS} = \sqrt{\frac{qE_{JBS}}{4\pi\varepsilon_S}} \quad [3.35]$$

The electric field at the Schottky contact varies with distance away from the P-N junction. The highest electric field is observed at the middle of the Schottky contact with a progressively smaller value closer to the P-N junction. In an analytical model with a worst case scenario, it is prudent to use the electric field at the middle of the contact to compute the leakage current. Until the depletion regions from the adjacent P-N junctions produce a potential barrier under the Schottky contact, the electric field at the metal-semiconductor interface in the middle of the contact increases with the applied reverse bias voltage as in the case of the Schottky rectifier. A potential barrier is established by the P-N junctions after depletion of the drift region below the Schottky contact. The voltage at which the depletion regions from the adjacent junctions intersect under the Schottky contact is referred to as the *pinch-off voltage*. The pinch-off voltage ($V_P$) can be obtained from the device cell parameters:

$$V_P = \frac{qN_D}{2\varepsilon_S}(p - s - x_J)^2 - V_{bi} \qquad [3.36]$$

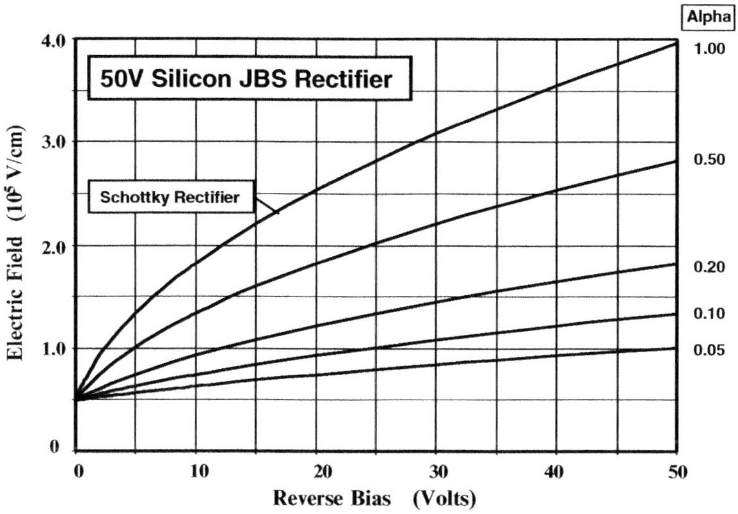


Fig. 3.11 Electric Field at the Schottky Contact for 50 V Silicon JBS Rectifiers.

Although a potential barrier begins to form after the reverse bias exceeds the pinch-off voltage, the electric field continues to rise at the Schottky contact due to encroachment of the potential to the Schottky contact. This problem is more acute for the silicon JBS rectifier because of the open shape of the planar junction. In order to analyze the impact of this on the reverse leakage current, the electric field $E_{JBS}$ can be related to the reverse bias voltage by:

$$E_{JBS} = \sqrt{\frac{2qN_D}{\varepsilon_S}(\alpha V_R + V_C)} \qquad [3.37]$$

where $\alpha$ is a coefficient used to account for the build up in the electric field after pinch-off and $V_C$ is the Schottky contact potential.

As an example, consider the case of the 50 volt silicon JBS rectifier discussed earlier in the chapter with a cell pitch (p) of 1.25 microns and a $P^+$ region with dimension 's' of 0.25 microns. The pinch-off voltage for this structure is only 1 volt for a drift region with doping concentration of $1 \times 10^{16}$ cm$^{-3}$. Due to the two-dimensional nature of the planar P-N junction in the silicon JBS rectifier structure, it is difficult to derive an analytical expression for alpha. However, the reduction of the electric field at the Schottky contact can be predicted by assuming various values for alpha in Eq. [3.37]. The results are shown in Fig. 3.11 for alpha values ranging between 0.05 and 1.00. An alpha of unity corresponds to the Schottky rectifier structure with no shielding. It can be observed that substantial reduction of the electric field at the Schottky contact is obtained as alpha is reduced.

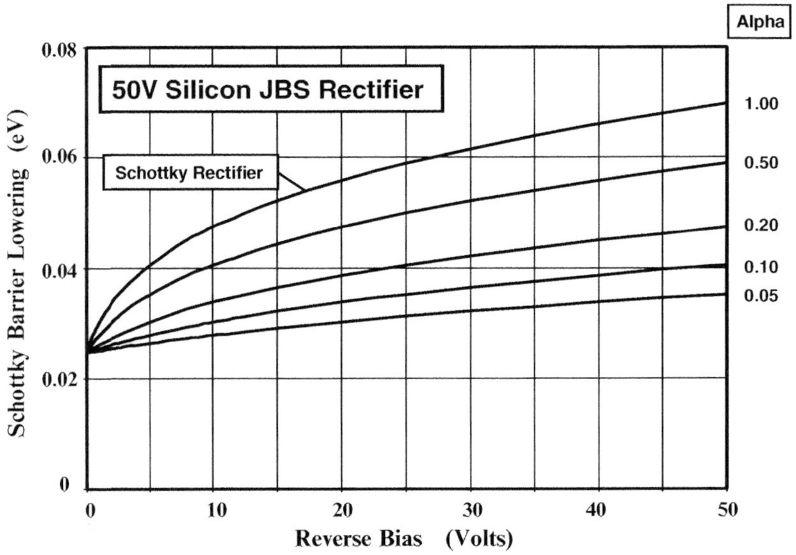


Fig. 3.12 Schottky Barrier Lowering in 50 V Silicon JBS Rectifiers with various Alpha Coefficients.

The impact of the reduction of the electric field at the Schottky contact on the Schottky barrier lowering is shown in Fig. 3.12. Without the shielding by the P-N junction, a barrier lowering of 0.07 eV occurs in the Schottky rectifier. The barrier lowering is reduced to 0.05 eV with an alpha of 0.2 in the JBS rectifier structure. Although this may appear to be a small change, it has a large impact on the reverse leakage current as shown in Fig. 3.13. For these plots, a value of 0.7 was assumed for the constant $\beta$ based upon the results of the numerical simulations discussed below. For the JBS structure with pitch of 1.25 microns, an implant window (2s) of 0.5 microns, and junction depth of 0.5 microns, the Schottky contact area is reduced to only 40 percent of the cell area. This results in a proportionate reduction of leakage current at low reverse bias voltages. The suppression of the Schottky barrier lowering and pre-breakdown multiplication, by the presence of the P-N junction, reduces the rate of increase in leakage current with increasing reverse bias. The net effect is a reduction in leakage current density by a factor of 35x when the reverse bias reaches 50 volts for the case of an alpha of 0.5. This demonstrates that a very large improvement in reverse power dissipation can be achieved with the JBS rectifier structure.

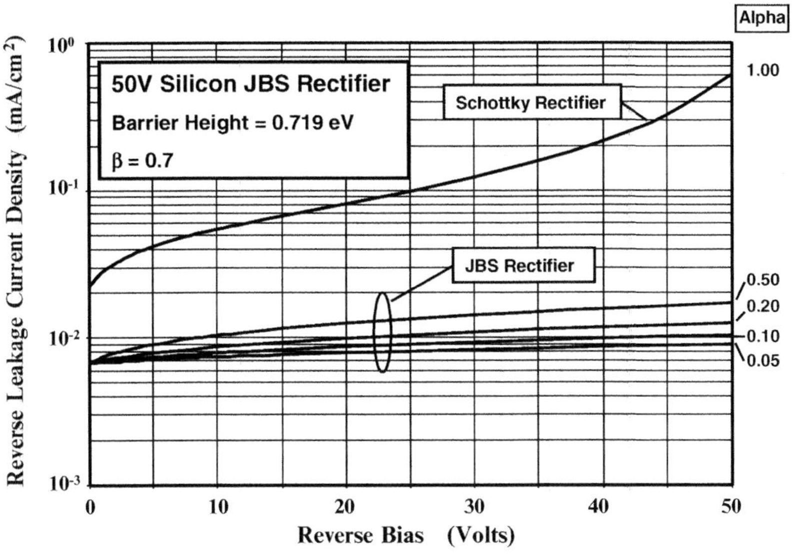


Fig. 3.13 Reverse Leakage Current for 50 V Silicon JBS Rectifiers with various Alpha coefficients.

## Simulation Example

In order to validate the above model for the reverse characteristics of the silicon JBS rectifier, the results of two-dimensional numerical simulations on a 50 V structure are described here. The structure had a drift region with a doping concentration of $8 \times 10^{15}$ $cm^{-3}$ and a thickness of 3 microns. The $P^+$ region had a depth of 0.5 microns with an ion-implant window (dimension s in Fig. 3.5) of 0.25 microns. The work function of the Schottky metal was chosen to obtain a barrier height of 0.65 eV.

A three dimensional view of the electric field distribution in the JBS rectifier cell is shown in Fig. 3.8E. The Schottky contact is located on the lower right-hand-side in the figure with the $P^+$ region located at the top of the figure. A high electric field ($4 \times 10^5$ V/cm) is observed at the P-N junction. However, the electric field at the middle of the Schottky contact is greatly reduced ($2.45 \times 10^5$ V/cm). It can also be seen that the electric field becomes smaller when proceeding towards the P-N junction. Consequently, a worst case analysis of the leakage current can be performed by using the electric field at the middle of the Schottky contact.

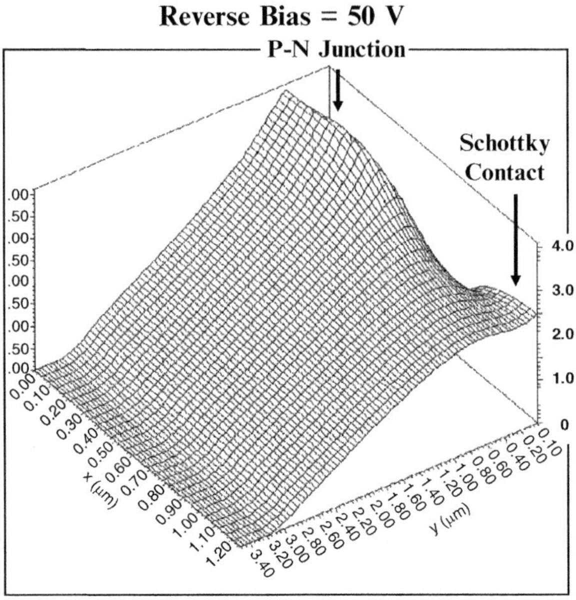


Fig. 3.8E Electric Field Distribution in a 50 V Silicon JBS Rectifier.

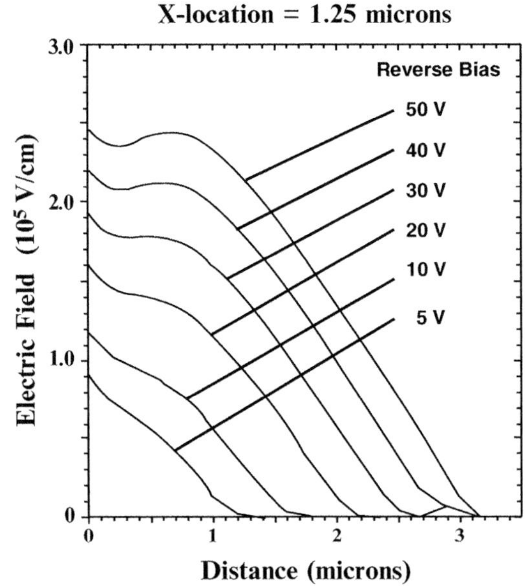


Fig. 3.9E Growth of the Electric Field at the Middle of the Schottky Contact in a 50 V Silicon JBS Rectifier.

The increase in the electric field at the middle of the Schottky contact in the JBS rectifier structure with cell pitch of 1.25 microns is shown in Fig. 3.9E. To contrast this behavior with the Schottky diode, the growth of the electric field for the Schottky rectifier is shown in Fig. 3.10E. From these figures, it is apparent that the electric field at the Schottky contact is suppressed in the JBS rectifier due to the incorporation of the P-N junction. An even greater suppression of the electric field at the Schottky contact can be obtained by reducing the cell pitch as shown in Fig. 3.11E for a structure with pitch of 1.00 microns.

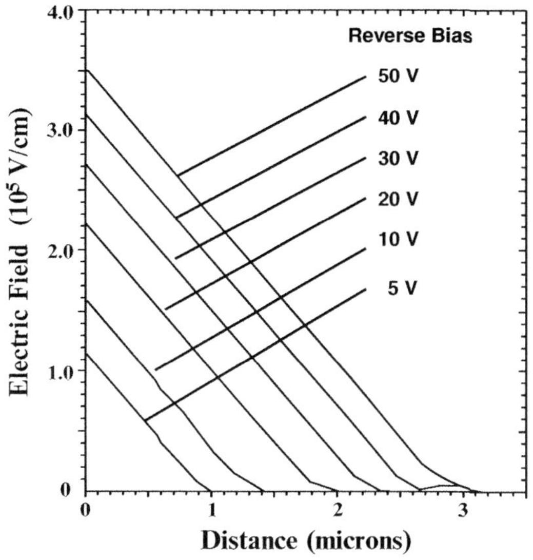


Fig. 3.10E Growth of the Electric Field in a 50 V Silicon Schottky Rectifier.

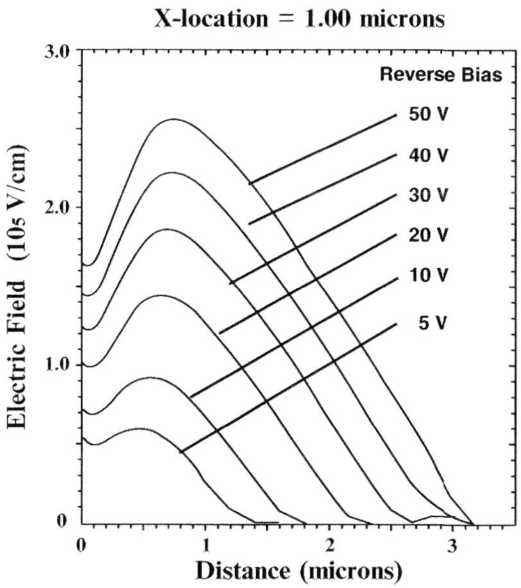


Fig. 3.11E Growth of the Electric Field at the Middle of the Schottky Contact in a 50 V Silicon JBS Rectifier.

The coefficient alpha, that governs the rate at which the electric field increases at the middle of the Schottky contact in the analytical model for the JBS rectifier structure, can be extracted from the results of the two-dimensional numerical simulations. The increase in the electric field at the middle of the Schottky contact, obtained from the numerical simulations, is shown in Fig. 3.12E for the JBS rectifiers with cell pitch (p) of 1.25 and 1.00 microns by the symbols. The results of calculations based upon using the analytical equation [3.37] are shown by the solid lines with the values fro alpha adjusted to fit the results of the numerical simulations. The case with alpha of unity fits the Schottky rectifier quite well as expected. The value for alpha for the JBS rectifier with pitch of 1.25 microns is found to be 0.45 while that for a pitch of 1.00 microns is 0.18. With these values of alpha, the analytical model accurately predicts the behavior of the electric field at the middle of the Schottky contact. It can therefore be used to compute the Schottky barrier lowering and leakage current in JBS rectifiers.

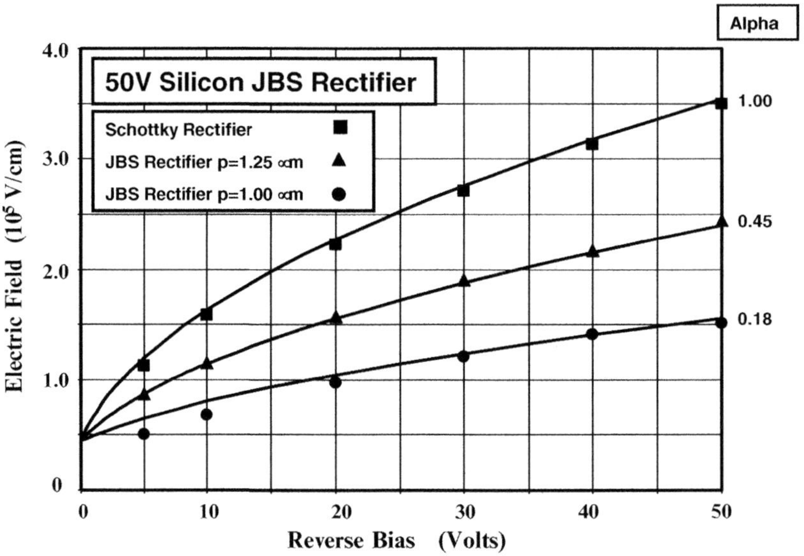


Fig. 3.12E Growth of the Electric Field at the Middle of the Schottky Contact for Silicon 50V JBS Rectifiers.

The reverse *i-v* characteristics of the 50 V Silicon JBS rectifier with a cell pitch of 1.25 microns obtained from the numerical simulations are shown in Fig. 3.13E together with those for the Schottky rectifier for comparison. Both devices had a cross-sectional width of 1.25 microns with a depth of 1 micron. The breakdown voltage for the JBS rectifier cell is observed to be 64 volts. This is consistent with an edge termination limited breakdown voltage of 50 volts at 80 percent of the parallel-plane breakdown voltage. The leakage current at small reverse bias voltages is 2.5x times smaller in the JBS rectifier structure. This is consistent with the reduction of the Schottky contact area by a factor of 2.5x in this JBS rectifier structure. The leakage current for the Schottky rectifier increases by a factor of 100x when the reverse voltage increases to 60 volts. In contrast, the reverse leakage current for this JBS rectifier increases by a factor of only 4x when the reverse voltage is increased to 60 volts. This increase is consistent with predictions of the analytical model. Based upon the simulation results, the leakage current for the JBS rectifier with pitch of 1.25 microns is 75x smaller than for the Schottky rectifier at a reverse bias of 60 volts.

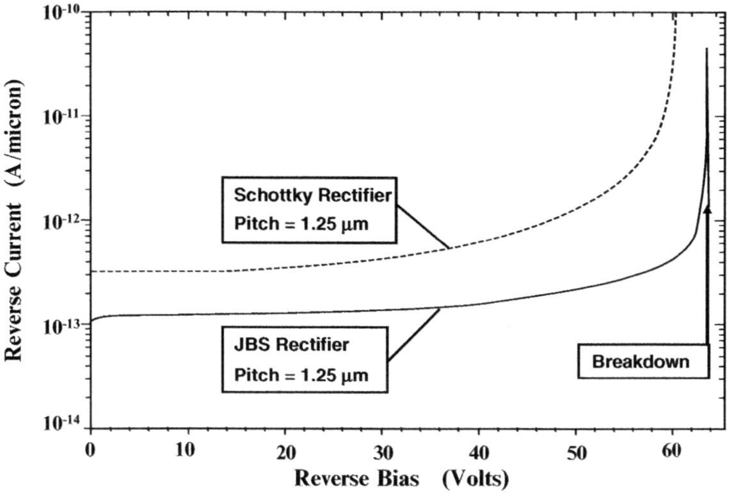


Fig. 3.13E Reverse Blocking Characteristics for a Silicon 50V JBS Rectifier.

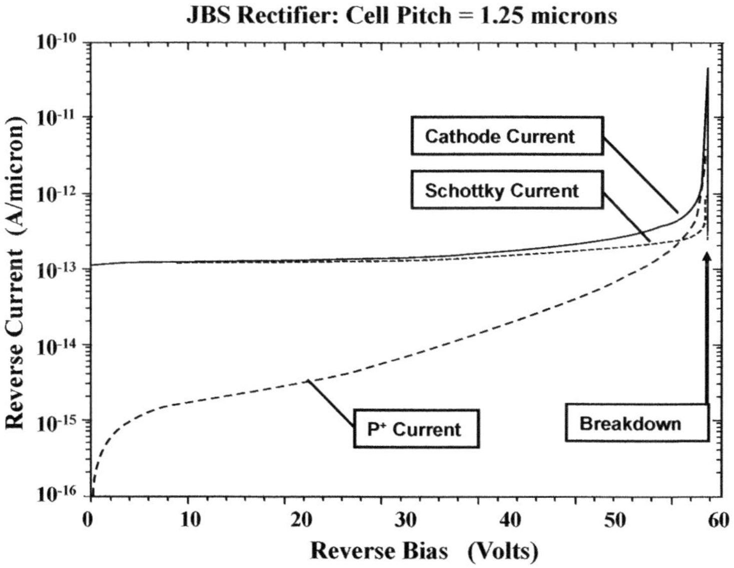


Fig. 3.14E Reverse Blocking Characteristics for a Silicon 50V JBS Rectifier.

The numerical simulations provide insight into the current flow within the JBS rectifier structure during the reverse blocking mode. The total reverse current flowing through the cathode electrode is compared with the currents flowing through the Schottky contact and the contact to the $P^+$ region in Fig. 3.14E. At reverse bias voltages below 50 volts, the cathode current is essentially equal to the current flowing through the Schottky contact. The current flowing through the contact to the $P^+$ region increases rapidly when the reverse voltage exceeds 30 volts and becomes comparable to the current in the Schottky contact at reverse bias voltages above 55 volts. It is worth pointing out that the breakdown occurs due to a very rapid increase in the current flowing through the contact to the $P^+$ region indicating that the avalanche multiplication occurs below the $P^+$ region due to the much larger local electric field as shown in Fig. 3.8E. This shields the Schottky contact against damage if the JBS rectifier is forced into breakdown during circuit operation.

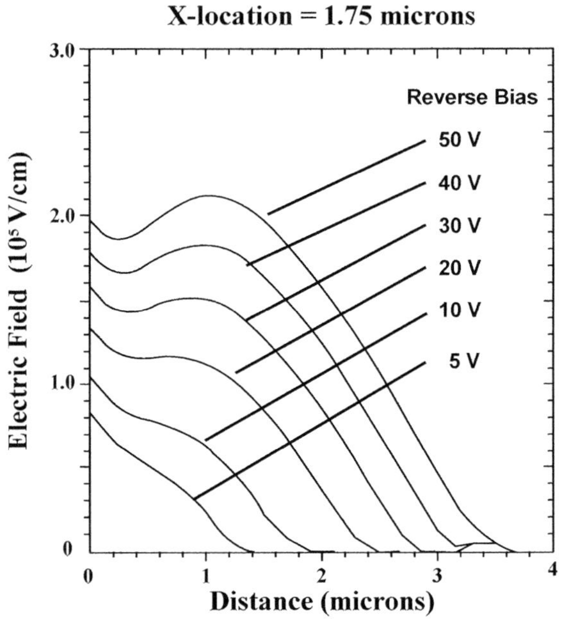


Fig. 3.15E Growth of the Electric Field at the Middle of the Schottky Contact in a 50 V Silicon JBS Rectifier.

The shielding of the Schottky contact in the JBS rectifier structure can be enhanced by using a larger junction depth. In order to illustrate this, the case of increasing the junction depth to 1.00 microns is considered here. For comparison purposes, the size of the Schottky contact for this structure was chosen to be equal to that for the JBS rectifier structure with a pitch of 1.25 microns and junction depth of 0.5 microns. The electric field profile for the structure with junction depth of 1.00 microns is shown in Fig. 3.15E for various reverse bias voltages. By comparison with the electric field profiles in Fig. 3.9E for the structure with junction depth of 0.5 microns, it can be observed that the electric field at the middle of the Schottky contact has been reduced. This indicates that the value for alpha has been reduced by the larger junction depth.

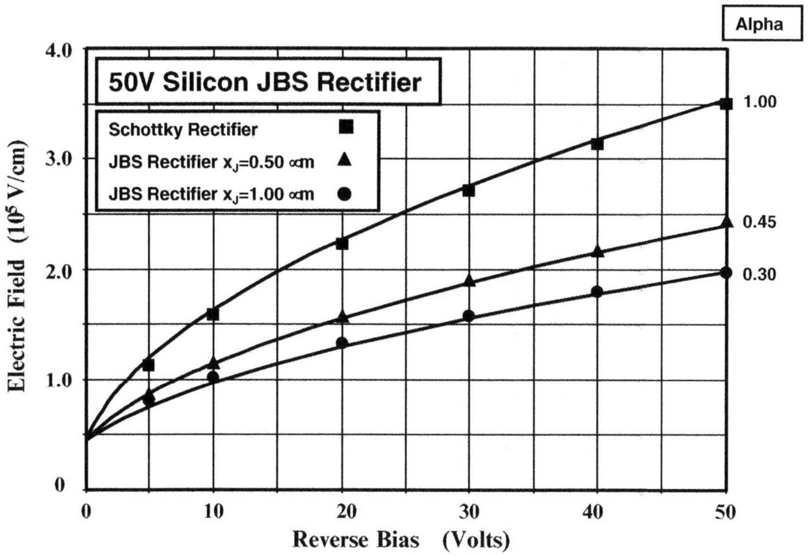


Fig. 3.16E Growth of the Electric Field at the Middle of the Schottky Contact for Silicon 50V JBS Rectifiers.

The growth of the electric field with reverse bias voltage for the JBS rectifier structures with the different junction depths is compared in Fig. 3.16E. In this figure, the value for alpha in the analytical model was adjusted to match the simulation data. It can be seen that the alpha is reduced from 0.45 to 0.3 with the larger junction depth due to the larger aspect ratio. This is beneficial for suppressing the Schottky barrier lowering and decreasing the leakage current. A deeper junction can also be utilized for the edge termination to enhance the breakdown voltage.

### 3.3.2 Silicon Carbide JBS Rectifier: Reverse Leakage Model

The leakage current in the silicon carbide JBS rectifier can be calculated using the same approach as for the silicon JBS rectifier structure. Firstly, it is important to account for the smaller Schottky contact area in the JBS rectifier cell. Secondly, it is necessary to include Schottky barrier lowering while accounting for the smaller electric field at the Schottky contact due to shielding by the P-N junction. Third, the thermionic field emission current must be included while accounting for the smaller electric field at the Schottky contact due to shielding by the P-N junction. After making these adjustments, the leakage current for the silicon carbide JBS rectifier can be calculated by using:

$$J_L = \left(\frac{p-s}{p}\right) AT^2 \exp\left(-\frac{q\phi_b}{kT}\right).\exp\left(\frac{q\Delta\phi_{bJBS}}{kT}\right).\exp\left(C_T E_{JBS}^2\right) \qquad [3.38]$$

where $C_T$ is a tunneling coefficient ($8 \times 10^{-13}$ $cm^2/V^2$ for 4H-SiC). In contrast to the Schottky rectifier, the barrier lowering for the JBS rectifier is determined by the reduced electric field $E_{JBS}$ at the contact:

$$\Delta\phi_{bJBS} = \sqrt{\frac{qE_{JBS}}{4\pi\varepsilon_S}} \qquad \textbf{[3.39]}$$

As in the case of the silicon JBS structure, the electric field at the Schottky contact varies with distance away from the P-N junction. The highest electric field is observed at the middle of the Schottky contact with a progressively smaller value closer to the P-N junction. When developing an analytical model with a worst case scenario, it is prudent to use the electric field at the middle of the contact to compute the leakage current.

Until the depletion regions from the adjacent P-N junctions produce a potential barrier under the Schottky contact, the electric field at the metal-semiconductor interface in the middle of the contact increases with the applied reverse bias voltage as in the case of the Schottky rectifier. A potential barrier is established by the P-N junctions after depletion of the drift region below the Schottky contact. As in the case of the silicon JBS rectifier structure, the pinch-off voltage ($V_P$) can be obtained from the device cell parameters:

$$V_P = \frac{qN_D}{2\varepsilon_S}(p-s)^2 - V_{bi} \qquad \textbf{[3.40]}$$

It is worth pointing out that the built-in potential for 4H-SiC is much larger than for silicon. Although the potential barrier begins to form after the reverse bias exceeds the pinch-off voltage, the electric field continues to rise at the Schottky contact due to encroachment of the potential to the Schottky contact. This problem is less acute for the silicon carbide structure than in the silicon JBS rectifier because of the rectangular shape of the P-N junction resulting from the very low diffusion coefficients for dopants in 4H-SiC. In order to analyze the impact of this on the reverse leakage current, the electric field $E_{JBS}$ can be related to the reverse bias voltage by:

$$E_{JBS} = \sqrt{\frac{2qN_D}{\varepsilon_S}(\alpha V_R + V_{bi})} \qquad \textbf{[3.41]}$$

where $\alpha$ is a coefficient used to account for the build up in the electric field after pinch-off.

As an example, consider the case of the 3kV silicon carbide JBS rectifier discussed earlier in the chapter with a cell pitch (p) of 1.25 microns and a $P^+$ region with dimension 's' of 0.5 microns. The pinch-off voltage for this structure is only 2 volts for a drift region with doping concentration of $1 \times 10^{16}$ $cm^{-3}$. Due to the two-dimensional nature of the P-N junction in the JBS rectifier structure, it is difficult to derive an analytical expression for alpha. However, the reduction of the electric field at the Schottky contact can be predicted by assuming various values for alpha in Eq. [3.41]. The results are shown in Fig. 3.14 for alpha values ranging between 0.05 and 1.00. An alpha of unity corresponds to the Schottky rectifier structure with no shielding. It can be observed that substantial reduction of the electric field at the Schottky contact is obtained as alpha is reduced.

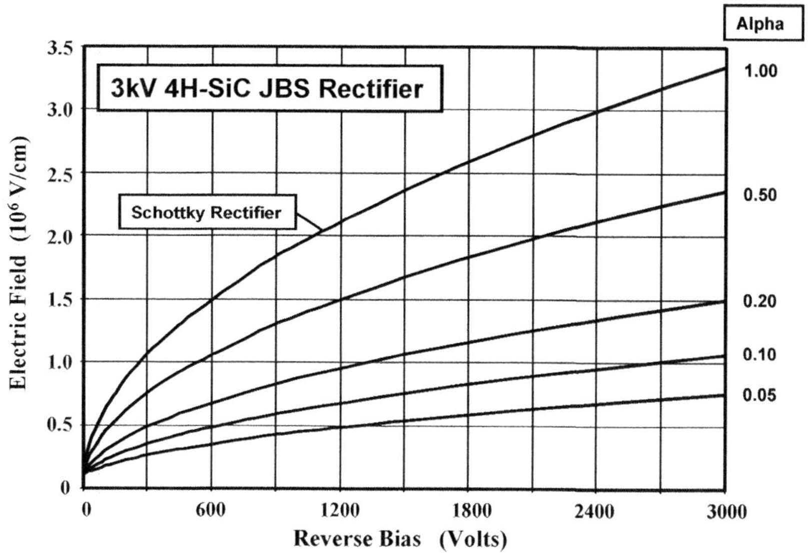


Fig. 3.14 Electric Field at the Schottky Contact for 3 kV Silicon Carbide JBS Rectifiers.

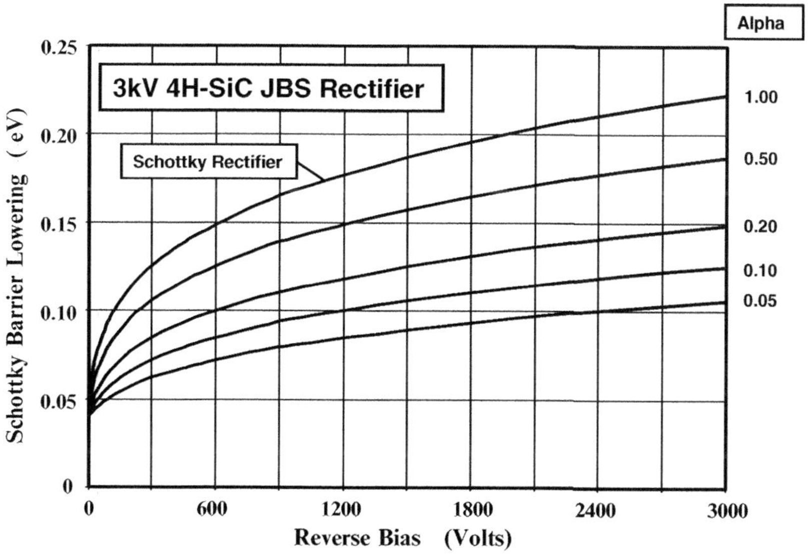


Fig. 3.15 Schottky Barrier Lowering in 3 kV Silicon Carbide JBS Rectifiers with various Alpha Coefficients.

The impact of the reduction of the electric field at the Schottky contact, due to the shielding by the P-N junction in the JBS structure, on the Schottky barrier lowering is shown in Fig. 3.15. Without the shielding by the P-N junction, a barrier lowering of 0.22 eV occurs in the Schottky rectifier. This is much greater than for silicon devices due to the larger electric field at the contact. The barrier lowering is reduced to 0.15 eV with an alpha of 0.2 in the 4H-SiC JBS rectifier structure. These smaller values for alpha are appropriate for the silicon carbide structure because the rectangular shape of the P-N junction favors a stronger shielding of the Schottky contact.

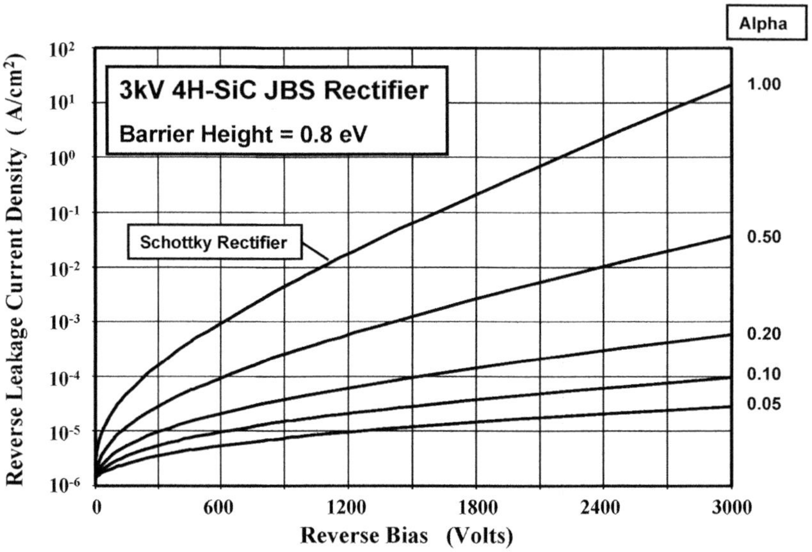


Fig. 3.16 Reverse Leakage Current for 3 kV Silicon Carbide JBS Rectifiers with various Alpha coefficients.

As discussed in Chapter 4, the larger barrier lowering for silicon carbide, in conjunction with the thermionic field emission current, results in an increase in leakage current by six-orders of magnitude when the voltage increases to 2500 volts for the 3 kV Schottky rectifier. This is reproduced in Fig. 3.16, as the plot with alpha of unity by using a barrier height of 0.8 eV. It can be seen that the leakage current is greatly reduced by the shielding in the JBS rectifier structure. For the 4H-SiC JBS rectifier structure with pitch of 1.25 microns, an implant window (s) of 0.5 microns, and junction depth of 0.5 microns, the Schottky contact area is reduced to 60 percent of the cell area. This results in a proportionate reduction of leakage current at low reverse bias voltages. More importantly, the suppression of the electric field at the Schottky contact, by the presence of the P-N junction, greatly reduces the rate of increase in leakage current with increasing reverse bias. The net effect is a reduction in leakage current density by a factor of 570x when the reverse bias reaches 3 kV for the case of an alpha of 0.5 and a factor of 36,000x for an alpha of 0.2. This demonstrates that a very large improvement in reverse power dissipation can be achieved with the 4H-SiC JBS rectifier structure with a modest increase in the on-state voltage drop.

However, the leakage current density observed with a Schottky barrier height of 0.8 eV leads to a power dissipation of 100 W/cm$^2$ at a reverse bias of 3000 volts for a JBS rectifier with alpha of 0.5 even at 300 °K. The power dissipation in the reverse blocking mode can be reduced by increasing the Schottky barrier height. As an example, if the barrier height is increased from 0.8 to 1.1 eV, the leakage current is reduced by 5-orders of magnitude as shown in Fig. 3.17. This is adequate to reduce the power dissipation in the reverse blocking mode to below that in the forward conduction mode even at elevated temperatures ensuring stable operation of the rectifier. This change in barrier height will increase the on-state voltage drop by 0.3 volts by shifting the *i-v* characteristics, shown earlier in Fig. 3.10, to larger voltages. The on-state voltage drop for the 4H-SiC JBS rectifiers with a barrier height of 1.1 eV is only 1 volt, which is a very good value for a rectifier designed to support 3000 volts.

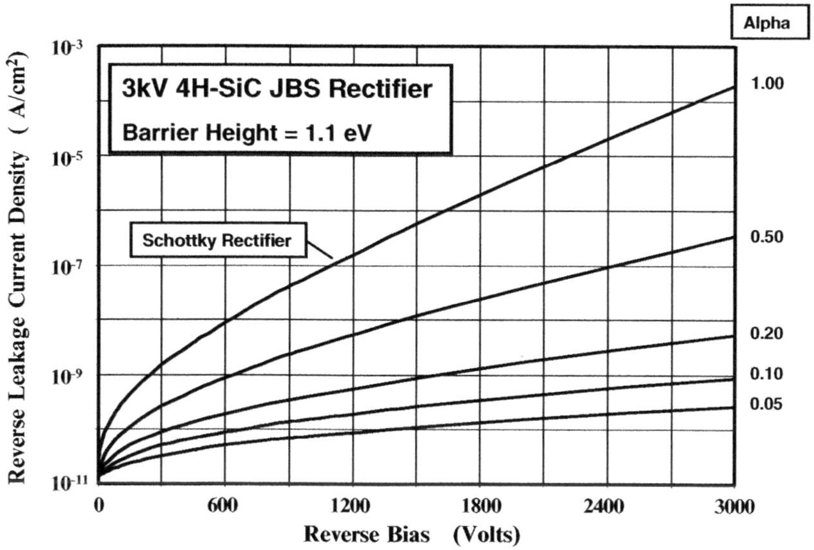


Fig. 3.17 Reverse Leakage Current for 3 kV Silicon Carbide JBS Rectifiers with various Alpha coefficients.

## Simulation Example

In order to validate the model for the reverse blocking characteristics for the silicon carbide JBS rectifier structure, a structure with breakdown voltage of 3000 volts, using a drift region with doping concentration of 1 x $10^{16}$ cm$^{-3}$ and thickness of 20 microns, will be considered here. The two-dimensional numerical simulations were performed with various spacing between the $P^+$ regions by varying the pitch (p) while maintaining an implant window (2s) of 1 micron. The depth of the $P^+$ region was also varied to examine its impact on the electric field reduction at the Schottky contact.

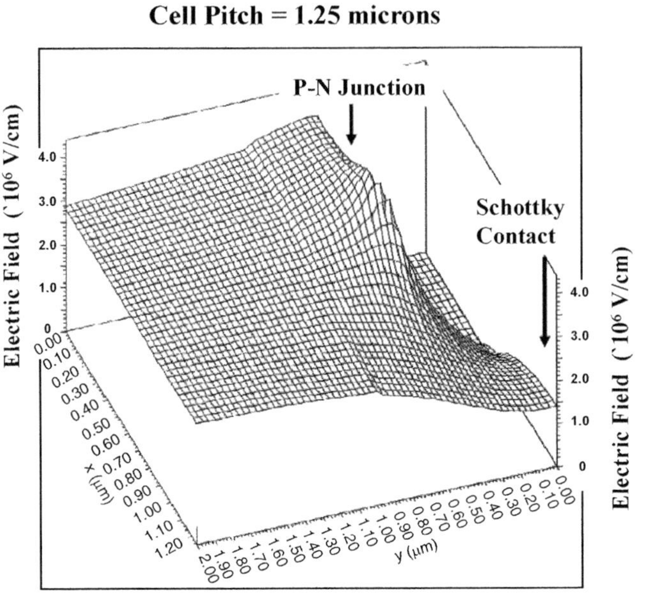


Fig. 3.17E Electric Field distribution in a 3 kV 4H-SiC JBS Rectifier.

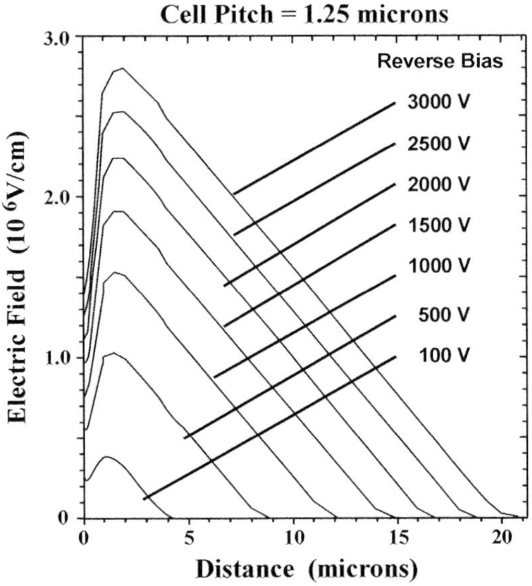


Fig. 3.18E Electric Field variation with Reverse Voltage in a 3000 V 4H-SiC JBS Rectifier.

A three-dimensional view of the electric field distribution in a 3000 volt 4H-SiC JBS rectifier structure with cell pitch of 1.25 microns is shown in Fig. 3.17E at a reverse bias of 3000 volts (just prior to breakdown). It can be seen that the highest electric field occurs at the P/N junction and that the electric field is suppressed at the Schottky contact. It is worth pointing out that the maximum electric field at the Schottky contact occurs at the location furthest away from the P-N junction (at x = 1.25 microns for the structure in Fig. 3.17E). Consequently, the largest leakage due to barrier lowering and tunneling components will occur at this location. For this reason, the highest electric field at the Schottky contact will be used to analyze the reverse leakage characteristics for the 4H-SiC JBS rectifiers.

The electric field profile at the center of the Schottky contact is shown in Fig. 3.18E for the case of a cell pitch of 1.25 microns. It can be observed that the electric field at the surface under the contact is significantly reduced when compared the peak electric field in the bulk. The peak of the electric field occurs at a depth of about 2 microns. At a reverse bias of 3000 volts, the electric field at the Schottky contact is only 1.4 x $10^6$ V/cm compared with 2.8 x $10^6$ V/cm at the electric field maxima which occurs in the bulk at a depth of 1.5 microns.

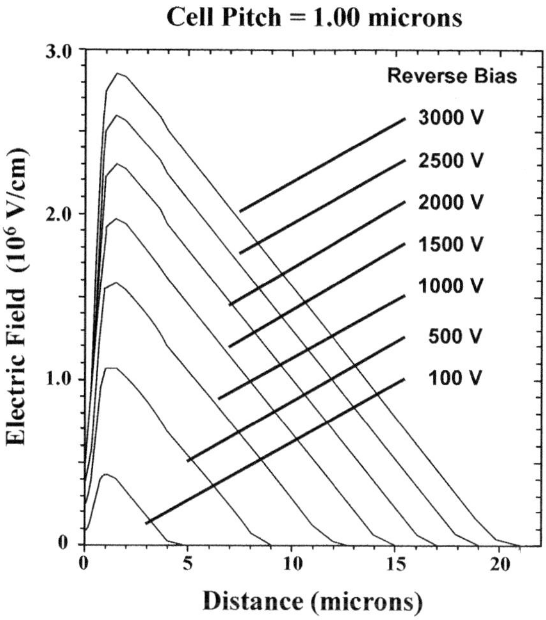


Fig. 3.19E Electric Field variation with Reverse Voltage in a 4H-SiC JBS Rectifier.

An even greater reduction of the electric field at the Schottky contact can be achieved by reducing the cell pitch while maintaining the same size for the window used to form the $P^+$ region. This is illustrated in Fig. 3.19E for the case of a cell pitch of 1.00 microns. Here, the electric field at the Schottky contact is only 7 x $10^5$ V/cm compared with 2.8 x $10^6$ V/cm at the maxima when the reverse bias voltage reaches 3000 volts. The larger reduction of the electric field at the Schottky contact occurs due to the depletion of the space between the $P^+$ regions at a smaller reverse bias voltage and the formation of a larger potential barrier under contact due to the larger channel aspect ratio. The larger potential barrier suppresses an increase in the electric field at the Schottky contact with increasing reverse bias voltage. This in turn has a very strong impact on reducing the increase in the leakage current with increasing reverse bias voltage.

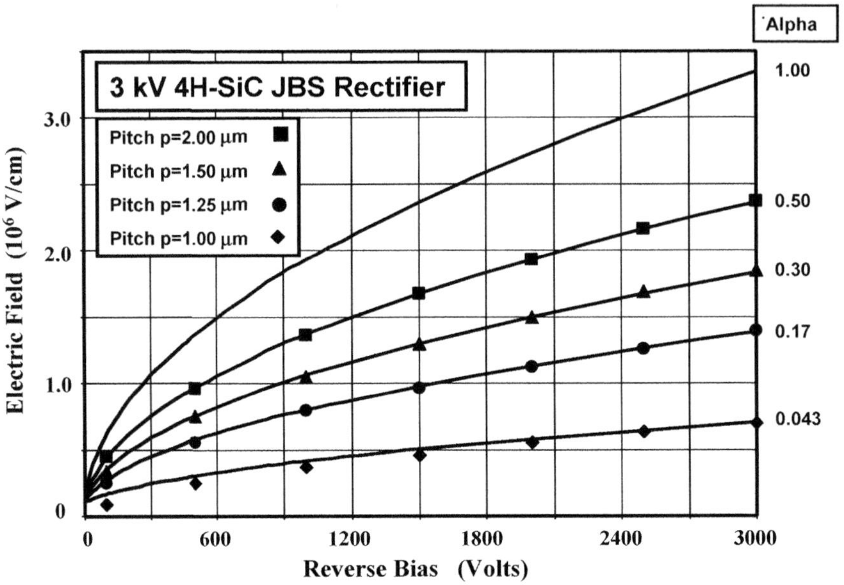


Fig. 3.20E Electric Field variation with Reverse Voltage in 4H-SiC JBS Rectifiers.

The increase in the electric field at the Schottky contact with increasing reverse bias voltage, obtained from the numerical simulations, is plotted in Fig. 3.20E for various cases of the cell pitch. These data points are compared with the calculated values obtained using the analytical model as shown by the solid lines. Here, the value for alpha was adjusted to obtain a good match to the simulation data for each cell pitch. The value for alpha is smaller for the silicon carbide JBS rectifier structure when compared with the silicon JBS rectifier structure with the same junction depth and spacing 'd'. This is due to the rectangular shape for the junction in the silicon carbide structure in comparison with a planar cylindrical shape for the junction in the silicon structure. The rectangular shape for the junction produces a stronger potential barrier under the Schottky contact which suppresses the electric field at the contact to a greater degree. This is accounted for in the analytical model by a smaller value for the alpha coefficient in Eq. [3.41].

For the silicon carbide structure, the electric field at the contact becomes close to the maximum value in the bulk when the pitch is over 2 microns. However, when the pitch is reduced to 1.25 microns, the electric field at the Schottky contact becomes less than half that for the normal Schottky rectifier structure. This reduced electric field at the contact is beneficial for suppressing the Schottky barrier lowering effect. The Schottky barrier lowering computed for the 4H-SiC JBS rectifiers by using the analytical model is compared with that in the normal Schottky rectifier structure in Fig. 3.21E. It can be observed that the Schottky barrier lowering for the JBS rectifier with a cell pitch of 1.25 microns is only 0.143 eV when compared with 0.223 eV for the normal Schottky rectifier.

The smaller electric field at the contact in the 4H-SiC JBS rectifier structures greatly reduces the leakage current because of not only suppressing the Schottky barrier lowering effect but also suppressing the tunneling current. This reduction of the leakage current is shown in Fig. 3.22E. With a cell pitch of 1.25

microns, the leakage current is reduced by a factor of 100,000 at reverse blocking voltages near the breakdown voltage. Since a pitch of 1.25 microns was demonstrated to produce excellent on-state characteristics (see Fig. 3.6E), this value is optimal for the 4H-SiC JBS rectifier with a blocking voltage of 3000 volts.

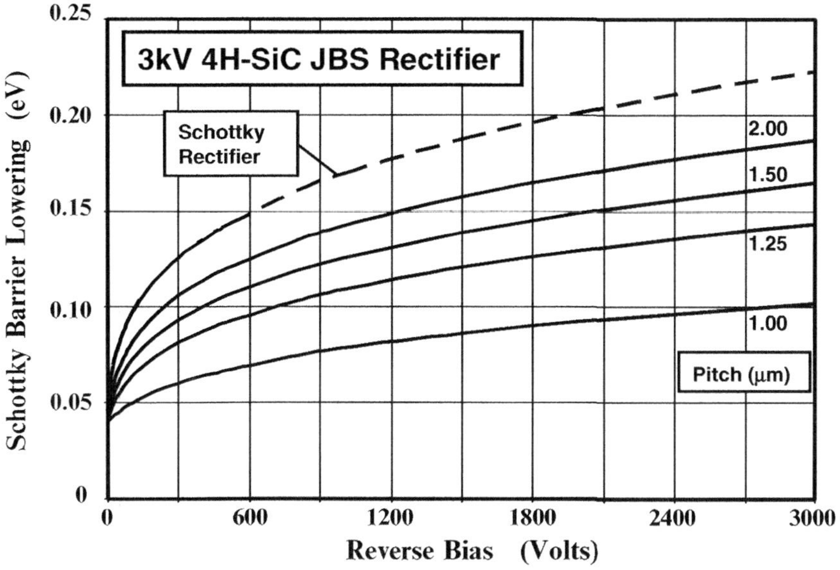


Fig. 3.21E Schottky Barrier Lowering in 4H-SiC JBS Rectifiers.

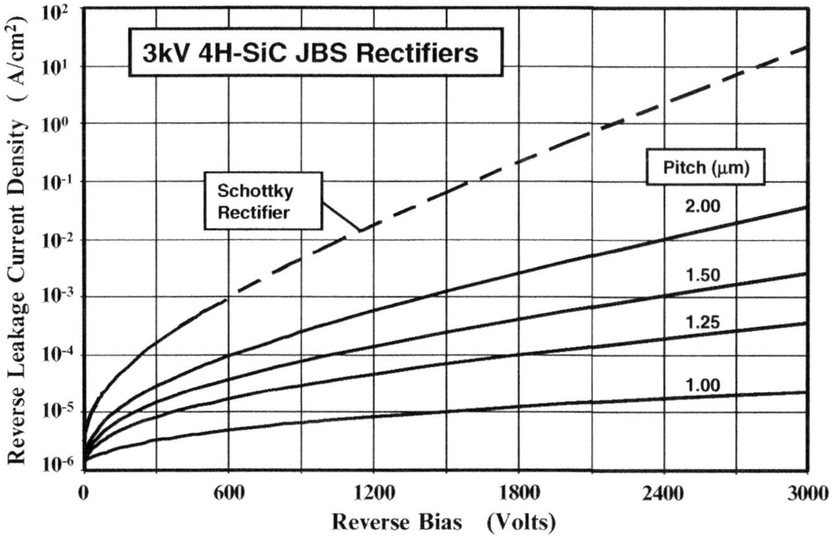


Fig. 3.22E Leakage Current Suppression in 4H-SiC JBS Rectifiers.

An even greater suppression of the electric field at the contact in the 4H-SiC JBS rectifier can be achieved by increasing the junction depth of the $P^+$ region. This can be accomplished by using boron ion implantation with various energies. As an example, the results of numerical simulations for the case of a junction depth of 0.9 microns are provided here. Unlike the case of silicon structures, the cell pitch does not have to be enlarged in the case of silicon carbide structure because of the lack of lateral diffusion during the annealing of the ion implanted layer. A three-dimensional view of the electric field distribution in a 3000 volt 4H-SiC JBS rectifier structure with cell pitch of 1.25 microns is shown in Fig. 3.23E at a reverse bias of 3000 volts (just prior to breakdown). It can be seen that the electric field is suppressed at the Schottky contact to a greater degree than for the junction depth of 0.5 microns. The maximum electric field at the Schottky contact has reduced from 1.4 x $10^6$ V/cm (see Fig. 3.17E) to 0.6 x $10^6$ V/cm.

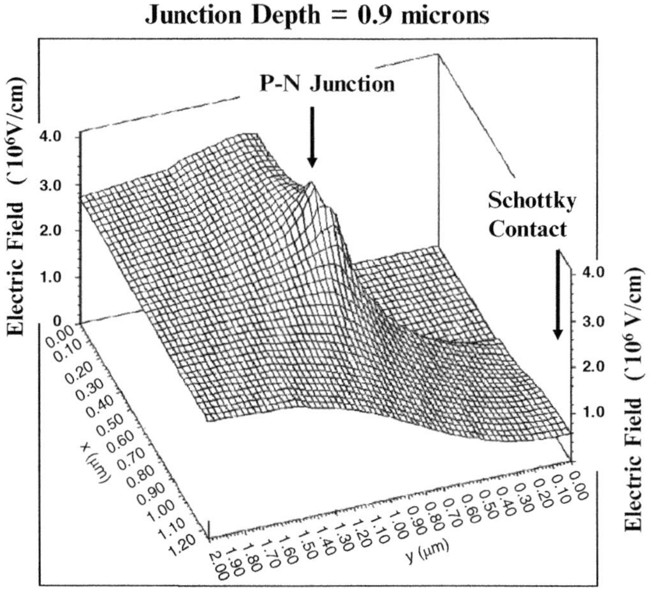


Fig. 3.23E Electric Field distribution in a 3 kV 4H-SiC JBS Rectifier.

The impact of the junction depth of the $P^+$ region on the degree of suppression of the electric field at the Schottky contact can be observed in Fig. 3.24E. In this figure, the cell pitch was maintained at 1.25 microns while the junction depth was increased from 0.1 to 0.9 microns. The data points provide the values extracted from the two dimensional numerical simulations. The solid lines provide the values obtained by using the analytical values with various values of alpha selected to match the data acquired from the numerical simulations. It can be observed from the figure that the value for alpha decreases with increasing junction depth because a stronger potential barrier is formed under the metal contact when the junction depth is made larger. Based up on these results, it can be concluded that a junction depth of 0.5 microns is adequate for suppressing the electric field under the Schottky contact for a JBS rectifier with a cell pitch of 1.25 microns.

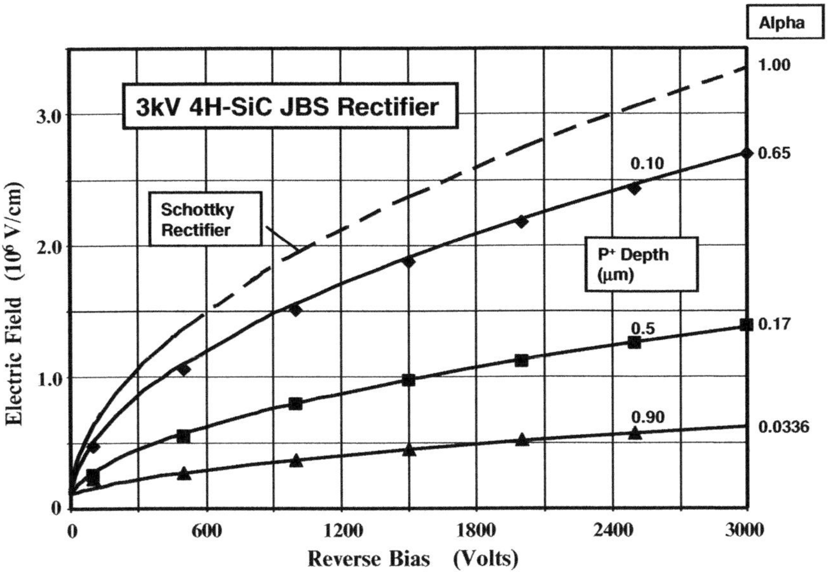


Fig. 3.24E Electric Field variation with Reverse Voltage in 4H-SiC JBS Rectifiers.

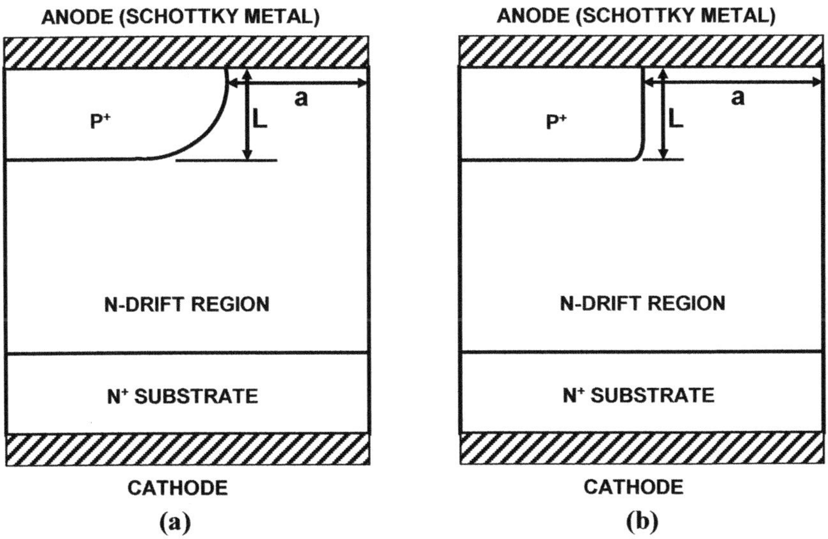


Fig. 3.25E Aspect Ratio Parameters for (a) the Silicon JBS Rectifier and (b) the Silicon Carbide JBS Rectifier.

The aspect ratio of the current conduction region located below the Schottky contact in the JBS rectifier has a strong influence on the suppression of the electric field at the contact, which is quantified by the coefficient alpha ($\alpha$) in Eq. (3.37) and Eq. (3.41). The aspect ratio is defined as:

$$AR = \frac{L}{2a} \quad [3.42]$$

where L is the length and 2a is the width of the channel in a JFET structure[8]. The equivalent dimensions are provided in Fig. 3.25E for (a) the silicon JBS rectifier structure and (b) the silicon carbide JBS rectifier structure. Based upon this diagram, the aspect ratio can be computed by dividing the junction depth ($x_J$) of the $P^+$ region to the space between the P-N junctions at the contact in the JBS rectifier. In the silicon JBS rectifier structure, the space between the junctions at the contact can be obtained by using $2(p - s - x_J)$ because the lateral diffusion of the $P^+$ region must be taken into account. In the case of the silicon carbide JBS rectifier structure, the space between the junctions at the contact can be obtained by using $2(p - s)$ because the lateral spread of the $P^+$ region is negligible.

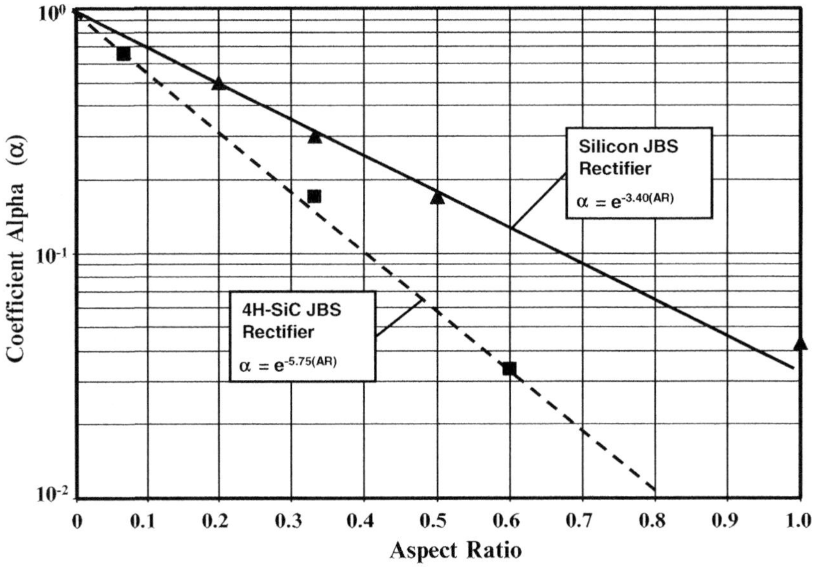


Fig. 3.26E Impact of Aspect Ratio on Alpha for JBS Rectifiers.

The variation of the coefficient alpha ($\alpha$) with aspect ratio, obtained from the numerical simulations, is shown in Fig. 3.26E for silicon and silicon carbide JBS rectifiers. It can be observed that the coefficient alpha ($\alpha$) varies exponentially with the aspect ratio. It is smaller for the case of silicon carbide devices when compared with silicon devices with the same aspect ratio because the electric field is not suppressed to the same degree in silicon devices due to the cylindrical shape of the $P^+$ region when compared with the rectangular shape for silicon carbide devices. The JBS structure is therefore particularly well suited for improving the performance of silicon carbide Schottky rectifiers.

## 3.4 Trade-Off Curve

In the textbook[1], it is demonstrated that the power dissipation can be minimized for a particular duty cycle and operating temperature by varying the Schottky barrier height during optimization of the Schottky rectifier structure. A smaller barrier height decreases the on-state voltage drop reducing conduction power losses while a large barrier height decreases the leakage current reducing reverse blocking power losses. Depending upon the duty cycle and the temperature, minimum power loss occurs at an optimum barrier height. A fundamental trade-off curve between on-state voltage drop and the leakage current was developed in the textbook based upon these considerations which does not depend upon the semiconductor material. However, the fundamental trade-off curve excludes the impact of the series resistance on the on-state voltage drop. More significantly, it excludes the strong influence of Schottky barrier lowering and pre-avalanche multiplication on the increase in leakage current for Schottky rectifiers.

When the voltage drop in the drift region is included for computing the on-state voltage drop of the silicon Schottky rectifier and the influence of Schottky barrier lowering and pre-breakdown multiplication are factored into the calculation of its leakage current, the trade-off curve degrades substantially as shown in Fig. 3.18 by dashed line corresponding to the triangular data points. In this figure, the trade-off curve for the silicon Schottky rectifier was generated by varying the Schottky barrier height. For an on-state voltage drop of 0.45 volts, the leakage current for the Schottky rectifier increases by two-orders of magnitude when compared with the fundamental trade-off curve.

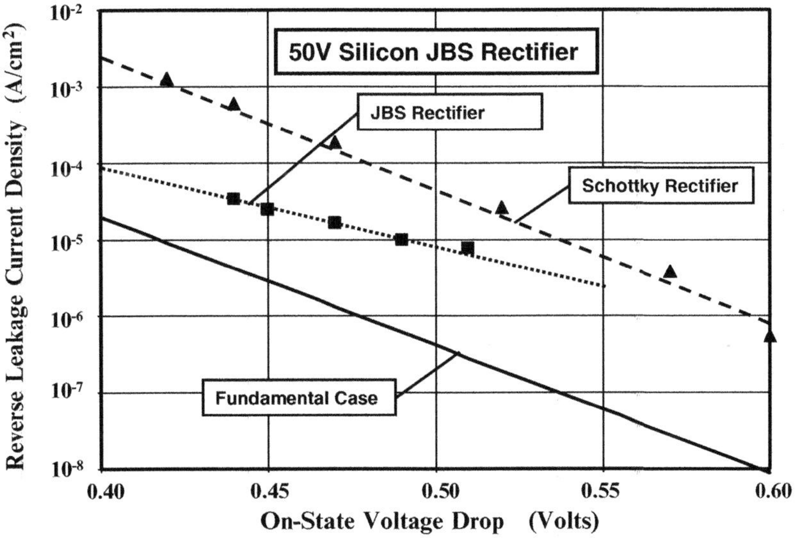


Fig. 3.18 Trade-Off Curve for the 50 V Silicon JBS Rectifier compared with that for a Schottky Rectifier.

In the silicon JBS rectifier, the Schottky barrier lowering effect is ameliorated by the reduced electric field at the contact. In addition, the pre-breakdown multiplication of the current flowing through the Schottky contact is suppressed by the much lower electric field at the vicinity of the Schottky contact. As pointed out earlier in the chapter, the breakdown shifts to the P-N junction in the JBS rectifier due to the larger electric field in this location. The trade-off curve for the JBS rectifier structure, computed by using the analytical models provided in the previous sections of this chapter in conjunction with the values for alpha extracted based upon the numerical simulations, is also shown in Fig. 3.18 by the dotted line corresponding to the square data points. For these silicon JBS rectifier structures, the pitch was varied while maintaining the same barrier height. For an on-state voltage drop of 0.45 volts, the leakage current for the JBS rectifier is reduced by an order of magnitude when compared with the Schottky rectifier. This implies that lower overall power dissipation can be achieved by using the JBS rectifier structure. In addition, it can be concluded that the JBS rectifier can be operated at a larger temperature before the on-set of thermal runaway.

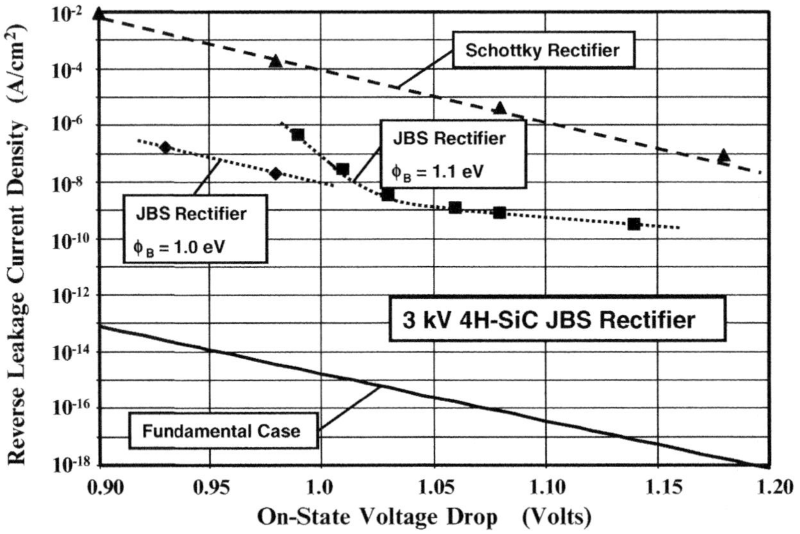


Fig. 3.19 Trade-Off Curve for the 3 kV 4H-SiC JBS Rectifier compared with that for a Schottky Rectifier.

A similar analysis can be performed for the silicon carbide JBS rectifier. In this case, a major improvement in reverse blocking characteristics occurs due to suppression of the Schottky barrier lowering and the thermionic field emission current due to the reduced electric field at the Schottky contact. When these phenomena are included in the analysis of the leakage current for the 4H-SiC Schottky rectifier, the trade-off curve for the case of devices designed to support 3000 volts obtained by varying the Schottky barrier height degrades significantly as shown in Fig. 3.19 by the dashed line corresponding to the triangular data points. It can be observed that the leakage current increases by more than 10 orders of magnitude when compared with the fundamental curve.

For the silicon carbide JBS rectifier structure, the Schottky barrier lowering and thermionic field emission phenomena are suppressed producing greatly reduced leakage currents at large reverse bias voltages. The trade-off curves obtained for these structures by varying the cell pitch are also shown in the Fig. 3.19 by the dotted lines. It can be observed that the leakage current is four orders of magnitude smaller than for the Schottky rectifier for the same on-state voltage drop in the range of 1.0 to 1.1 volts. For a barrier height of 1.1 eV, the minimum on-state voltage drop that can be obtained for the JBS rectifier by increasing the cell pitch is about 0.97 volts. Consequently, its performance begins to approach that for the Schottky rectifier when the cell pitch is increased beyond 1.5 microns. However, if the Schottky barrier height is reduced to 1.0 eV while keeping a small cell pitch to suppress the leakage current, the trade-off curve for the JBS rectifier remains four orders of magnitude better than for the Schottky rectifier even for on-state voltage drops below 1 volt as shown in the figure.

## 3.5 Summary

In this chapter, it has been demonstrated that a significant improvement in the leakage current of Schottky rectifiers can be achieved by incorporation of a P-N junction to shield the Schottky contact against high electric fields generated in the semiconductor. In the case of silicon devices, the reduction of the electric field at the contact suppresses the Schottky barrier lowering and pre-breakdown multiplication phenomena. This reduces the leakage current by an order of magnitude while the increase in on-state voltage due to loss of Schottky contact area is small. In the case of silicon carbide devices, the reduction of the electric field at the Schottky contact suppresses the Schottky barrier lowering and the thermionic field emission current leading to a reduction of leakage current by four orders of magnitude. Consequently, this approach has become quite popular for the development of Schottky rectifiers from silicon carbide[9,10,11].

The use of a P-N junction to improve the performance of Schottky rectifiers has also been utilized for creating high performance anti-parallel diodes within silicon power MOSFET structures[12] that are used as synchronous rectifiers in switch mode power supply circuits. In this application, the power MOSFET is used in place of a diode in the buck converter circuit topology. During on-state current conduction, gate voltage is applied to the power MOSFET to turn-on the channel resulting in current flow with an on-state voltage drop that is smaller than that of Schottky rectifiers. However, to avoid a shoot-through problem during the switching transients, it is necessary to carry the load current for a short time through an alternate path. If this current flows through the body diode in the power MOSFET structure, circuit operation is disrupted by the slow reverse recovery time for this bipolar diode. It is therefore preferable to incorporate a Schottky rectifier in parallel with the P-N body diode. The JBS structure has been found to be well suited for the integration of the Schottky rectifier with the power MOSFET because it enables the use of the contact metal for the source region of the MOSFET as the Schottky contact metal in the JBS diode. Although the barrier height for this contact is relatively small making the leakage current for the normal Schottky rectifier prohibitively large, the leakage current in the JBS structure is suppressed by using the shielding from the P-N junctions. The P-base region within the power MOSFET structure can be utilized for forming these P-N junctions with no additional processing steps.

## References


[1] B.J. Baliga, "Fundamentals of Power Semiconductor Devices", Springer Scientific, New York, 2008.
[2] B.J. Baliga, "The Pinch Rectifier: A Low Forward Drop High Speed Power Diode", IEEE Electron Device Letters, Vol. 5, pp. 194-196, 1984.
[3] B.J. Baliga, "Pinch Rectifier", U. S. Patent 4,641,174, Issued February 3, 1987.
[4] M. Mehrotra and B.J. Baliga, "Very Low Forward Drop JBS Rectifiers Fabricated using Sub-micron Technology", IEEE Transactions on Electron Devices, Vol. 41, pp. 1655-1660, 1994.
[5] B.J. Baliga, "Silicon Carbide Power Devices", World Scientific Publishing Company, 2005.
[6] B.J. Baliga, "Analysis of Junction Barrier controlled Schottky Rectifier Characteristics", Solid State Electronics, Vol. 28, pp. 1089-1093, 1985.
[7] B.J. Baliga, "Modern Power Devices", Chapter 4, John Wiley and Sons, 1987.
[8] B.J. Baliga, "Modern Power Devices", Krieger Publishing Company, Malabar, FL, 1992.
[9] R. Held, N. Kaminski, and E. Niemann, "SiC Merged P-N/Schottky Rectifiers for High Voltage Applications", Silicon Carbide and Related materials – 1997, Material Science Forum, Vol. 264-268, pp. 1057-1060, 1998.
[10] F. Dahlquist, et al, "A 2.8 kV JBS Diode with Low Leakage", Silicon Carbide and Related materials – 1999, Material Science Forum, Vol. 338-342, pp. 1179-1182, 2000.
[11] J. Wu, "A 4308 V, 20.9 mO-cm$^2$ 4H-SiC MPS Diodes based on a 30 micron Drift Layer", Silicon Carbide and Related materials – 2003, Material Science Forum, Vol. 457-460, pp. 1109-1112, 2004.
[12] B.J. Baliga and D. Alok, "Paradigm Shift in Planar Power MOSFET Technology", Power Electronics Technology Magazine, pp. 24-32, November 2003.

# Chapter 4

# Trench Schottky Barrier Controlled Schottky Rectifiers

As discussed in the previous chapter, the leakage current for silicon and silicon carbide Schottky rectifiers can be greatly reduced at high reverse bias voltages by shielding the metal contact from the high electric field generated within the semiconductor. The approach utilized in the previous chapter is based up on creating P-N junctions located under the Schottky contact with carefully chosen spacing between them to create a potential barrier under the metal contact during the reverse blocking mode. One of the short-comings of this approach is the need to anneal the ion implanted P-type regions at very high temperatures in order to activate the dopant and remove the lattice damage. At these high temperatures, some dissociation can occur at the semiconductor surface which degrades the quality of the metal-semiconductor junction that must be subsequently formed. Although an issue for silicon devices, this problem is particularly severe for silicon carbide due to the very high (~1600 °C) annealing temperature for activation of ion implanted regions.

Consequently, a second method has been proposed to ameliorate the barrier lowering effect in vertical silicon Schottky rectifiers by shielding the Schottky contact utilizing a second Schottky contact with a larger barrier height[1,2]. The basic concept was to create a potential barrier to shield the main current carrying low barrier height Schottky contact against high electric fields generated in the semiconductor by using closely spaced Schottky contacts with a large barrier height. In order to produce a strong potential barrier under the main Schottky contact, it is preferable to locate the second high barrier Schottky metal within a vertically walled trench. This device was therefore named the *'Trench-Schottky-Barrier controlled Schottky (TSBS) rectifier' structure*.

B.J. Baliga, *Advanced Power Rectifier Concepts*, DOI: 10.1007/978-0-387-75589-2_4, 

In the TSBS rectifier structure, the on-state current is designed to flow in the un-depleted gaps between the trenches when the diode is forward biased. When a reverse bias is applied, a potential barrier forms under the main Schottky contact which suppresses the electric field at this contact. This prevents a large increase in the leakage current with reverse bias and keeps the leakage current low even if the barrier height of the main contact is small. However, a high electric field is produced at the second Schottky contact located in the trenches. This produces a rapid increase in the leakage current with reverse bias at this contact. However, by using a large Schottky barrier height for this second contact, the absolute value for the leakage current at large reverse bias voltages can be kept much smaller than at the main low barrier height Schottky contact.

This chapter provides analytical models for the TSBS rectifier structure. The physics of operation of these structures is also elucidated by using two-dimensional numerical analysis. It is demonstrated that the TSBS concept improves the performance of silicon and silicon carbide devices[3]. These device structures can be fabricated without the need for high temperature process steps that can degrade the silicon carbide surface so that the quality of the Schottky contact is improved.

## 4.1 Trench Schottky Barrier controlled Schottky (TSBS) Rectifier Structure

The Trench Schottky Barrier controlled Schottky (TSBS) rectifier structure is illustrated in Fig. 4.1. It consists of a high barrier height metal placed in a trench region to generate a potential barrier that can shield the main low barrier height Schottky contact (at location B) in the reverse blocking mode. The magnitude of

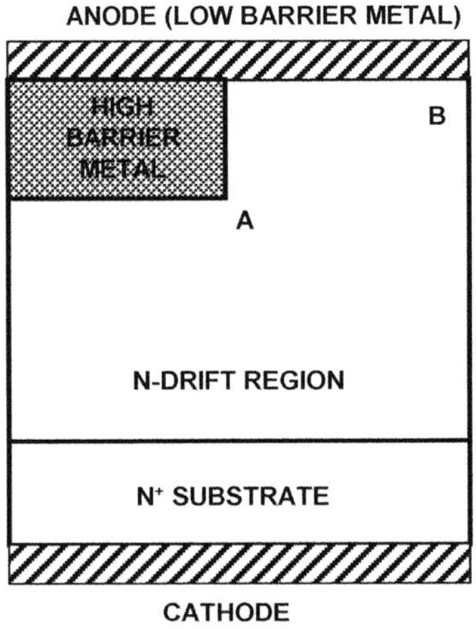


Fig. 4.1 Trench Schottky Barrier controlled Schottky (TSBS) Rectifier Structure.

the potential barrier depends upon the separation between the trenches and the trench depth. A smaller separation and larger trench depth favors an increase in the magnitude of the potential barrier leading to a greater reduction of the electric field at the Schottky contact. A reduction of the electric field at the Schottky contact produces a smaller barrier lowering and field emission effect, which is beneficial for reducing the leakage current at high reverse bias voltages. However, a high electric field can be generated at the sharp corner of the metal in the trenches leading to degradation in the blocking voltage within the device cell structure. Proper optimization of the cell structure can maintain a cell breakdown voltage above that at the edge termination.

The space between the trenches is also chosen so that there is an un-depleted region below the main low barrier height Schottky contact during on-state operation to enable unipolar conduction with a low on-state voltage drop. In the TSBS rectifier, the current flow through the metal in the trenches is relatively small due to its large barrier height. Due to small band gap for silicon, it is difficult to make the difference in barrier heights between the main contact and the metal in the trench more than 0.3 eV. The larger band gap for silicon carbide provides the opportunity to utilize a greater difference between the barrier heights of the main contact and the metal in the trench. Consequently, the TSBS concept is well suited for development of silicon carbide structures with very high breakdown voltages.

The TSBS structure can be fabricated by first depositing the main low barrier height metal on a pristine semiconductor surface to obtain a high quality interface. The metal layer is then patterned to open windows where the trenches are formed. The metal layer can serve as a barrier during the etching of the semiconductor to form self-aligned trenches if the appropriate reactive-ion-etching chemistry is used. The second high barrier height metal can then be evaporated to fill the trenches and cover the first metal layer to complete the cell structure. The second metal layer must of course be patterned to terminate the device at the edges. The same device structure, shown in Fig. 4.1, can be produced for both silicon and silicon carbide devices. Consequently, a single basic model can be created for TSBS rectifiers made from both these semiconductors. However, the difference in the Schottky barrier heights that are appropriate for silicon and 4H-SiC result in some differences during device optimization as discussed below.

In silicon JBS rectifiers, the P-N junction is formed with an annealing cycle that creates a planar junction with its extension in the lateral direction as shown in Fig. 3.1. The additional area consumed by the lateral diffusion degrades the on-state characteristics. In addition, the cylindrical shape of the junction allows encroachment of the cathode potential towards the Schottky contact producing an enhancement of the electric field at the contact with increasing reverse bias voltage. The rectangular shape for the metal contact in the trenches in the TSBS rectifier structure produces superior on-state and reverse blocking characteristics when compared with the silicon JBS rectifier structure.

As in the case of the JBS rectifier structure, it will be assumed that the breakdown voltage is reduced to about 80 percent of the ideal parallel-plane value due to the edge termination. The doping concentration for the drift region must be computed after accounting for this reduction in the breakdown voltage:

$$N_D = \left(\frac{5.34x10^{13}}{BV_{PP}}\right)^{4/3} \quad \textbf{[4.1]}$$

where $BV_{PP}$ is the breakdown voltage for the parallel-plane case after accounting for the edge termination. The maximum depletion width in the TSBS structure is limited to that associated with the breakdown voltage (BV) of the structure, as given by:

$$t = W_D(BV) = \sqrt{\frac{2\varepsilon_S BV}{qN_D}} \quad \textbf{[4.2]}$$

The thickness of the drift region required below the bottom of the trenches is therefore less than the depletion width for the ideal parallel-plane junction with the above doping concentration. The resistance of the drift region below the low barrier height main Schottky contact is consequently enhanced above that for the ideal parallel-plane case due to current transport between the trenches and the lower doping concentration of the drift region.

## 4.2 Forward Conduction Model

Analysis of the on-state voltage drop of the TSBS rectifier requires taking into consideration the current constriction at the main low barrier height Schottky contact due to the presence of the trenches and the enhanced resistance of the drift region due to current spreading from the main Schottky contact to the $N^+$ substrate. Several models were developed for the spreading resistance for the JBS rectifier structure in the previous chapter. It was found that model C with a 45 degree current spreading angle from the bottom of the P-N junction is the most suitable. In addition, this model was applied to the silicon carbide JBS rectifier by assuming a rectangular shape for the junction. It is therefore appropriate to utilize this model for the on-state current flow in both silicon and silicon carbide TSBS rectifier structures.

The current flow pattern in the forward conduction mode for the TSBS rectifier is illustrated in Fig. 4.2 with the shaded area. The model takes into account the increase in current density at the main Schottky contact due to the space taken up by the trenches and the presence of a depletion layer at the large barrier height metal. The current through the main Schottky contact flows only within the un-depleted portion (with dimension 'd') of the drift region at the top surface. Consequently, the current density at the main Schottky contact ($J_{FS}$) is related to the cell (or cathode) current density ($J_{FC}$) by:

$$J_{FS} = \left(\frac{p}{d}\right) J_{FC} \qquad [4.3]$$

where p is the cell pitch. The dimension 'd' is determined by the cell pitch (p), the size of the window (2s) for etching the trenches, and the on-state depletion width ($W_{D,ON}$):

$$d = p - s - W_{D,ON} \qquad [4.4]$$

Depending up on the lithography used for device fabrication to minimize the space (dimension 's') taken up by the trenches, the current density at the main Schottky contact can be enhanced by a factor of two or more. This must be taken into account when computing the voltage drop across the Schottky contact given by:

$$V_{FS} = \phi_B + \frac{kT}{q} \ln\left(\frac{J_{FS}}{AT^2}\right) \qquad [4.5]$$

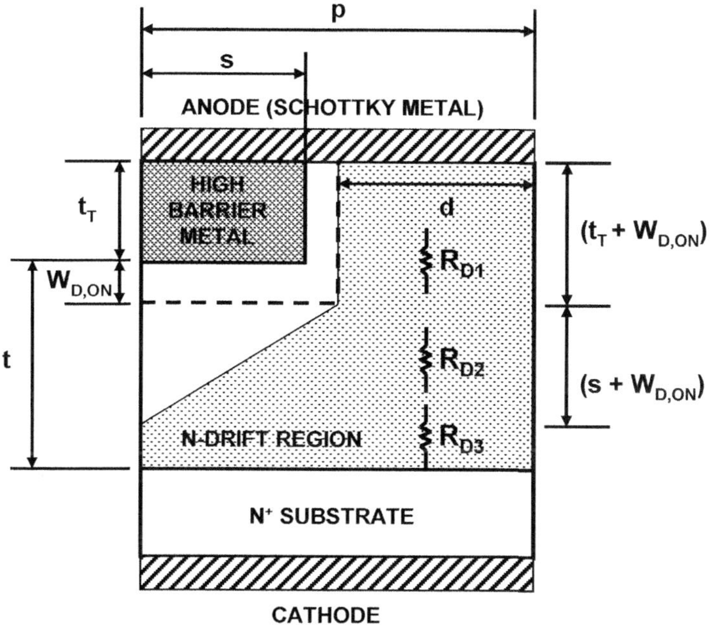


Fig. 4.2 Current Flow Pattern in the TSBS Rectifier Structure during Operation in the On-State.

After flowing across the main Schottky contact, the current flows through the un-depleted portion of the drift region. In the model, it is assumed that the current flows through a region with a uniform width 'd' until it reaches the bottom of the depletion region and then spreads to the entire cell pitch (p) at a 45 degree spreading angle. The current paths overlap at a distance ($s + W_{D,ON}$) from the bottom of the depletion region. The current then flows through a uniform cross-sectional area.

The net resistance to current flow can be calculated by adding the resistance of the three segments. The resistance of the first segment of uniform width 'd' is given by:

$$R_{D1} = \frac{\rho_D.(t_T + W_{D,ON})}{d.Z} \quad \text{[4.6]}$$

The resistance of the second segment can be derived by using the same approach used for Model C for the silicon JBS rectifier:

$$R_{D2} = \frac{\rho_D}{Z}\ln\left(\frac{p}{d}\right) \quad \text{[4.7]}$$

The resistance of the third segment with a uniform cross-section of width p is given by:

$$R_{D3} = \frac{\rho_D.(t - s - 2W_{D,ON})}{p.Z} \quad \text{[4.8]}$$

The specific resistance for the drift region can be calculated by multiplying the cell resistance ($R_{D1}$ + $R_{D2}$ + $R_{D3}$) by the cell-area (p.Z):

$$R_{sp,drift} = \frac{\rho_D.p.(t_T + W_{D,ON})}{d} + \rho_D.p.\ln\left(\frac{p}{d}\right) + \rho_D.(t - s - 2W_{D,ON}) \quad \text{[4.9]}$$

In addition, it is important to include the resistance associated with the thick, highly doped $N^+$ substrate. In the case of silicon devices, the contribution from the substrate is typically 2 x $10^{-5}$ $\Omega$-cm$^2$. For 4H-SiC, the specific resistance contributed by the $N^+$ substrate is typically 4 x $10^{-4}$ $\Omega$-cm$^2$.

The on-state voltage drop for the TSBS rectifier at a forward cell current density $J_{FC}$, after including the substrate contribution, is given by:

$$V_F = \phi_B + \frac{kT}{q}\ln\left(\frac{J_{FS}}{AT^2}\right) + (R_{sp,drift} + R_{sp,subs}).J_{FC} \quad \text{[4.10]}$$

The on-state depletion layer width for the silicon and silicon carbide structures is determined by the contact potential for the metal in the trenches and the on-state voltage drop. The contact potential for a metal-semiconductor junction is given by[4]:

$$qV_C = \phi_M - (\chi_S + E_C - E_{FS}) \quad \text{[4.11]}$$

where $\phi_M$ is the work function of the metal in the trenches, $\chi_S$ is the electron affinity of the semiconductor, $E_C$ is the conduction band energy, and $E_{FS}$ is the

Fermi level energy in the semiconductor. The Fermi level position in the semiconductor can be computed by using:

$$E_{FS} = E_i + \frac{kT}{q} \ln\left(\frac{n_0}{n_i}\right) \quad [4.12]$$

where $E_i$ is the intrinsic energy level, k is Boltzmann's constant, T is the absolute temperature, q is the charge of the electron, $n_0$ is the equilibrium concentration for electrons and $n_i$ is the intrinsic carrier concentration. For simplicity, the equilibrium concentration of electrons in the semiconductor can be assumed to be equal to the doping concentration for the analysis of TSBS rectifiers.

When computing the on-state voltage drop using the above equations, it is satisfactory to make the approximation that the depletion layer width can be computed by subtracting an on-state voltage drop from the contact potential:

$$W_{D,ON} = \sqrt{\frac{2\varepsilon_S (V_C - V_{ON})}{qN_D}} \quad [4.13]$$

Due to the abrupt junctions formed with the metal-semiconductor contacts, it is appropriate to assume that the entire depletion occurs in the semiconductor for the TSBS rectifier structure.

### 4.2.1 Silicon TSBS Rectifier: Example

In order to understand the operation of the silicon TSBS rectifier structure, it is instructive to consider a specific example of a device with breakdown voltage of 50 volts. If the edge termination limits the breakdown to 80 percent of the ideal value, the parallel-plane breakdown voltage is 62.5 volts. This voltage can be supported by a depletion region width of 2.85 microns in the drift region with doping concentration of 8 x $10^{15}$ $cm^{-3}$. For this example, it will be assumed that the barrier height for the main Schottky contact is 0.60 eV while that for the metal in the trenches is 0.85 eV. These values are representative of using a low barrier height metal such as Chromium for the main Schottky contact and Platinum as the metal in the trenches.

The contact potential at the metal in the trenches in this case is found to be 0.639 volts. Using this value and an on-state voltage drop of 0.45 volts, the depletion region width in the N-drift region at the metal in the trenches is 0.17 microns. If the TSBS structure has a cell pitch (p) of 1.00 microns, and the trenches are created using a window (2s) of 0.5 microns, the dimension 'd' is found to be 0.58 microns. Consequently, the current density at the main Schottky contact region where the current is transported is enhanced by a factor about 2 times when compared with the cathode (or average cell) current density. This enhancement in current density is smaller than that for the silicon JBS rectifier structure because extra space is consumed in the silicon JBS rectifier structure by the lateral diffusion of the junction.

The impact of changing the cell pitch (p) on the on-state characteristics can be predicted by using the above model for the TSBS rectifier structure. The results obtained for the case of a trench depth of 0.5 microns are shown in Fig. 4.3. The *i-v* characteristic of a normal Schottky rectifier with the same drift region parameters is included for comparison. The impact on the on-state voltage drop is small as long as the pitch is more than 1.00 microns. The on-state voltage drop of the TSBS rectifier structure at an on-state current density of 100 A/cm$^2$ is 0.343 volts compared with 0.322 volts for the normal Schottky rectifier structure with the same barrier height as the main Schottky contact. The ability to obtain a low on-state voltage drop in the silicon TSBS structure with a smaller pitch than for the JBS rectifier structure produces superior on-state characteristics.

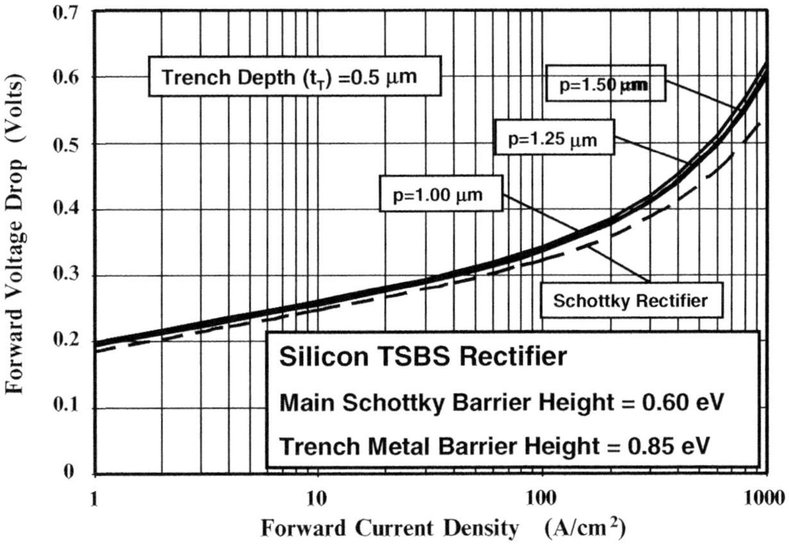


Fig. 4.3 Forward Characteristics of 50V Silicon TSBS Rectifier Structures.

The degree of shielding of the main Schottky contact in the TSBS rectifier structure against the high electric fields generated in the drift region during the reverse blocking mode depends upon the depth of the trenches. A deeper trench improves the *aspect ratio* of the region between the metal in the trenches. As discussed in the previous chapter, a larger aspect ratio has been found to provide greater shielding in the case of the vertical junction field effect transistor leading to a larger blocking gain[5]. However, the resistance of the first segment in the model for the drift region resistance increases with trench depth leading to a larger on-state voltage drop. This is illustrated for the TSBS rectifier structure with a fixed cell pitch of 1 micron in Fig. 4.4 where the analytically calculated forward *i-v* characteristics are shown for various trench depths. Increasing the trench depth from 0.5 to 1 micron increases the on-state voltage drop by a small amount from 0.343 to 0.348 volts. In this figure, the case of a structure with zero depth for the trench is also included. In this case, the on-state voltage for the TSBS rectifier structure is reduced to 0.338 volts. This is still larger than the on-state voltage drop of 0.322 volts for the normal Schottky rectifier because of the space consumed by the large barrier height metal which enhances the current density at the main Schottky contact.

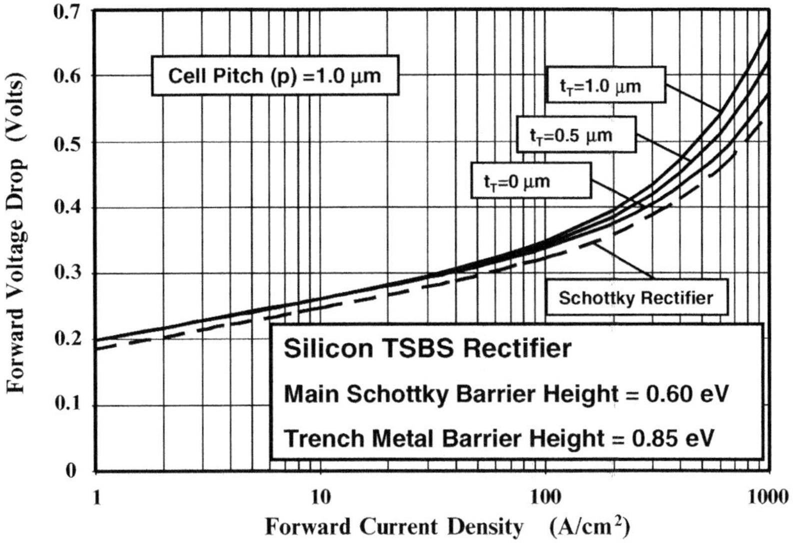


Fig. 4.4 Forward Characteristics of 50V Silicon TSBS Rectifier Structures.

## Simulation Example

In order to validate the above model for the on-state characteristics of the silicon TSBS rectifier, the results of two-dimensional numerical simulations of a 50 V structure are described here. All the devices had a drift region thickness of 3 microns with a doping concentration of 8 x $10^{15}$ $cm^{-3}$. For all the TSBS rectifier structures, a work function of 4.80 eV was used for the main Schottky contact and 5.05 eV for the metal in the trenches. In all cases, the window for etching the trenches (twice the dimension 's' in Fig. 4.2) was kept at 0.5 microns.

The on-state *i-v* characteristic for the TSBS rectifier with a cell pitch 'p' of 1.00 microns and a trench depth of 0.5 microns is shown in Fig. 4.1E. This plot includes the total current (cathode current) flowing through the structure together with the current flowing through the main Schottky contact (dotted line) and the metal in the trenches (dashed line). It can be observed that the current flowing through the metal in the trenches is 3 orders of magnitude smaller than the current through the main Schottky contact because of its larger barrier height. This justifies using an analytical model for the TSBS rectifier with only the current flow through the main low barrier height Schottky contact. At an on-state current density of 100 A/$cm^2$, the on-state voltage drop obtained from the numerical simulations for this structure is 0.35 volts. The value obtained from the numerical simulations is in good agreement with the value obtained by using in the analytical model for this structure with a barrier height of 0.6 eV for the main Schottky contact (see Fig. 4.3). The saturation current density for the main Schottky contact can be determined to be 8 x $10^{-4}$ A/cm$^2$ by extrapolation of the linear portion of the *i-v* characteristics obtained from the numerical simulations (see Fig. 4.1E). In computing this saturation current density, it is important to recognize that the area of the main Schottky contact is only half the cell area. Using this value for the saturation current density, the effective Schottky barrier height for the main Schottky contact in the simulations is found to be 0.60 eV, which is the same value used for the analytical models.

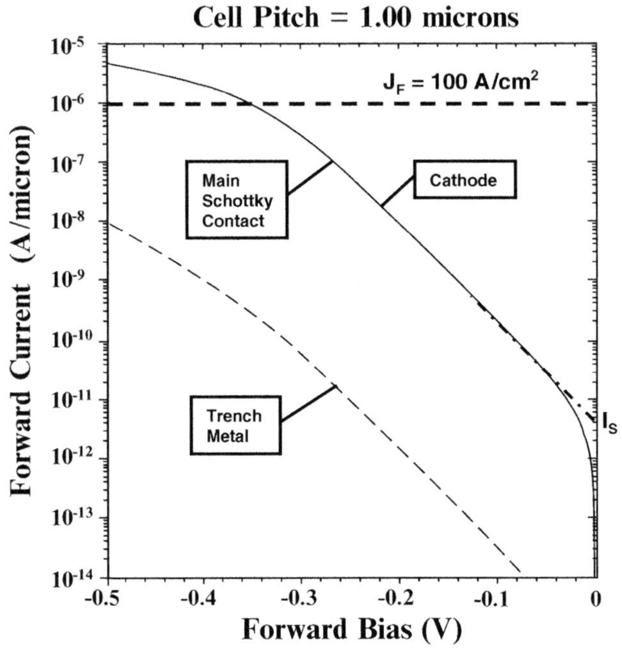


Fig. 4.1E On-State Characteristics for a 50V Silicon TSBS Rectifier.

The current flow-lines for the above TSBS rectifier structure with cell pitch (p) of 1 micron and trench depth of 0.5 microns are shown in Fig. 4.2E at an on-state voltage drop of 0.4 volts. It can be seen that all the current flow-lines converge to the main Schottky contact located on the upper right-hand-side of the structure demonstrating that very little current flows through the metal contact in the trenches. The current flow pattern is consistent with pattern used to develop the analytical model with an approximately uniform cross-section between the trenches followed by a spreading of the current at an angle of about 45 degrees (see Fig. 4.2). The current then becomes uniformly distributed below a depth of 1.5 microns from the surface. This justifies the use of a three region analytical model for the series resistance of the drift region with no current flow through the metal electrode located in the trenches.

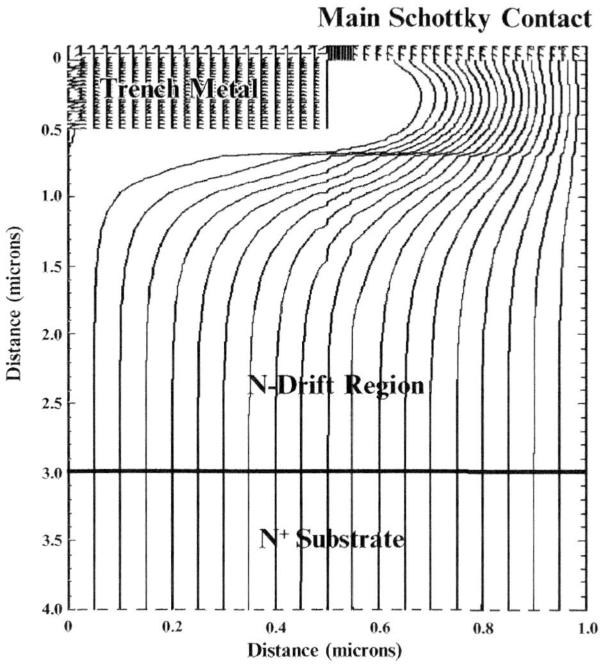


Fig. 4.2E On-State Current Distribution in a 50V Silicon TSBS Rectifier.

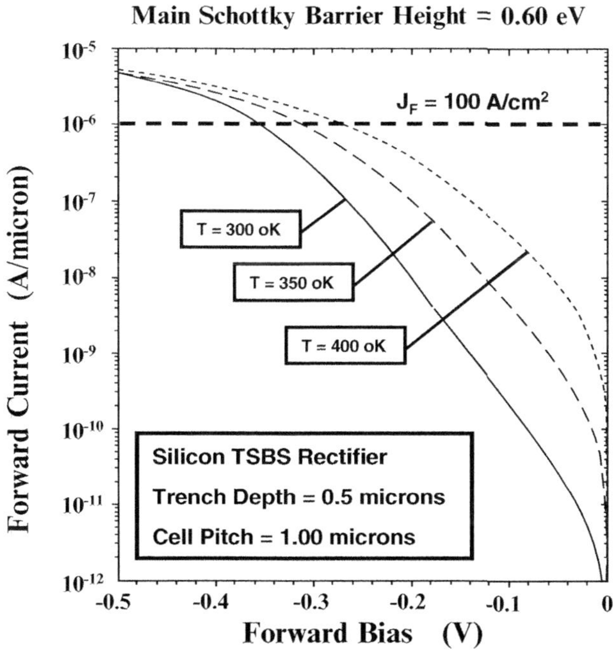


Fig. 4.3E On-State Characteristics for a Typical 50V Silicon TSBS Rectifier.

The on-state characteristics for the silicon TSBS rectifier have the same temperature behavior as the normal Schottky rectifier. The on-state voltage drop decreases with increasing temperature[4] as shown in Fig. 4.3E. This produces a reduction in the on-state power losses as the temperature increases. The reverse blocking power loss becomes dominant at high temperatures because the leakage current density increases rapidly with increasing temperature. The power loss exhibits a minima at a temperature beyond which the device can undergo thermal runaway.

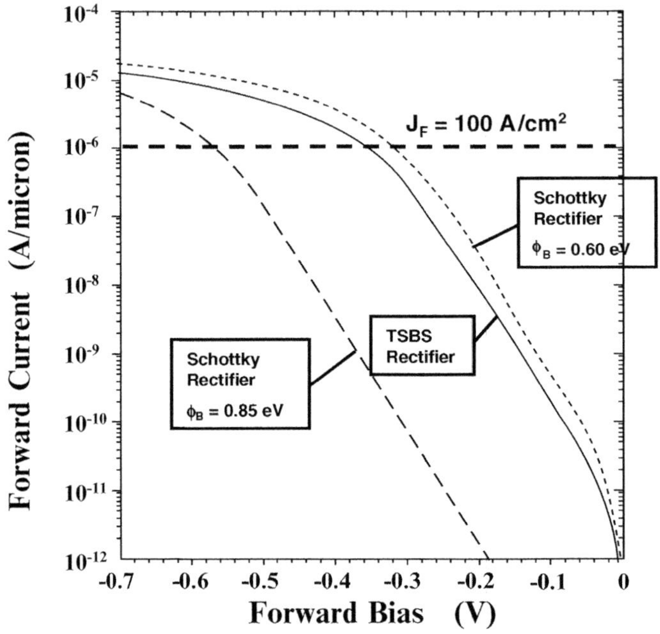


Fig. 4.4E Comparison of the On-State Characteristics for a Typical 50V Silicon TSBS Rectifier with Schottky Rectifiers.

In the TSBS rectifier structure, the area for the main Schottky contact must be sacrificed to introduce the second Schottky contact for providing the shielding during reverse bias operation. This produces an increase in current density at the main Schottky contact and a larger spreading resistance of the drift region. Both of these phenomena can be expected to increase the on-state voltage drop of the TSBS rectifier structure when compared with the normal Schottky rectifier. This impact on the on-state characteristics can be illustrated by using the *i-v* characteristics of the typical TSBS rectifier structure as an example. The *i-v* characteristics of the typical TSBS rectifier with a cell pitch of 1 micron and a trench depth of 0.5 microns are compared with the normal Schottky rectifier with a cell pitch of 1 micron in Fig. 4.4E. It can be observed that the on-state voltage drop for the TSBS rectifier structure is 0.03 volts larger than for the normal Schottky rectifier with same barrier height of 0.60 eV as the main Schottky contact in the TSBS rectifier. For completeness, the i-v characteristics for the normal Schottky rectifier with the same barrier height of 0.85 eV as the metal in the trenches of the TSBS rectifier are also displayed in the figure. The on-state voltage drop for this normal Schottky rectifier is 0.21 volts larger than that of the TSBS rectifier structure.

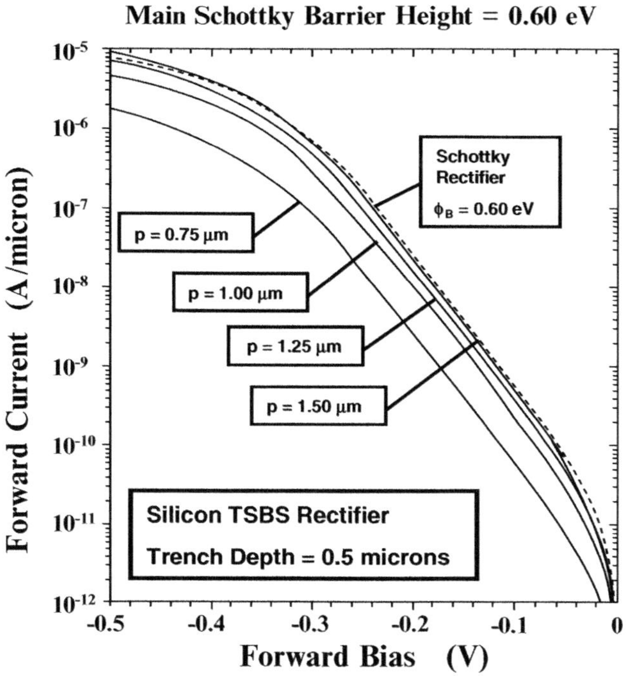


Fig. 4.5E On-State Characteristics for 50V Silicon TSBS Rectifiers.

The on-state characteristics of 50 V TSBS rectifiers with a cell pitch of 0.75, 1.00, 1.25 and 1.50 microns are compared with that of the TSBS rectifier and a normal Schottky rectifier with cell pitch of 1.00 microns in Fig. 4.5E. All the structures had a main Schottky contact with barrier height of 0.60 eV. In the TSBS rectifiers, the trench had a width (2s) of 1 micron and a depth of 0.5 microns. From the graph, the on-state voltage drop may appear to be reducing with increasing cell pitch. However, after accounting for the difference in the areas of the structures, the on-state voltage drop at an on-state current density of 100 A/cm$^2$ is found to be approximately equal for all the TSBS rectifier structures with cell pitch greater than 1 micron. These results are consistent with those predicted by the analytical model (see Fig. 4.3) confirming its utility for analysis of the JBS rectifier structure in the on-state. However, when the pitch is reduced to 0.75 microns, the on-state voltage drop increases to 0.41 volts compared with 0.35 volts for the other TSBS rectifier structures. Based upon these results, it is better to use a cell pitch of 1 micron for the TSBS rectifier in order to suppress the electric field at the main Schottky contact to a greater degree while obtained a low on-state voltage drop.

In the TSBS rectifier structure, an increase in the trench depth improves the shielding of the main Schottky contact. However, this also increases the resistance of the current path from the Schottky contact to the cathode. The impact of increasing the trench depth from 0 to 1.0 micron in the TSBS rectifier structure can be demonstrated using numerical simulations for the case of a cell pitch of 1 micron. The on-state *i-v* characteristics for these structures are compared with the normal Schottky rectifier in Fig. 4.6E. The i-v characteristics of the structures nearly coincide for the cases of a trench depth of 0.25 and 0.50 microns. The on-state voltage drop at a current density of 100 A/cm$^2$ increases slightly for the cases of deeper trench regions. These results are in good agreement with the predictions of the analytical model (see Fig. 4.4). The case of zero trench depth is also shown in the figure by the *i-v* characteristics with the dashed line. The on-state voltage drop for this case is still larger than that of the normal Schottky rectifier due to the space consumed by the metal with the larger barrier height.

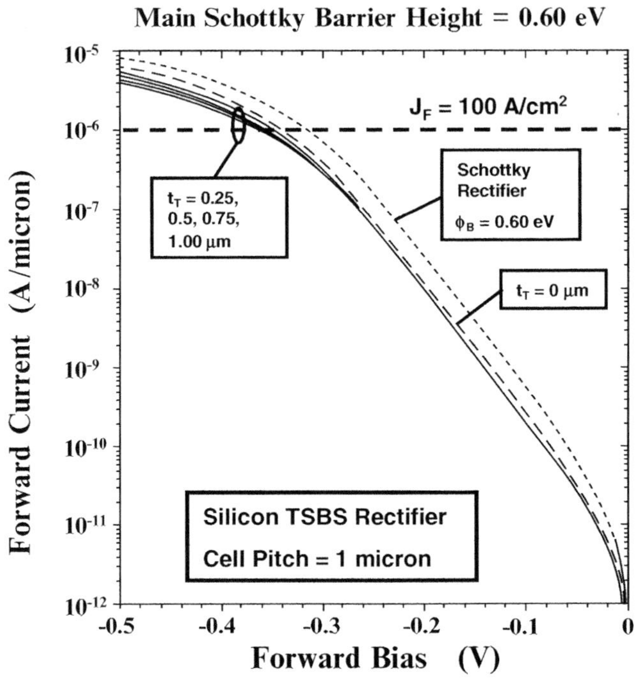


Fig. 4.6E On-State Characteristics for 50V Silicon TSBS Rectifiers.

### 4.2.2 Silicon Carbide TSBS Rectifier: Example

As mentioned earlier, the same on-state model can be used for the silicon carbide TSBS rectifier and the silicon structure because the cell structure is identical. However, in the case of silicon carbide devices, it is possible to utilize metal layers with larger Schottky barrier heights because of its larger energy band gap than silicon. Although this produces a larger zero bias depletion width at the metal in the trenches, the depletion width under on-state current flow is similar to the silicon structure because the silicon carbide device is operated at a larger on-state voltage drop. One significant difference between the silicon and silicon carbide structures is that the trench depth is a smaller fraction of the total drift region thickness for the case of high voltage silicon carbide rectifiers. This reduces the impact of current spreading within the silicon carbide cell structure. However, it is important to include the resistance associated with the thick, highly doped $N^+$ substrate because this is substantially larger than for silicon devices. The specific resistance of the $N^+$ substrate can be determined by taking the product of its resistivity and thickness. For 4H-SiC, the lowest available resistivity for $N^+$ substrates is 20 mΩ-cm. If the thickness of the substrate is 200 microns, the specific resistance contributed by the $N^+$ substrate is $4 \times 10^{-4}$ Ω-cm$^2$.

The on-state voltage drop for the TSBS rectifier at a forward cell current density $J_{FC}$, including the substrate contribution, is given by Eq. (4.10). The Richardson's constant for 4H-SiC is 146 A°K$^{-2}$cm$^{-2}$. As an example, 4H-SiC TSBS rectifiers with a blocking voltage capability of 3000 volts will be analyzed in this section. This reverse blocking capability can be obtained by using a drift region with a doping concentration of $8.5 \times 10^{15}$ cm$^{-3}$ and thickness of 20 microns after accounting for loss in voltage at the edge termination. A typical 4H-SiC TSBS rectifier structure has a cell pitch of 1 micron, a trench width (2s) of 1 micron, and a trench depth of 0.5 microns.

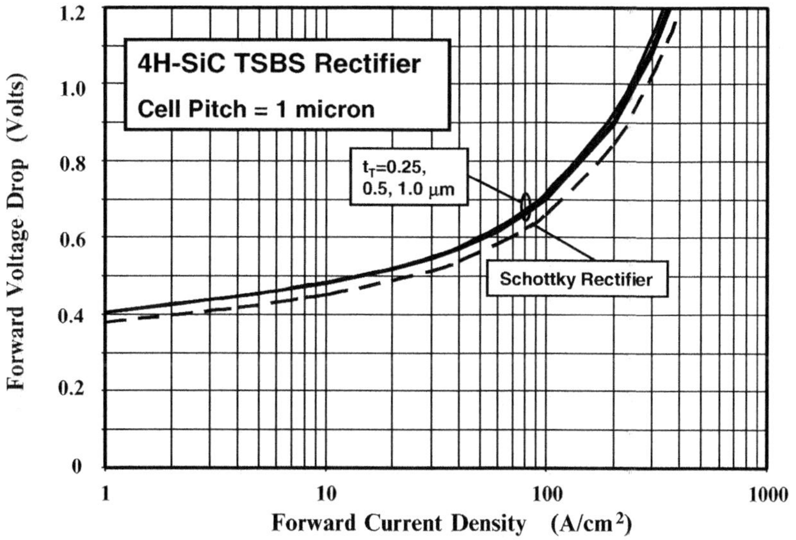


Fig. 4.5 Forward Characteristics of 3kV 4H-SiC TSBS Rectifiers.

The forward characteristics of 3kV TSBS rectifiers, calculated using the analytical model with a barrier height of 0.8 eV for the main Schottky contact, are shown in Fig. 4.5 with the trench depth as a parameter. It can be observed that the on-state characteristics are not strongly affected by the trench depth. This is due to the relatively small trench depth when compared with the total thickness of 20 microns for drift region. The current spreading in these high voltage silicon carbide

TSBS rectifier structures occurs within the top 2 microns making the contribution from the rest of the drift region nearly equal for all the cases. The trench depth can therefore be selected for sufficiently reducing the electric field at the main Schottky contact during the reverse blocking mode.

It is worth pointing out that the area for the main Schottky contact is reduced in half by the presence of the trenches in the TSBS rectifier structures. This results in an increase in the current density at the main Schottky contact which produces the observed shift in the *i-v* characteristics to larger voltages when compared with the normal Schottky rectifier. In comparison with the Schottky rectifier characteristics (shown by the dashed line in the figure), the increase in on-state voltage drop at a forward current density of 100 A/cm$^2$ is small (0.04 volts) for these TSBS rectifier structures.

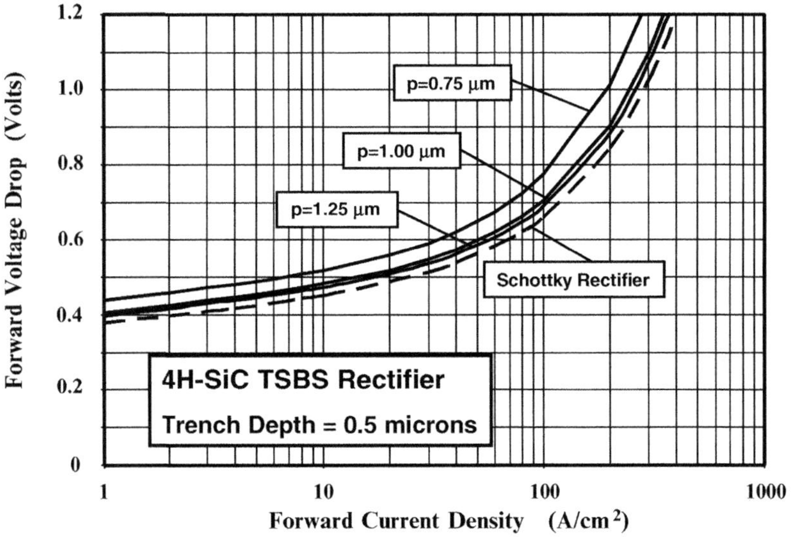


Fig. 4.6 Forward Characteristics of 3kV 4H-SiC TSBS Rectifiers.

The impact of changing the cell pitch on the on-state characteristics of the 4H-SiC TSBS rectifier is illustrated in Fig. 4.6. In this case, the trench depth was kept at the same value of 0.5 microns for the structures. It can be observed that increasing the cell pitch from 1.00 to 1.25 microns produces only a small improvement in the on-state voltage drop. In contrast, reducing the cell pitch to 0.75 microns produces a significant increase in the on-state voltage drop. It is therefore necessary to maintain a cell pitch of at least 1 micron to obtain a low on-state voltage drop. This value is sufficient to suppress the electric field at the main Schottky contact as shown later in the chapter.

**Simulation Example**

In order to validate the above model for the on-state characteristics of the silicon carbide TSBS rectifier, the results of two-dimensional numerical simulations on a 3000 V structure are described here. The structure had a drift region thickness of 20 microns with a doping concentration of 8.5 x $10^{15}$ $cm^{-3}$. The trench region had a depth of 0.5 microns with an etching window (dimension 2s in Fig. 4.2) of 1.0 microns. A work-function of 4.5 eV was used for the main Schottky contact and 5.0 eV for the metal in the trenches. This corresponds to a Schottky barrier height of about 0.8 eV for the main Schottky metal and 1.3 eV for the metal in the trenches based upon an electron affinity of 3.7 eV for 4H-SiC.

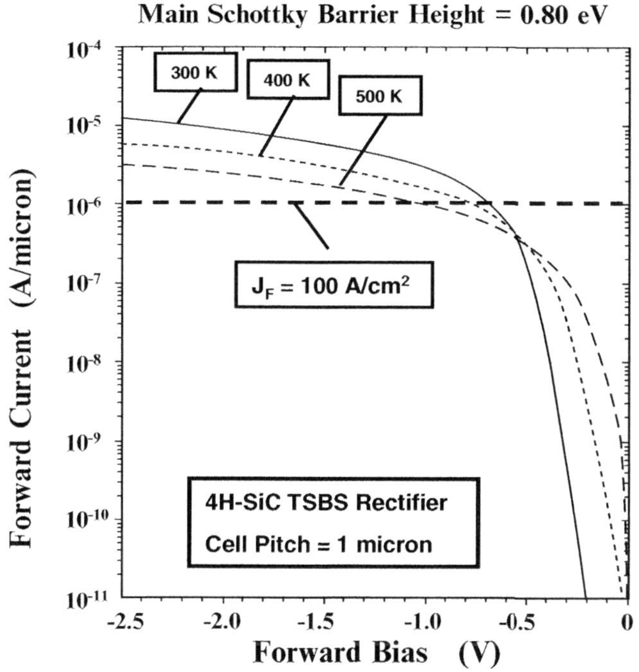


Fig. 4.7E Forward Characteristics of a Typical 3 kV 4H-SiC TSBS Rectifier.

The forward *i-v* characteristics of the typical 3000 V 4H-SiC TSBS rectifier obtained from the numerical simulations are shown in Fig. 4.7E for the case of a cell pitch of 1 micron and a trench depth of 0.5 microns. The on-state voltage drop at a current density of 100 A/cm$^2$ is 0.7 volts at 300 $^o$K which is identical to the value obtained by using the analytical model providing validation for the model. This 4H-SiC rectifier exhibits the desirable positive temperature coefficient for the on-state voltage drop because of the substantial contribution from the drift region resistance. The drift region resistance increases with temperature due to a reduction of the mobility for electrons.

In the TSBS structure, the analytical model for the on-state characteristics is based upon the assumption that the current through the metal in the trenches can be neglected. The current flow at the high barrier metal in the trenches in the typical TSBS structure can be observed in Fig. 4.8E for the case of a cell pitch of 1 micron. The current through the high barrier metal is about 7 orders of magnitude less than that through the low barrier metal at the on-state operating point. This validates the assumptions used when developing the analytical model for the 4H-SiC TSBS rectifier and demonstrates that the on-state voltage drop for the silicon carbide TSBS rectifier is controlled by current transported through the main Schottky contact metal.

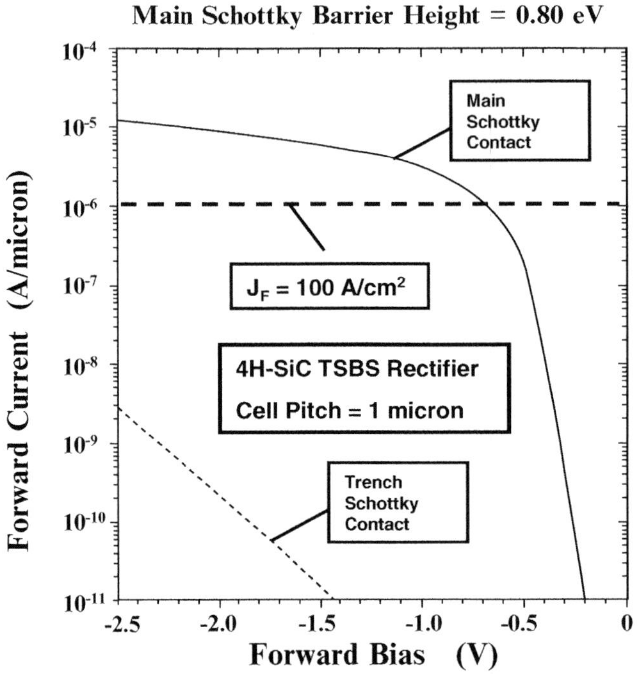


Fig. 4.8E Forward Characteristics of a Typical 3 kV 4H-SiC TSBS Rectifier.

With the analytical model for the silicon carbide TSBS structure, it was found that the on-state voltage drop is not strongly dependent on the depth of the trenches but is sensitive to the cell pitch. The influence of the cell pitch can be demonstrated by performing numerical simulations of the TSBS structure with the same trench depth. It was found that the on-state voltage drop increases substantially when the cell pitch is reduced to 0.75 microns as shown in Fig. 4.9E. The on-state voltage drop increases from 0.7 volts at a cell pitch of 1 micron to 0.9 volts at a cell pitch of 0.75 microns. This increase is similar to the results obtained by using the analytical model (see Fig. 4.6). The reason for the increase in the on-state voltage drop is the greater constriction of the current at the Schottky contact for the smaller cell pitch because the dimension 'd' is reduced from 0.32 microns at a cell pitch of 1 micron to only 0.07 microns for the pitch of 0.75 microns. It will be shown later in this chapter that a cell pitch of 1 micron is adequate to suppress the electric field at the main Schottky contact in the reverse blocking mode.

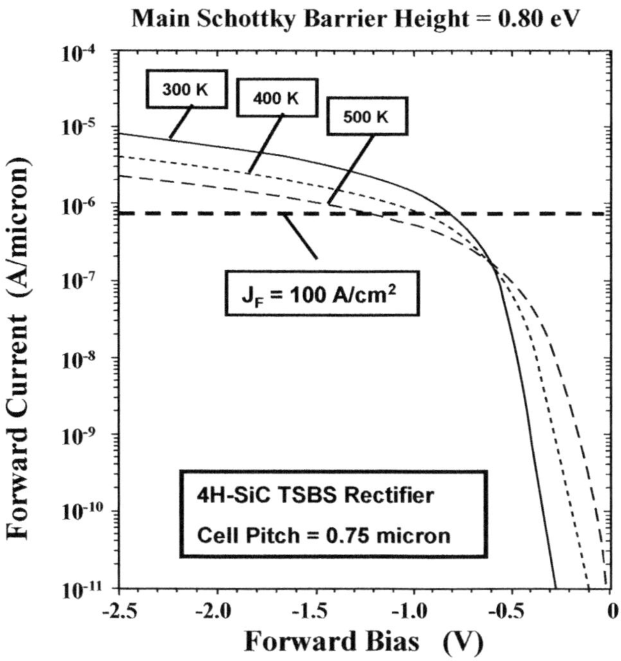


Fig. 4.9E Forward Characteristics of a 3 kV 4H-SiC TSBS Rectifier.

## 4.3 TSBS Rectifier Structure: Reverse Leakage Model

The reverse leakage current in the TSBS rectifier is reduced when compared with the Schottky rectifier due to the smaller electric field at the metal-semiconductor interface at the main Schottky contact. In addition, the area of the main Schottky contact is a fraction of the total cell area resulting in a smaller reverse current contribution. In the case of the silicon TSBS rectifier, the reduced electric field at the Schottky contact suppresses the barrier lowering effect. In the case of the silicon carbide TSBS rectifier, the reduced electric field not only decreases the barrier lowering but also mitigates the influence of thermionic field emission. The impact of the reduction of the electric field at the Schottky contact in the TSBS rectifier structure on the leakage current is analyzed in this section.

### 4.3.1 Silicon TSBS Rectifier: Reverse Leakage Model

As in the case of the JBS rectifier structure, the leakage current model for the TSBS rectifier must take into account the smaller main Schottky contact area within the cell and the influence of the smaller electric field generated at the Schottky contact due to the shielding effect. The leakage current contribution from the main Schottky contact in the silicon TSBS rectifier can be obtained by using:

$$J_L = \left(\frac{p-s}{p}\right) AT^2 \exp\left(-\frac{q\phi_b}{kT}\right).\exp\left(\frac{q\beta\Delta\phi_{bTSBS}}{kT}\right) \qquad [4.14]$$

where $\beta$ is a constant to account for a smaller barrier lowering closer to the trenches as discussed previously for the JBS rectifier structure. In the previous chapter on Schottky rectifiers, it was demonstrated that the high electric field at the Schottky contact produces a reduction of the effective barrier height due to the image force lowering phenomenon. In contrast with the Schottky rectifier, the barrier lowering for the TSBS rectifier is determined by the reduced electric field $E_{TSBS}$ at the main Schottky contact:

$$\Delta\phi_{bTSBS} = \sqrt{\frac{qE_{TSBS}}{4\pi\varepsilon_S}} \qquad [4.15]$$

In the TSBS rectifier structure, the electric field at the Schottky contact varies with distance away from the trenches. The highest electric field is observed at the middle of the main Schottky contact with a progressively smaller value closer to the trenches. In an analytical model with a worst case scenario, it is prudent to use the electric field at the middle of the main Schottky contact to compute the leakage current. Until the depletion regions from the adjacent trenches produce a potential barrier under the main Schottky contact, the electric field at the metal-semiconductor interface in the middle of the contact increases with the applied reverse bias voltage as in the case of the normal Schottky rectifier. A potential barrier is established by the trenches after depletion of the drift region below the main Schottky contact. The voltage at which the depletion regions from the adjacent trenches intersect under the main Schottky contact is referred to as the *pinch-off voltage*. The pinch-off voltage ($V_P$) can be obtained from the device cell parameters:

$$V_P = \frac{qN_D}{2\varepsilon_S}(p-s)^2 - V_{CT} \qquad [4.16]$$

where $V_{CT}$ is the contact potential for the metal in the trenches.

Although a potential barrier begins to form after the reverse bias exceeds the pinch-off voltage, the electric field continues to rise at the main Schottky contact due to encroachment of the potential to the main Schottky contact. In order to analyze the impact of this on the reverse leakage current, the electric field $E_{TSBS}$ can be related to the reverse bias voltage by:

$$E_{TSBS} = \sqrt{\frac{2qN_D}{\varepsilon_S}(\alpha V_R + V_{CM})} \qquad [4.17]$$

where $\alpha$ is a coefficient used to account for the build up in the electric field after pinch-off and $V_{CM}$ is the contact potential for the main Schottky contact.

The above analysis for the electric field at the main Schottky contact is identical to that for the silicon and silicon carbide JBS rectifiers developed in the previous chapter. Consequently, the same graph relation the electric field to the reverse bias for various values of alpha shown in Fig. 3.11 is also applicable to the TSBS structure. The value for alpha for the silicon TSBS rectifier structure is different from that for the silicon JBS rectifier structure because the rectangular shape of the trenches in the TSBS structure produces a greater suppression of the electric field.

The impact of the reduction of the electric field at the Schottky contact on the Schottky barrier lowering in the silicon TSBS rectifier structure is also the same as that shown in Fig. 3.12. Without the shielding by the metal in the trenches, a barrier lowering of 0.07 eV occurs in the normal silicon Schottky rectifier. The barrier lowering is reduced to 0.05 eV with an alpha of 0.2 in the TSBS rectifier structure. Although this may appear to be a small change, it has a large impact on the reverse leakage current as was already shown in Fig. 3.13.

The analysis of the leakage current for the TSBS rectifier structure requires inclusion of the contribution from the trench metal contact. The leakage current flowing through this contact can in principle be made much smaller than that flowing through the main contact metal by making the barrier height for the trench contact large. For silicon devices, the largest practical value for the Schottky barrier height is only 0.85 volts by using platinum silicide. The leakage current flowing through the trench contact metal increases rapidly with increasing reverse bias voltage because of the Schottky barrier lowering phenomenon. This increase is exacerbated by the larger electric field generated at the trench corners in the TSBS rectifier structure.

The leakage current contribution from the trench contact in the silicon TSBS rectifier can be modeled by using the planar Schottky barrier theory:

$$J_{LTS} = \left(\frac{s}{p}\right) AT^2 \exp\left(-\frac{q\phi_{bTS}}{kT}\right).\exp\left(\frac{q\Delta\phi_{bTS}}{kT}\right) \qquad [4.18]$$

The barrier lowering for the trench contact in the TSBS rectifier is determined by the electric field $E_{TS}$ at this contact:

$$\Delta\phi_{bTS} = \sqrt{\frac{qE_{TS}}{4\pi\varepsilon_S}} \qquad [4.19]$$

where the electric field ($E_{TS}$) at the trench contact can be computed using parallel-plane analysis. However, this methodology produces a leakage current contribution from the trench metal contact that is much smaller than actually prevalent in the TSBS rectifier structure. The leakage current contribution from the trench metal contact is significantly enhanced by the high electric field generated at the sharp corners of the trenches. The larger electric field at the trench corners aggravates the Schottky barrier lowering phenomenon producing a very rapid increase in leakage current at high reverse bias voltages. Since the electric field at the trench corners is determined by two-dimensional effects, it is not amenable to simple analytical modeling.

### Simulation Example

In order to validate the above model for the reverse characteristics of the silicon TSBS rectifier, the results of two-dimensional numerical simulations on a 50 V structure are described here. The structure had a drift region with a doping concentration of 8.5 x $10^{15}$ $cm^{-3}$ and a thickness of 3 microns. The typical silicon TSBS rectifier structure had a trench depth of 0.5 microns with trench etching window (dimension 2s in Fig. 4.2) of 1 micron. The work function of the main Schottky metal was chosen to obtain a barrier height of 0.60 eV and that for the trench metal was chosen to obtain a barrier height of 0.85 eV.

A three dimensional view of the electric field distribution in the typical TSBS rectifier cell is shown in Fig. 4.10E. The main Schottky contact is located on the lower right-hand-side in the figure with the trench region located at the top of the figure. A high electric field (4 x $10^5$ V/cm) is observed at the metal interface at the bottom of the trench. However, the electric field at the middle of the main Schottky contact is greatly reduced (1.5 x $10^5$ V/cm). It can also be seen that the electric field becomes smaller when proceeding towards the P-N junction. Consequently, a worst case analysis of the leakage current can be performed by using the electric field at the middle of the main Schottky contact. It is worth pointing out that a large electric field (6 x $10^5$ V/cm) is generated at the sharp corner of the trench. This can degrade the breakdown voltage within the cell. A reduction of this electric field can be achieved by rounding the bottom of the trenches.

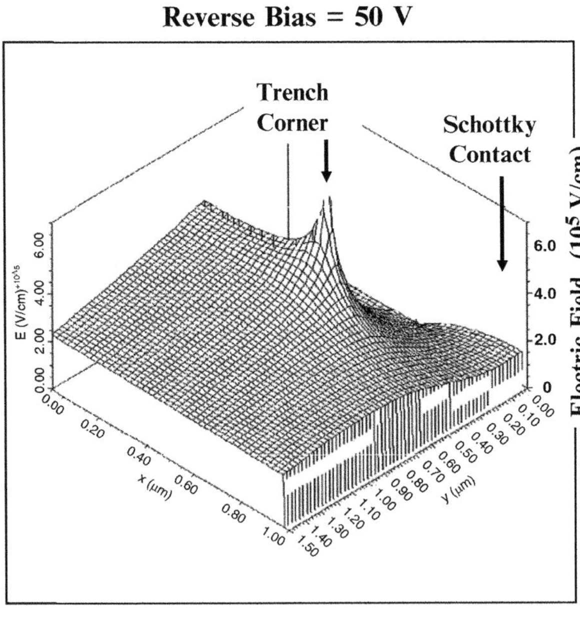


Fig. 4.10E Electric Field Distribution in a Typical 50 V Silicon TSBS Rectifier.

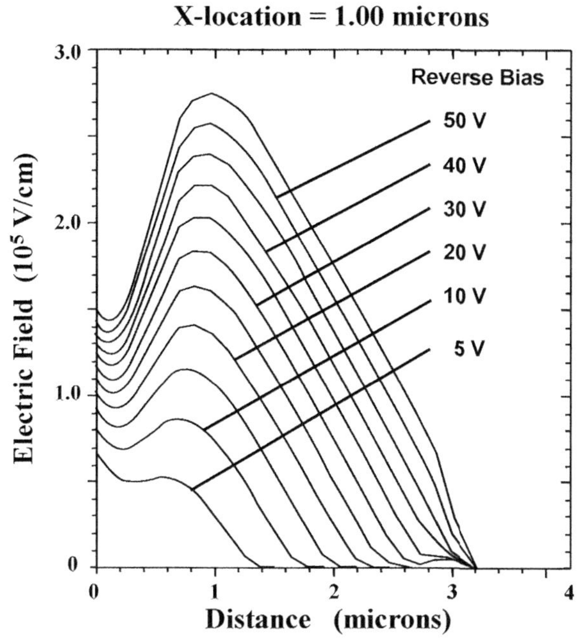


Fig. 4.11E Growth of the Electric Field at the Middle of the main Schottky Contact in a Typical 50 V Silicon TSBS Rectifier.

The growth in the electric field at the middle of the main Schottky contact with increasing reverse bias voltage for the typical TSBS rectifier structure with cell pitch of 1.00 microns is shown in Fig. 4.11E. In comparison with the growth of the electric field for the normal Schottky rectifier previously shown in Fig. 3.10E, it is apparent that the electric field at the main Schottky contact is suppressed in the TSBS rectifier. Although the potential barrier in this TSBS rectifier structure occurs at a depth of about 0.5 microns, the peak electric field occurs at a depth of about 1 micron. Consequently, a sufficient thickness for the drift region is required below the bottom of the trenches to prevent punch-through of the electric field to the $N^+$ substrate at high reverse bias voltages.

An even greater suppression of the electric field at the main Schottky contact can be obtained by either increasing the trench depth or reducing the cell pitch in the TSBS rectifier structure. The improvement obtained with a trench depth of 0.75 microns is shown in Fig. 4.12E for a TSBS rectifier structure with cell pitch of 1.00 microns. The electric field at the center of the main Schottky contact is now reduced to 0.9 x $10^5$ V/cm compared with 1.5 x $10^5$ V/cm for the typical TSBS rectifier structure with trench depth of 0.5 microns. The location of the peak electric also shifts to a greater depth for the structure with the deeper trench resulting in strong punch-through of the electric field profile to the $N^+$ substrate at reverse bias voltages above 40 volts. These simulations are intended to illustrate the impact of changing the trench depth while maintaining the same total thickness for the drift region. In practice, it is preferable to increase the total thickness of the drift region when the trench depth is increased to avoid the punch-through of the electric field to the $N^+$ substrate.

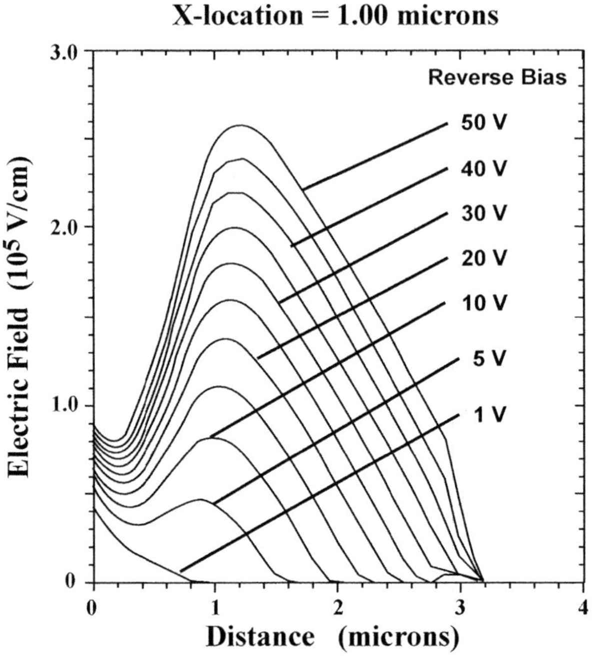


Fig. 4.12E Growth of the Electric Field at the Middle of the main Schottky Contact in a 50 V Silicon TSBS Rectifier.

The greater reduction of the electric field at the main Schottky contact obtained by reducing the cell pitch is illustrated in Fig. 4.13E for the a TSBS rectifier structure with cell pitch of 0.75 microns and trench depth of 0.5 microns. The electric field at the center of the main Schottky contact is now reduced to only 0.35 x $10^5$ V/cm compared with 1.5 x $10^5$ V/cm for the typical TSBS rectifier structure with trench depth of 0.5 microns. However, as pointed out earlier, the increase in the on-state voltage drop does not warrant such a drastic reduction of the electric field at the main Schottky contact.

The electric field generated at the corners of the trenches is also dependent up on the space between the trenches. On the one hand, when the trenches are located close together, the depletion of the space between them occurs at a small reverse bias voltage. The electric field at the trench corners is not significantly enhanced in this case. However, the area of the main Schottky contact is reduced when the space between the trenches is reduced producing a larger on-state voltage drop. On the other hand, if the space between the trenches is enlarged, the electric field enhancement at the trench corners becomes severe.

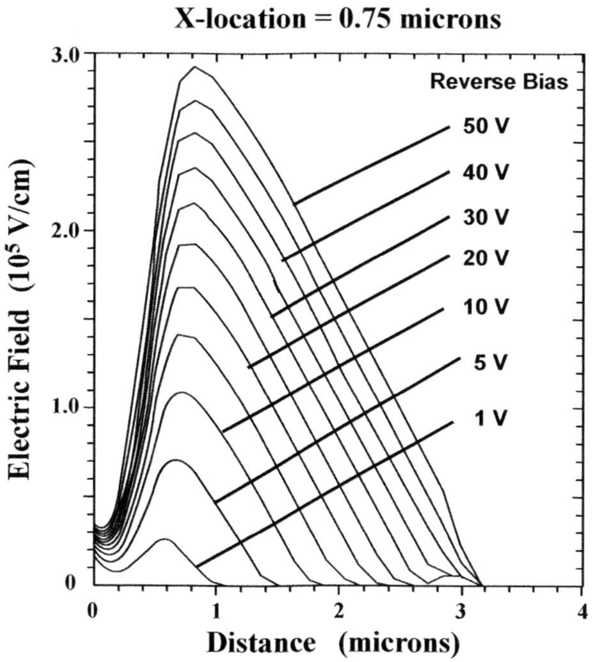


Fig. 4.13E Growth of the Electric Field at the Middle of the main Schottky Contact in a 50 V Silicon TSBS Rectifier.

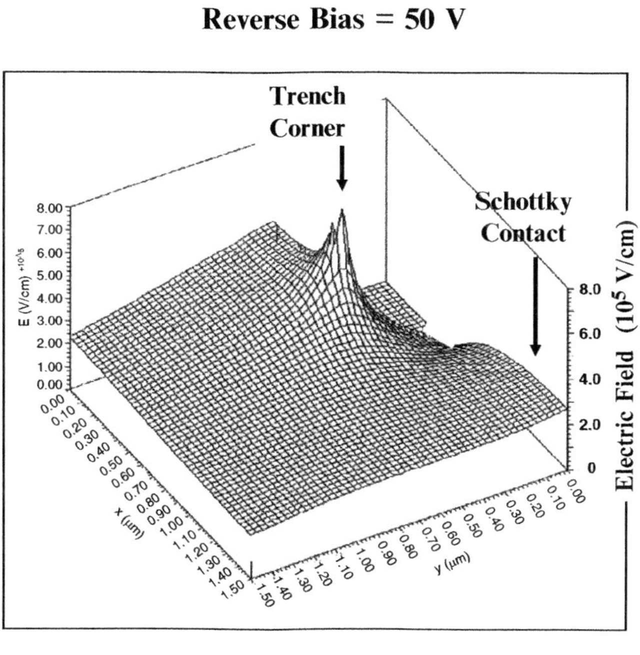


Fig. 4.14E Electric Field Distribution in a 50 V Silicon TSBS Rectifier.

This can be observed in the three-dimensional view of the electric field for a TSBS rectifier structure with a cell pitch of 1.5 microns and a trench depth of 0.5 microns shown in Fig. 4.14E. The electric field at the corner of the trenches has increased to 8 x $10^5$ V/cm at a reverse bias of 50 volts when compared with 6.0 x $10^5$ V/cm for the typical TSBS rectifier structure with a cell pitch of 1 micron. The enhanced electric field at the trench corners can reduce the breakdown voltage of the TSBS rectifier within the cell. It is important to maintain this breakdown voltage above the breakdown voltage at the edge termination. It is also worth pointing out that the larger space between the trenches results in a larger electric field at the main Schottky contact (compare Fig. 4.14E with Fig. 4.10E).

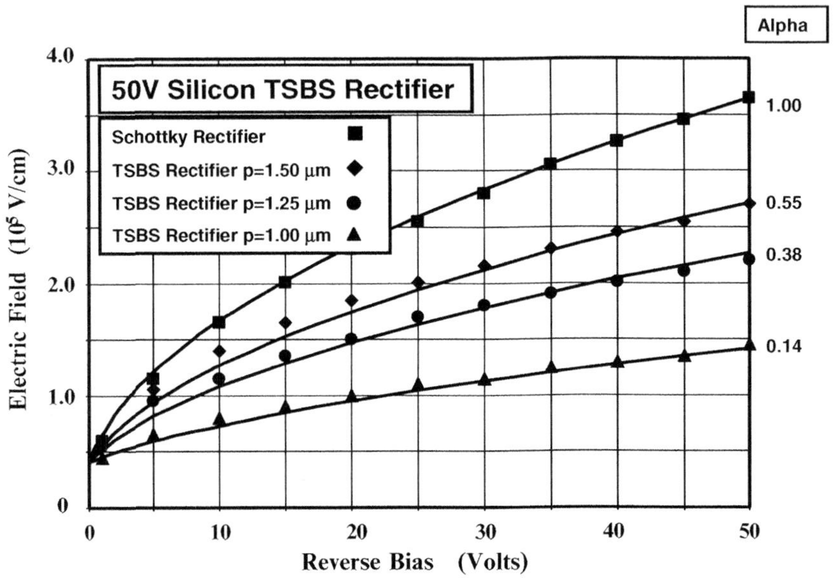


Fig. 4.15E Growth of the Electric Field at the Middle of the Schottky Contact for Silicon 50V TSBS Rectifiers.

The coefficient alpha, that governs the rate at which the electric field increases at the middle of the main Schottky contact in the analytical model for the TSBS rectifier structure, can be extracted from the results of the two-dimensional numerical simulations. The increase in the electric field at the middle of the main Schottky contact, obtained from the numerical simulations, is shown in Fig. 4.15E for the TSBS rectifiers with cell pitch (p) ranging from 1.00 to 1.50 microns by the symbols. The results of calculations based upon using the analytical Eq. [4.17] are shown by the solid lines with the values fro alpha adjusted to fit the results of the numerical simulations. The case with alpha of unity fits the Schottky rectifier quite well as expected. The value for alpha for the TSBS rectifier with pitch of 1.00 microns is found to be 0.14. It increases to 0.38 for a cell pitch of 1.25 microns while the alpha for a pitch of 1.50 microns is 0.55. With these values of alpha, the analytical model accurately predicts the behavior of the electric field at the middle of the Schottky contact. It can therefore be used to compute the Schottky barrier lowering and leakage current in TSBS rectifiers.

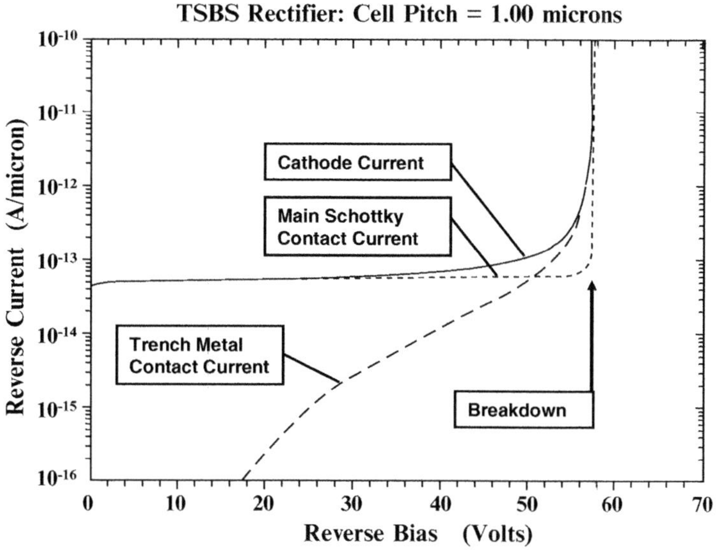


Fig. 4.16E Reverse Blocking Characteristics for a Silicon 50V TSBS Rectifier.

The numerical simulations provide insight into the current flow within the TSBS rectifier structure during the reverse blocking mode. The total reverse current flowing through the cathode electrode is compared with the currents flowing through the main Schottky contact and the trench metal contact in Fig. 4.16E. At reverse bias voltages below 30 volts, the cathode current is equal to the current flowing through the main Schottky contact. The current flow through the main Schottky contact remains essentially constant with increasing voltage due to suppression of growth in the electric field at the main Schottky contact. However, the current flowing through the trench metal contact increases rapidly when the reverse voltage exceeds 30 volts and becomes comparable to the current in the main Schottky contact at reverse bias voltages above 50 volts. It is necessary to use a sufficiently large barrier height for the trench metal contact to reduce this contribution. The leakage current from the trench contact can also be reduced by rounding the bottom of the trenches to reduce the electric field at the corners. It is worth pointing out that the breakdown occurs due to a very rapid increase in the current flowing through both the contacts.

The reverse *i-v* characteristic of the 50 V Silicon TSBS rectifier with a cell pitch of 1.00 microns obtained from the numerical simulations is shown in Fig. 4.17E together with that for the Schottky rectifiers using barrier heights corresponding to the main contact metal (0.60 eV) and the metal in the trenches (0.85 eV). All the devices had a cross-sectional width (cell pitch) of 1 micron. The breakdown voltage for the TSBS rectifier cell is observed to be 57 volts. This is consistent with an edge termination limited breakdown voltage of 50 volts at 80 percent of the parallel-plane breakdown voltage. The leakage current at small reverse bias voltages is 2x times smaller in the TSBS rectifier structure than for the Schottky rectifier with a barrier height of 0.60 eV. This is consistent with the reduction of the Schottky contact area by a factor of 2x in this TSBS rectifier structure. The leakage current for the Schottky rectifier increases by a factor of 7x when the reverse voltage increases to 50 volts. In contrast, the reverse leakage current for this TSBS rectifier increases by a factor of only 2x when the reverse voltage is increased to 50 volts. This increase is consistent with predictions of the analytical model. The leakage current contributed by a planar Schottky rectifier with a barrier height of 0.85 eV, corresponding to the trench contact, is also shown in the figure. This contribution remains well below that for the other devices. From this observation, it can be concluded that the enhancement of the electric field at the trench corners has a significant adverse impact on the leakage current contributed by the trench metal in the TSBS rectifier structure.

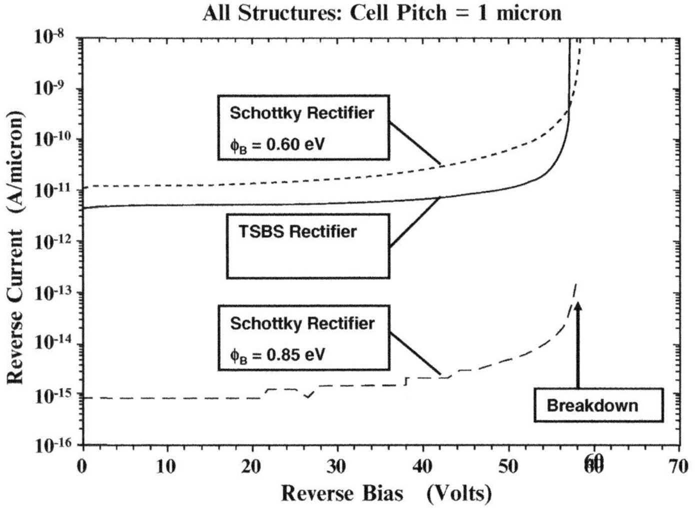


Fig. 4.17E Reverse Blocking Characteristics for a Typical Silicon 50V TSBS Rectifier compared with Schottky Rectifiers.

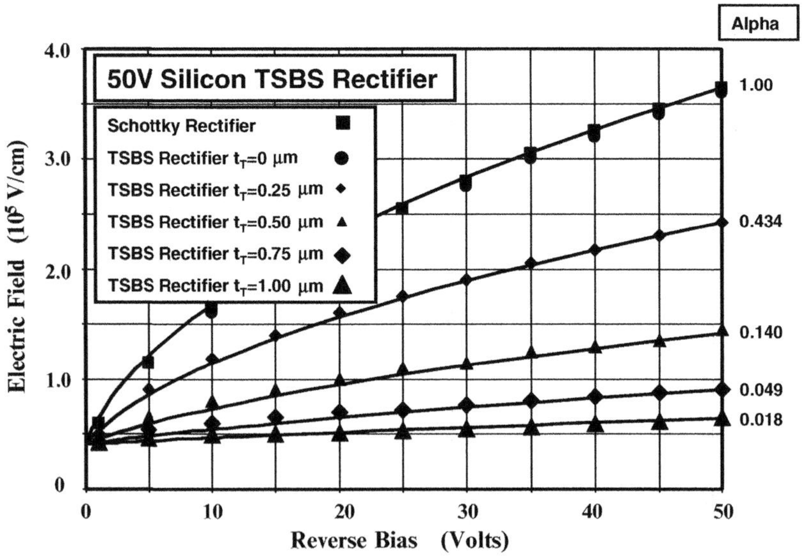


Fig. 4.18E Growth of the Electric Field at the Middle of the Schottky Contact for Silicon 50V TSBS Rectifiers.

The growth of the electric field with reverse bias voltage for the TSBS rectifier structures with the different trench depths is compared in Fig. 4.18E. In this figure, the value for alpha in the analytical model was adjusted to match the simulation data. It can be seen that the alpha is reduced from 1.00 to 0.018 with the increasing trench depth due to the larger channel aspect ratio. This is beneficial for suppressing the Schottky barrier lowering and decreasing the leakage current.

### 4.3.2 Silicon Carbide TSBS Rectifier: Reverse Leakage Model

The leakage current in the silicon carbide TSBS rectifier can be calculated using the same approach as for the silicon TSBS rectifier structure. Firstly, it is important to account for the smaller main Schottky contact area in the TSBS rectifier cell. Secondly, it is necessary to include Schottky barrier lowering while accounting for the smaller electric field at the main Schottky contact due to shielding by the trench metal electrode. Third, the thermionic field emission current must be included while accounting for the smaller electric field at the main Schottky contact due to shielding by the trench metal contact. After making these adjustments, the leakage current for the silicon carbide TSBS rectifier can be calculated by using:

$$J_L = \left(\frac{p-s}{p}\right) AT^2 \exp\left(-\frac{q\phi_b}{kT}\right).\exp\left(\frac{q\Delta\phi_{bTSBS}}{kT}\right).\exp\left(C_T E_{TSBS}^2\right) \quad \textbf{[4.20]}$$

where $C_T$ is a tunneling coefficient (8 x $10^{-13}$ $cm^2/V^2$ for 4H-SiC). In contrast to the Schottky rectifier, the barrier lowering for the TSBS rectifier is determined by the reduced electric field $E_{TSBS}$ at the contact:

$$\Delta\phi_{bTSBS} = \sqrt{\frac{qE_{TSBS}}{4\pi\varepsilon_S}} \quad \textbf{[4.21]}$$

As in the case of the silicon TSBS structure, the electric field at the main Schottky contact varies with distance away from the trenches. The highest electric field is observed at the middle of the main Schottky contact with a progressively smaller value closer to the trenches. When developing an analytical model with a worst case scenario, it is prudent to use the electric field at the middle of the main Schottky contact to compute the leakage current.

Until the depletion regions from the adjacent trenches produce a potential barrier under the main Schottky contact, the electric field at the metal-semiconductor interface in the middle of the main Schottky contact increases with the applied reverse bias voltage as in the case of the normal Schottky rectifier. A potential barrier is established by the trench metal contacts after depletion of the drift region below the Schottky contact. As in the case of the silicon TSBS rectifier structure, the pinch-off voltage ($V_P$) can be obtained from the device cell parameters:

$$V_P = \frac{qN_D}{2\varepsilon_S}(p-s)^2 - V_{CT} \quad \textbf{[4.22]}$$

where $V_{CT}$ the contact potential for the metal in the trenches. It is worth pointing out that the contact potential for 4H-SiC is larger than for silicon because it is advantageous to utilize a larger barrier height for the metal in the trenches. Although the potential barrier begins to form after the reverse bias exceeds the pinch-off voltage, the electric field continues to rise at the main Schottky contact due to encroachment of the potential to the main Schottky contact. In order to analyze the impact of this on the reverse leakage current, the electric field $E_{TSBS}$ can be related to the reverse bias voltage by:

$$E_{TSBS} = \sqrt{\frac{2qN_D}{\varepsilon_S}(\alpha V_R + V_{CT})} \quad [4.22]$$

where $\alpha$ is a coefficient used to account for the build up in the electric field after pinch-off.

Since the TSBS structure in silicon carbide has the same rectangular shape for the trenches as the P-N junction in the silicon carbide JBS structure, the plots provided in Chapter 3 are also applicable to the TSBS rectifier structure. The reader can refer to Fig. 3.14 for the reduction of electric field at the main Schottky contact for various values of alpha; Fig. 3.15 for the reduction of Schottky barrier lowering at the main Schottky contact for various values of alpha; and Fig. 3.16 for the reduction of leakage current at the main Schottky contact for various values of alpha. However, the values for alpha extracted from the simulations of TSBS rectifier structures must be utilized for the analysis of specific TSBS rectifier structures.

### Simulation Example

In order to validate the above model for the reverse leakage characteristics of the silicon carbide TSBS rectifier, the results of two-dimensional numerical simulations on a 3000 V structure are described here. The structure had a drift region thickness of 20 microns with a doping concentration of $8.5 \times 10^{15}$ $cm^{-3}$. The trench region had a depth of 0.5 microns with an etching window (dimension 2s in Fig. 4.2) of 1.0 microns. A work-function of 4.5 eV was used for the main Schottky contact and 5.0 eV for the metal in the trenches. This corresponds to a Schottky barrier height of about 0.8 eV for the main Schottky metal and 1.3 eV for the metal in the trenches based upon an electron affinity of 3.7 eV for 4H-SiC.

The reverse blocking characteristics of the TSBS structure were also studied by using two-dimensional numerical simulations. The breakdown voltage of the TSBS structure was found to be the same (~ 3000 volts) as that for the Schottky rectifiers for the same drift region parameters. The currents flowing through the main Schottky contact and trench metal are shown in Fig. 4.19E as a function of the reverse bias voltage. It can be seen that the leakage current contribution from the main Schottky contact dominates at reverse bias voltages up to 2500 volts due to its low barrier height. However, the exponential increase in leakage current with reverse bias on the unprotected trench metal contact allows its leakage current to catch-up and then exceed the leakage current of the main contact at a reverse bias of 2900 volts.

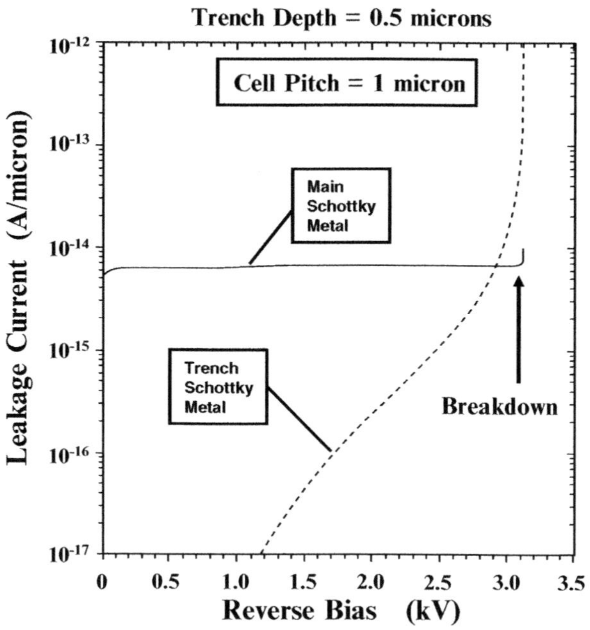


Fig. 4.19E Reverse Blocking Characteristics of a 3 kV 4H-SiC TSBS Rectifier.

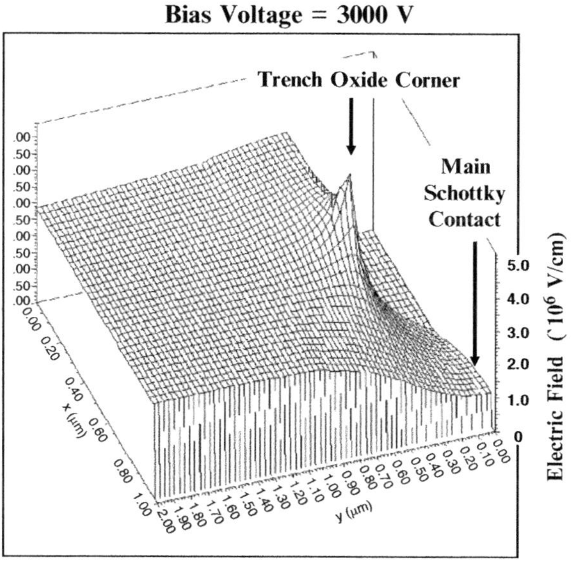


Fig. 4.20E Electric Field distribution in a 4H-SiC TSBS Rectifier.

The increase in the leakage current from the trench contact metal is exacerbated by the enhanced electric field generated at the sharp corners of the trenches. This enhanced electric field can be observed in the three-dimensional view of the electric field in a TSBS structure with cell pitch of 1.00 micron shown in Fig. 4.20E at a reverse bias of 3000 volts (just prior to breakdown). This graph also shows the expected reduction of the electric field at the main Schottky contact by the potential barrier created using the high Schottky barrier metal located in the trenches. As in the case of the JBS rectifier, the maximum electric field at the main Schottky contact occurs at the location furthest away from the trench (at x = 1.00 micron for the structure in Fig. 4.20E). Consequently, the largest barrier lowering and tunneling components will occur at this location. For this reason, the highest electric field at the Schottky contact will be used to analyze the reverse leakage characteristics for the TSBS rectifiers. The degree of field reduction is dependent on the spacing between the trench regions as well as the depth of the trench.

The electric field profile at the center of the main Schottky contact is shown in Fig. 4.21E for the case of a cell pitch of 1 micron. It can be seen that the electric field at the surface under the main Schottky contact is significantly reduced when compared the peak electric field in the bulk. The peak of the electric field occurs at a depth of about 1 micron. At a reverse bias of 3000 volts, the electric field at the Schottky contact is only 1 x $10^6$ V/cm compared with 2.85 x $10^6$ V/cm at the maxima in the bulk. An even greater reduction of the electric field at the main Schottky contact can be achieved by reducing the cell pitch. This is illustrated in Fig. 4.22E for the case of a cell pitch of 0.75 microns. Here, the electric field at the Schottky contact is only 2 x $10^5$ V/cm compared with 2.85 x $10^6$ V/cm at the maxima in the bulk.

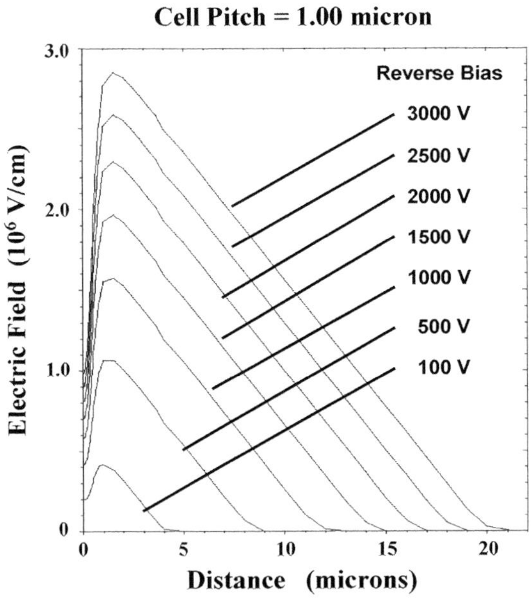


Fig. 4.21E Electric Field variation with Reverse Voltage in a 4H-SiC TSBS Rectifier.

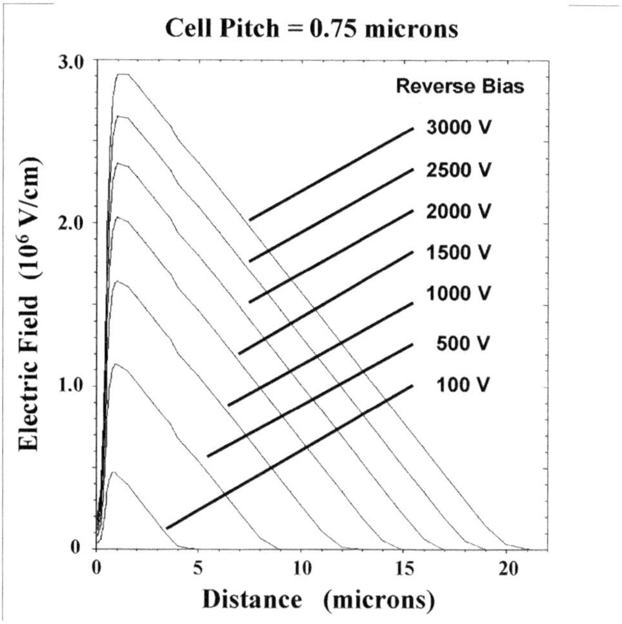


Fig. 4.22E Electric Field variation with Reverse Voltage in a 4H-SiC TSBS Rectifier.

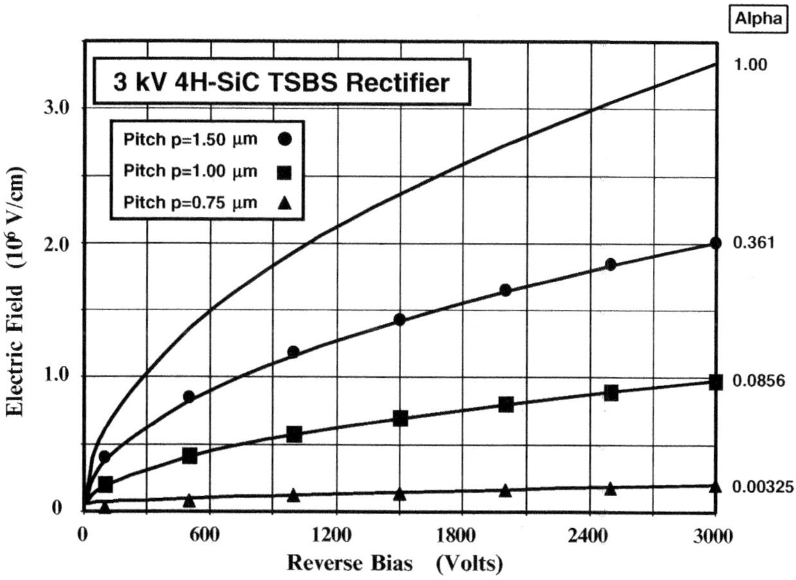


Fig. 4.23E Electric Field variation with Reverse Voltage in 4H-SiC TSBS Rectifiers.

The increase in the electric field at the main Schottky contact in the TSBS rectifier structure with increasing reverse bias voltage is charted in Fig. 4.23E for various cases of the cell pitch with the trench depth held constant at 0.5 microns. The electric field at the middle of the main Schottky contact becomes significantly less than for a normal Schottky rectifier when the cell pitch is made less than 2 microns. A greater suppression of the electric field is apparent as the cell pitch is reduced. In order to extract the alpha coefficient for the silicon carbide TSBS structure, the data obtained from the numerical simulations is shown by the symbols in Fig. 4.23E while the analytically calculated electric field using various alpha values are shown by the solid lines. Thus, the benefits of using the TSBS concept to suppress the electric field at the main Schottky contact can be obtained only with carefully optimized spacing which is slightly smaller than for the silicon carbide JBS rectifier[6]. With proper spacing, the reduction of the electric field provides significant benefits due to reducing the Schottky barrier lowering and tunneling currents.

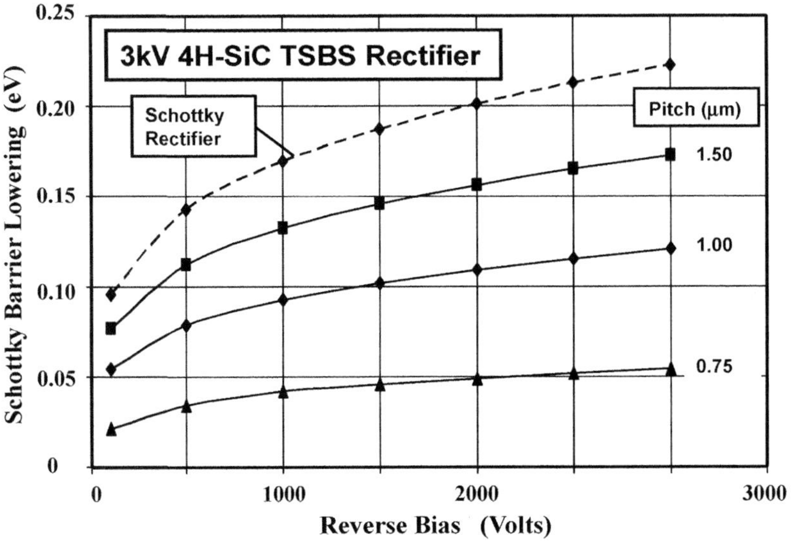


Fig. 4.24E Schottky Barrier Lowering in 4H-SiC TSBS Rectifiers.

The protection of the main Schottky contact against barrier lowering is quantified in Fig. 4.24E, where the values were determined using the analytical formulae with the electric field extracted from the simulations. With a pitch of 1 micron, the Schottky barrier lowering is reduced from 0.22 eV to 0.12 eV. The impact of ameliorating the Schottky barrier lowering phenomenon on the leakage current is quite dramatic for silicon carbide Schottky rectifiers because of the strong dependence of the tunneling component on the electric field. Even for a cell pitch of 1 micron, a reduction in the leakage current by five orders of magnitude is observed at high reverse bias voltages, as shown in Fig. 4.25E, demonstrating the advantage of utilizing the potential barrier to suppress the leakage current from the main contact. (Note that the leakage current from the trench metal was neglected in these plots.) An even greater reduction of the leakage current is possible by shrinking the cell pitch to 0.75 microns but this is accompanied by an increase in the on-state voltage drop. The TSBS structure is therefore very effective for improving the reverse blocking characteristics of high voltage silicon carbide rectifiers.

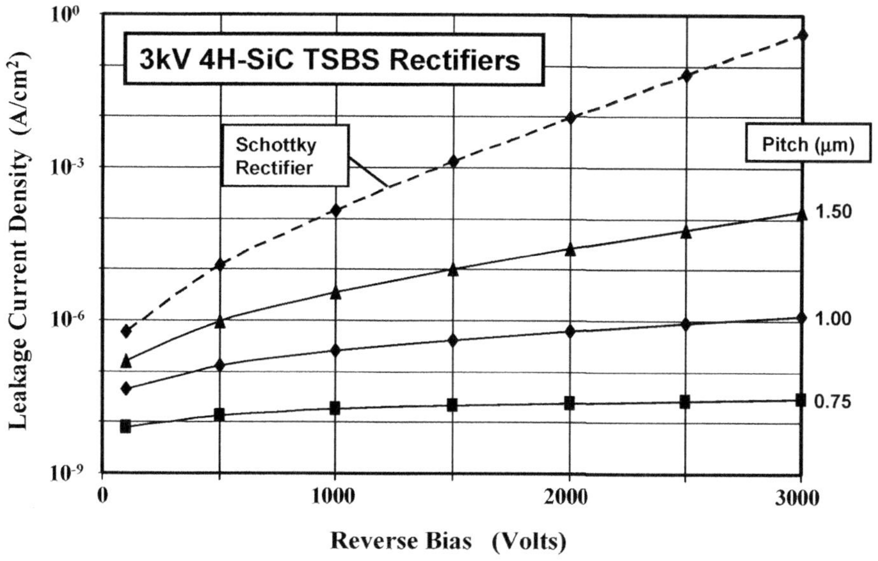


Fig. 4.25E Leakage Current in 4H-SiC TSBS Rectifiers.

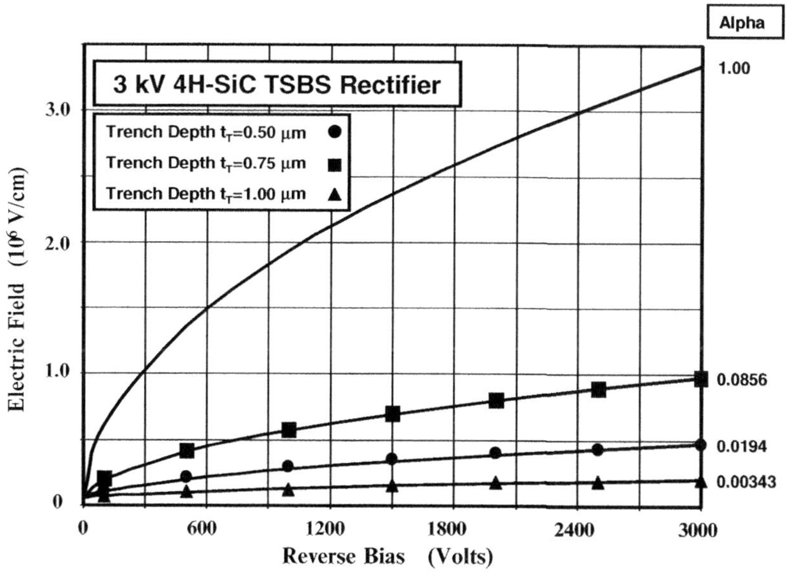


Fig. 4.26E Electric Field variation with Reverse Voltage in 4H-SiC TSBS Rectifiers.

The electric field at the main Schottky contact for the different trench depths is compared in Fig. 4.26E for the case of a fixed cell pitch of 1 micron. The electric field is reduced by a factor of about 3x when a trench depth of 0.5 microns is utilized. An even greater reduction of the electric field by a factor of 5x occurs for a trench depth of 1 micron. In order to extract the alpha coefficient for the silicon carbide TSBS structure, the data obtained from the numerical simulations is shown by the symbols in Fig. 4.26E while the analytically calculated electric field using various alpha values are shown by the solid lines.

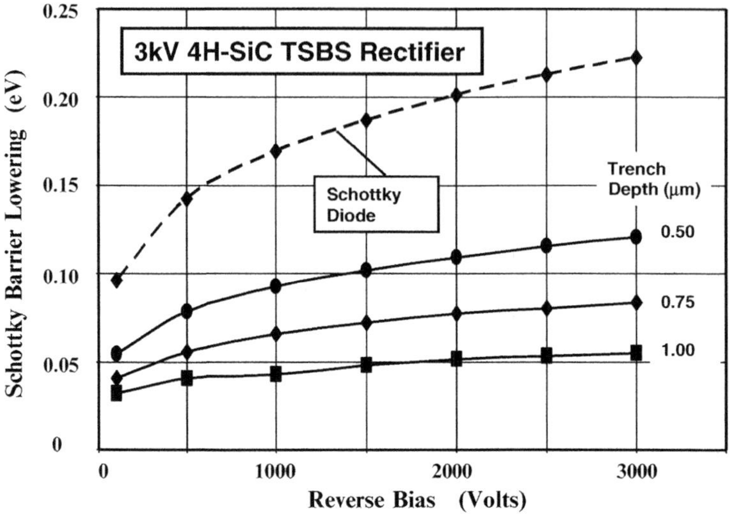


Fig. 4.27E Schottky Barrier Lowering in 4H-SiC TSBS Rectifiers.

The impact of the reduction of the electric field at the main Schottky contact on the Schottky barrier lowering is quantified in Fig. 4.27E. With a trench depth of 0.5 microns, the barrier lowering is reduced from 0.22 eV for a normal silicon carbide Schottky rectifier to 0.125 eV. When the trench depth is increased to 0.75 microns, the barrier lowering is reduced to only 0.08 eV while it becomes only 0.055 eV for a trench depth of 1.00 microns.

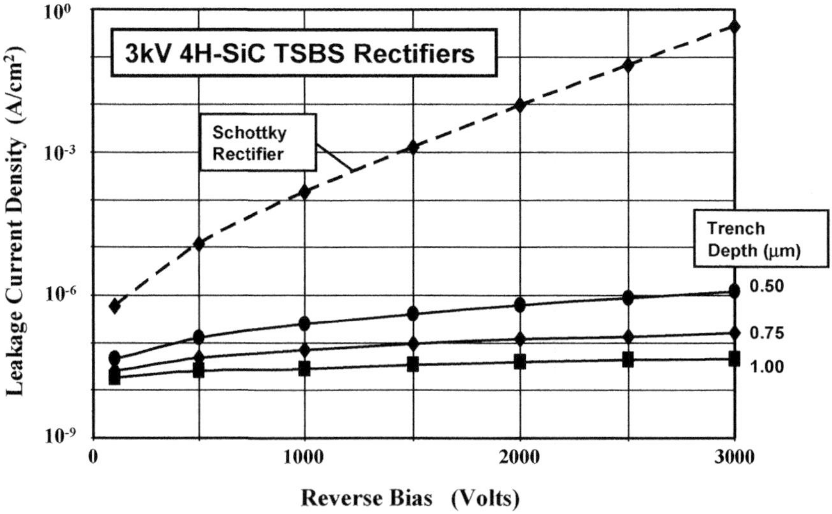


Fig. 4.28E Leakage Current in 4H-SiC TSBS Rectifiers.

This reduction in the Schottky barrier lowering reduces the leakage current in the silicon carbide TSBS Schottky rectifier structure as demonstrated in Fig. 4.28E. The reduction of the tunneling current in silicon carbide Schottky contacts was included when computing the leakage current during the analysis. It can be observed that the leakage current is reduced by 5 orders of magnitude for a trench depth of 0.5 microns. Further increase in the trench depth reduces the leakage current only by another order of magnitude. Consequently, it is sufficient to utilize a trench depth of 0.5 microns for the silicon carbide TSBS rectifier structure to suppress the leakage current.

The aspect ratio of the current conduction region located below the main Schottky contact in the TSBS rectifier has a strong influence on the suppression of the electric field at the contact, which is quantified by the coefficient alpha ($\alpha$) in Eq. (4.17) and Eq. (4.22). The aspect ratio for both the silicon and silicon carbide TSBS rectifier structures can be computed by dividing the trench depth by the space between the trenches:

$$AR = \frac{t_T}{2(p-s)} \qquad [4.23]$$

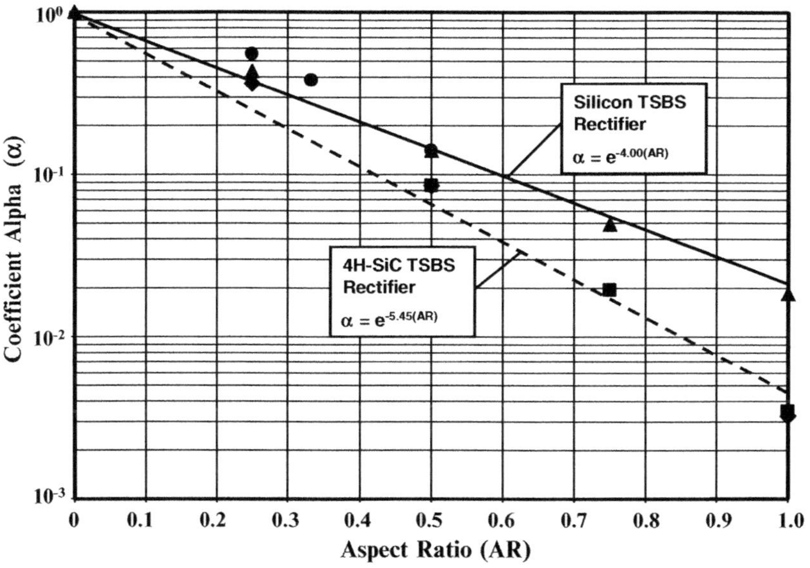


Fig. 4.29E Impact of Aspect Ratio on Alpha for TSBS Rectifiers.

The variation of the coefficient alpha ($\alpha$) with aspect ratio, obtained from the numerical simulations, is shown in Fig. 4.29E for silicon and silicon carbide TSBS rectifiers. It can be observed that the coefficient alpha ($\alpha$) varies exponentially with the aspect ratio. A single line fits the data obtained for silicon devices (and another single line fits the data for 4H-SiC devices) with different trench depths and cell pitches demonstrating that the aspect ratio is determining the alpha. It is smaller for the case of silicon carbide devices when compared with silicon devices with the same aspect ratio indicating that the electric field is not suppressed to the same degree in silicon devices. The TSBS structure is therefore particularly well suited for improving the performance of silicon carbide Schottky rectifiers.

## 4.4 Trade-Off Curve

In the textbook[4], it is demonstrated that the power dissipation can be minimized for a particular duty cycle and operating temperature by varying the Schottky barrier height during optimization of the Schottky rectifier structure. A smaller barrier height decreases the on-state voltage drop reducing conduction power losses while a large barrier height decreases the leakage current reducing reverse blocking power losses. Depending upon the duty cycle and the temperature, minimum power loss occurs at an optimum barrier height. A fundamental trade-off curve between on-state voltage drop and the leakage current was developed in the textbook based upon these considerations which does not depend upon the semiconductor material. However, the fundamental trade-off curve excludes the impact of the series resistance on the on-state voltage drop. More significantly, it excludes the strong influence of Schottky barrier lowering and pre-avalanche multiplication on the increase in leakage current for silicon Schottky rectifiers.

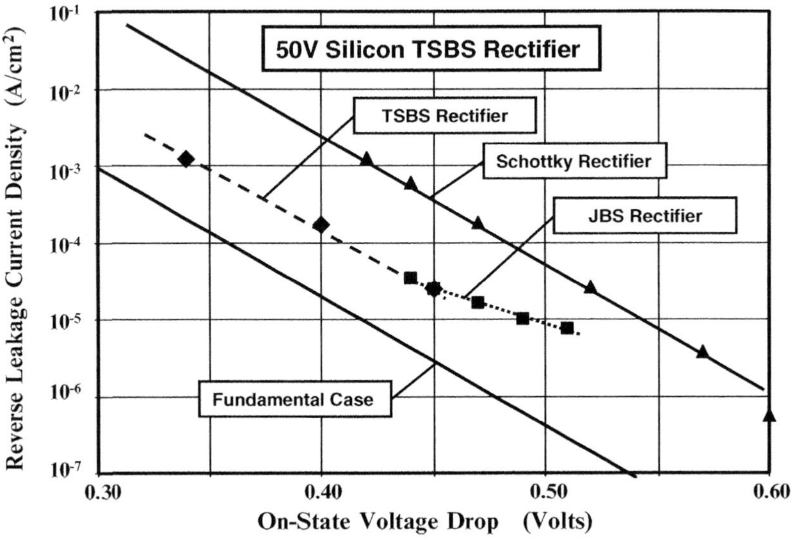


Fig. 4.7 Trade-Off Curve for the 50 V Silicon TSBS Rectifier compared with that for a Schottky Rectifier.

When the voltage drop in the drift region is included for computing the on-state voltage drop of the silicon Schottky rectifier and the influence of Schottky barrier lowering and pre-breakdown multiplication are factored into the calculation of its leakage current, the trade-off curve degrades substantially as shown in Fig.

4.7 by the solid line corresponding to the triangular data points. In this figure, the trade-off curve for the silicon Schottky rectifier was generated by varying the Schottky barrier height. For an on-state voltage drop of 0.45 volts, the leakage current for the Schottky rectifier increases by two-orders of magnitude when compared with the fundamental trade-off curve.

In the silicon TSBS rectifier, the Schottky barrier lowering effect is ameliorated by the reduced electric field at the main Schottky contact. In addition, the pre-breakdown multiplication of the current flowing through the main Schottky contact is suppressed by the much lower electric field at this contact. The trade-off curve obtained from the numerical simulations of the TSBS rectifier structure with a cell pitch of 1 micron and trench depth of 0.5 microns is also shown in Fig. 4.7 by the dashed line corresponding to the diamond data points. For these silicon TSBS rectifier structures, the barrier height for the main Schottky contact was varied while maintaining the same device structure. For an on-state voltage drop of 0.40 volts, the leakage current for the TSBS rectifier is reduced by an order of magnitude when compared with the normal Schottky rectifier. It is worth pointing out that the TSBS rectifier has the same leakage current as the JBS rectifier at an on-state voltage drop of 0.45 volts indicating that both structures are equally effective in reducing the leakage current.

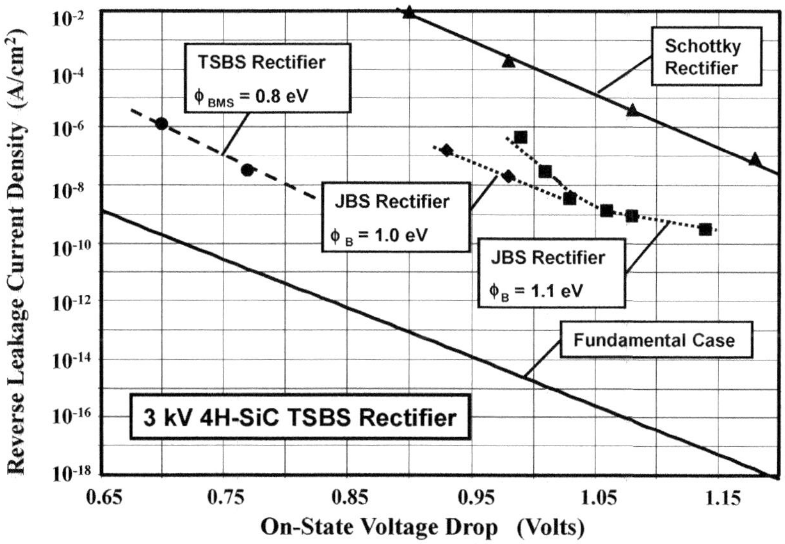


Fig. 4.8 Trade-Off Curve for the 3 kV 4H-SiC TSBS Rectifier compared with that for a Schottky Rectifier.

A similar analysis can be performed for the silicon carbide TSBS rectifier. In this case, a major improvement in reverse blocking characteristics occurs due to suppression of the Schottky barrier lowering and the thermionic field emission current due to the reduced electric field at the main Schottky contact. When these phenomena are included in the analysis of the leakage current for the 4H-SiC Schottky rectifier, the trade-off curve for the case of devices designed to support 3000 volts obtained by varying the Schottky barrier height degrades significantly as shown in Fig. 4.8 by the solid line corresponding to the triangular data points. It can be observed that the leakage current increases by more than 10 orders of magnitude when compared with the fundamental curve.

For the silicon carbide TSBS rectifier structure, the Schottky barrier lowering and thermionic field emission phenomena are suppressed producing greatly reduced leakage currents at large reverse bias voltages. The trade-off curves obtained for these structures by varying the cell pitch is shown in the Fig. 4.8 by the dashed line. It can be observed that the leakage current is seven orders of magnitude smaller than for the Schottky rectifier for the same on-state voltage drop. For the same leakage current density of 1 x $10^{-8}$ A/cm$^2$, the TSBS rectifier structure has a 0.2 V lower on-state voltage drop when compared with the JBS rectifier structure. This difference can be attributed to the larger depletion width under on-state operation in the JBS rectifier structure in silicon carbide because the built-in potential for the P-N junction in 4H-SiC is much larger than the contact potential for the metal is the trenches of the TSBS rectifier structure.

## 4.5 Summary

In this chapter, it has been demonstrated that a significant improvement in the leakage current of Schottky rectifiers can be achieved by incorporation of a trench region with high barrier contact metal to shield the main current carrying Schottky contact against high electric fields generated in the semiconductor. In the case of silicon devices, the reduction of the electric field at the contact suppresses the Schottky barrier lowering and pre-breakdown multiplication phenomena. This reduces the leakage current by an order of magnitude while the increase in on-state voltage due to loss of Schottky contact area is small. In the case of silicon carbide devices, the reduction of the electric field at the Schottky contact suppresses the Schottky barrier lowering and the thermionic field emission current leading to a reduction of leakage current by seven orders of magnitude. In addition, the TSBS rectifier structure does not require the high temperature annealing process required for formation of the P-N junctions in the JBS rectifier structure. This circumvents the degradation of the silicon carbide surface allowing formation of superior main Schottky contacts.

The application of the TSBS rectifier concept to silicon carbide devices was first explored at PSRC[2] for 1000 volt devices. Subsequently, experimental results were reported on 4H-SiC devices fabricated with blocking voltages of 300 volts using titanium as the main Schottky contact and nickel as the metal in the trenches[7]. A reduction of the leakage current by two-orders of magnitude was observed. A reduction in the leakage current by three-orders of magnitude has also been reported for 6H-SiC devices with blocking voltage of 100 volts[8].

**References**

[1] L. Tu and B.J. Baliga, "Schottky Barrier Rectifier including Schottky Barrier Regions of Differing Barrier Heights", U. S. Patent 5,262,668, Issued November 16, 1993.

[2] M. Praveen, S. Mahalingam, and B.J. Baliga, "Silicon Carbide Dual Metal Schottky Rectifiers", PSRC Technical Working Group Meeting Report, TW-97-002-C, 1997.

[3] B.J. Baliga, "Silicon Carbide Power Devices", World Scientific Publishing Company, 2005.

[4] B.J. Baliga, "Fundamentals of Power Semiconductor Devices", Springer Scientific, New York, 2008.

[5] B.J. Baliga, "Modern Power Devices", Chapter 4, John Wiley and Sons, 1987.

[6] B.J. Baliga, "High Voltage Silicon Carbide Devices", Material Research Society Symposium, Vol. 512, pp. 77-88, 1998.

[7] K.J. Schoen, et al, "A Dual Metal Trench Schottky Pinch-Rectifier in 4H-SiC", IEEE Electron Device Letters, Vol. 19, pp. 97-99, 1998.

[8] F. Roccaforte, et al, "Silicon Carbide Pinch Rectifiers using a Dual-Metal Ti-NiSi Schottky Barrier", IEEE transactions on Electron Devices, Vol. 50, pp. 1741-1747, 2003.

# Chapter 5

# Trench MOS Barrier Controlled Schottky Rectifiers

Device structures for reducing the electric field at the Schottky contact were discussed in the previous chapters by incorporation of either a P-N junction or a metal located within trenches. The suppression of the electric field enables significant reduction of the leakage current under reverse blocking operation. In this chapter, yet another approach to suppressing the electric field at the Schottky contact will be described. This approach utilizes a metal-oxide-semiconductor (MOS) structure incorporated within trenches etched around the Schottky contact. The depletion of the region between the trenches under reverse bias operation creates a potential barrier under the Schottky contact and screens it against high electric field in the bulk of the semiconductor drift region. This device structure is referred to as the *'Trench-MOS-Barrier controlled Schottky (TMBS) rectifier'* structure[1]. As in the case of the JBS and TSBS rectifier structures, the depth of the MOS structure used to form the potential barrier is small when compared with the total drift region thickness. In this device concept, the MOS structure is used to suppress the electric field at the Schottky contact but not used to shape the electric field profile in the drift region.

In the TMBS rectifier structure, the on-state current is designed to flow in the un-depleted gaps between the trenches when the diode is forward biased. With an MOS structure, very little depletion of the space between the trenches occurs resulting in a smaller resistance for this region when compared with the JBS and TSBS structures. When a reverse bias is applied, deep-depletion regions form at the MOS structure, which extend between the trenches creating a potential barrier under the Schottky contact. This suppresses the electric field at the Schottky contact preventing the large increase in the leakage current with reverse bias observed in the normal Schottky rectifier structure. In designing the TMBS

B.J. Baliga, *Advanced Power Rectifier Concepts*, DOI: 10.1007/978-0-387-75589-2_5,

rectifier structure, it important to monitor the electric field in the oxide to ensure it is within the range for reliable operation for silicon devices. In the case of silicon carbide devices, the electric field in the oxide can exceed the rupture strength leading to catastrophic failure. The TMBS concept is therefore not suitable for development of devices from silicon carbide.

This chapter provides analytical models for the TMBS rectifier structure. The physics of operation of these structures is also elucidated by using two-dimensional numerical analysis. It is demonstrated that the TMBS concept improves the performance of silicon devices but is not suitable for development of silicon carbide devices[2]. The silicon structures can be fabricated by first etching the trenches using a silicon nitride covered surface. This allows selective oxidation of the trench surfaces using the silicon nitride as an oxidation mask. The silicon nitride can then be selectively removed leaving the oxide in the trenches. The deposition of the anode metal simultaneously creates the Schottky contact on the top surface between the trenches and the metal for the MOS structure within the trenches. This produces a self-aligned device design than allows small dimensions for the space between the trenches. For this reason, the simulations described in this chapter have been performed using the same work function for the Schottky contact and the metal in the trenches.

## 5.1 Trench MOS Barrier controlled Schottky (TMBS) Rectifier Structure

The Trench MOS Barrier controlled Schottky (TMBS) rectifier structure is illustrated in Fig. 5.1. It consists of a trench region containing an MOS structure to generate a potential barrier that can shield the Schottky contact (at location B) in

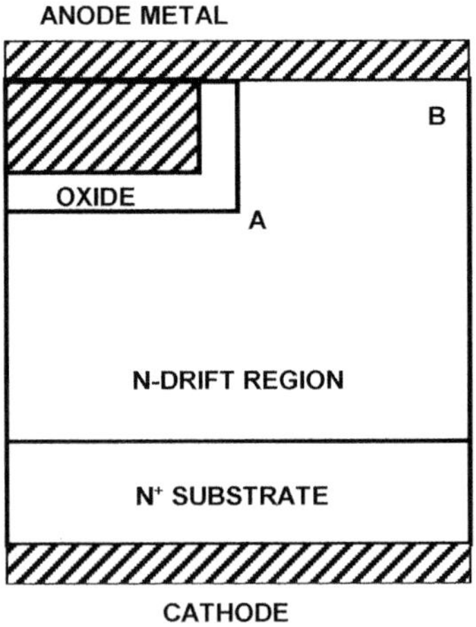


Fig. 5.1 Trench MOS Barrier controlled Schottky (TMBS) Rectifier Structure.

the reverse blocking mode. The magnitude of the potential barrier depends upon the separation between the trenches and the trench depth. A smaller separation and larger trench depth favors an increase in the magnitude of the potential barrier leading to a greater reduction of the electric field at the Schottky contact. A reduction of the electric field at the Schottky contact produces a smaller barrier lowering and field emission effect, which is beneficial for reducing the leakage current at high reverse bias voltages. However, a high electric field can be generated at the sharp corner of the trenches leading to high local electric fields in the oxide that can degrade reliability. In addition, the potential barrier created at location B under the Schottky contact is also dependent on the thickness of the oxide.

The space between the trenches is chosen so that there is an un-depleted region below the Schottky contact during on-state operation to enable unipolar conduction with a low on-state voltage drop. Very little depletion of the space between the trenches is observed with the MOS structure during on-state current flow. This is favorable for reducing the resistance of the drift region between the trenches when compared with the JBS and TSBS rectifier structures.

The TMBS structure can be fabricated by first depositing a layer of silicon nitride on the semiconductor surface. This silicon nitride is patterned to open windows for etching the trenches. The trenches can be formed by using reactive-ion-etching with the silicon nitride as a masking layer. An oxide is then formed on the silicon surfaces at the bottom and side-walls of the trenches by thermal oxidation. During this process step, the silicon nitride layer on the upper silicon surface acts as a mask against oxidation of the silicon on the upper surface. The silicon nitride layer on the upper surface of the silicon is now selectively removed while retaining the oxide on the sidewalls and bottom of the trenches. The anode metal is then evaporated to fill the trenches and cover the upper surface to complete the cell structure.

The anode metal layer must of course be patterned to terminate the device at the edges. The same device structure, shown in Fig. 5.1, can be produced for both silicon and silicon carbide devices. Consequently, a single basic model can be created for TMBS rectifiers made from both these semiconductors. However, the high electric field generated in the oxide in the case of silicon carbide devices precludes the development of viable devices.

As in the case of the JBS rectifier structure, it will be assumed that the breakdown voltage is reduced to about 80 percent of the ideal parallel-plane value due to the edge termination. The doping concentration for the drift region must be computed after accounting for this reduction in the breakdown voltage:

$$N_D = \left( \frac{5.34x10^{13}}{BV_{PP}} \right)^{4/3} \qquad [5.1]$$

where $BV_{PP}$ is the breakdown voltage for the parallel-plane case.

The maximum depletion width in the TMBS structure at the Schottky contact is limited to that associated with the breakdown voltage (BV) of the structure, as given by:

$$t = W_D(BV) = \sqrt{\frac{2\varepsilon_S BV}{qN_D}} \quad \textbf{[5.2]}$$

However, the depletion width under the MOS-trenches must be computed after accounting for the voltage supported across the oxide with the semiconductor operating under deep-depletion conditions[3].

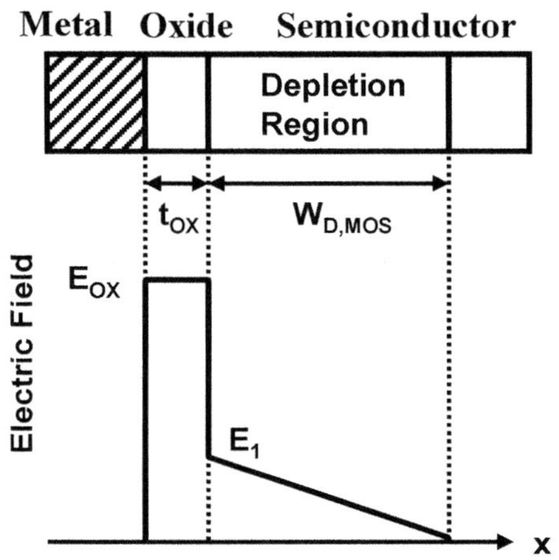


Fig. 5.2 Electric Field distribution under the MOS region of the TMBS Rectifier Structure during Operation in the Blocking Mode.

The distribution of the electric field under the MOS-trench regions of the TMBS rectifier structure is shown in Fig. 5.2 in the reverse blocking mode of operation. The positive voltage applied to the cathode is shared between the oxide and the semiconductor. Due to the presence of the Schottky contact in the vicinity of the MOS-structure, an inversion layer cannot form at the oxide-semiconductor interface. The semiconductor then operates in the deep-depletion mode with a depletion width decided by the electric field at the oxide-semiconductor interface (indicated as $E_1$ in the figure). Since the applied reverse bias voltage ($V_R$) is shared between the oxide and the semiconductor:

$$V_R = V_{OX} + V_S = E_{OX} t_{OX} + \frac{1}{2} E_1 W_{D,MOS} \quad \textbf{[5.3]}$$

where $V_{OX}$ is the voltage supported by the oxide and $V_S$ is the voltage supported within the semiconductor.

The electric field ($E_1$) in the semiconductor at the oxide interface and the electric field ($E_{OX}$) in the oxide are inter-related via Gausses Law:

$$E_{OX} = \frac{\varepsilon_S}{\varepsilon_{OX}} E_1 \tag{5.4}$$

In addition, the electric field in the semiconductor ($E_1$) is related to the depletion layer width by:

$$E_1 = \frac{qN_D}{\varepsilon_S} W_{D,MOS} \tag{5.5}$$

Combining these relationships:

$$V_R = \frac{qN_D}{C_{OX}} W_{D,MOS} + \frac{qN_D}{2\varepsilon_S} W^2_{D,MOS} \tag{5.6}$$

where $C_{OX}$ is the specific capacitance of the trench oxide ($\varepsilon_{OX}/t_{OX}$). The solution for this quadratic equation provides the depletion layer width in the semiconductor under the trench oxide:

$$W_{D,MOS} = \frac{\varepsilon_S}{C_{OX}} \left\{ \sqrt{1 + \frac{2V_R C^2_{OX}}{q\varepsilon_S N_D}} - 1 \right\} \tag{5.7}$$

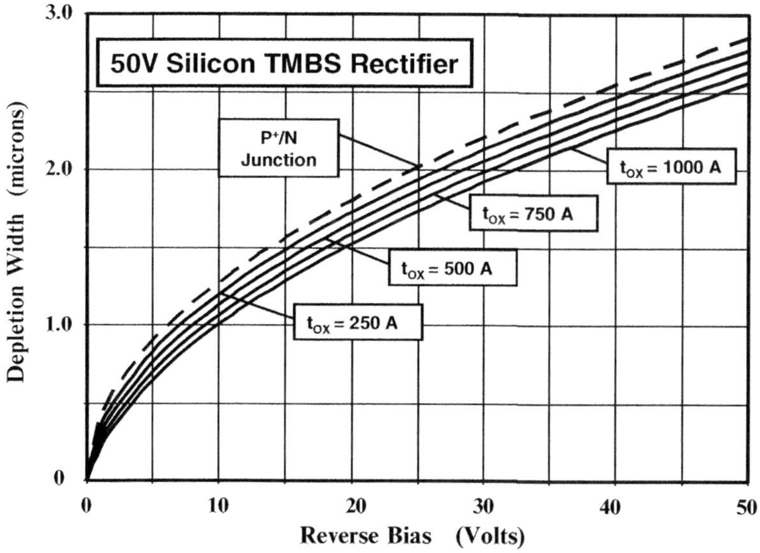


Fig. 5.3 Depletion Layer Width under the MOS-Trench Region in a 50V Silicon TMBS Rectifier Structures.

As an example, the depletion width under the MOS-trench region is provided in Fig. 5.3 for the case of a drift region with doping concentration of $8 \times 10^{15}$ cm$^{-3}$. The oxide thickness was varied from 250 to 1000 angstroms. The depletion layer width for the case of an abrupt $P^+$/N junction has also been included in this figure using dashed lines for comparison. It can be observed that the depletion layer width for the MOS structure is smaller than for the P-N junction because some of the applied reverse bias is supported across the oxide. However, the difference in depletion layer width is only about 10%.

## 5.2 Forward Conduction Model

Analysis of the on-state voltage drop of the TMBS rectifier requires taking into consideration the current constriction at the main low barrier height Schottky contact due to space consumed by the presence of the trenches and the enhanced resistance of the drift region due to current spreading from the Schottky contact to the $N^+$ substrate. Several models were developed for the spreading resistance for the JBS rectifier structure in the previous chapter. It was found that model C with a 45 degree current spreading angle from the bottom of the P-N junction is the most suitable. In addition, this model was applied to the silicon carbide JBS rectifier by assuming a rectangular shape for the junction. It is therefore appropriate to utilize this model for the on-state current flow in both silicon and silicon carbide TMBS rectifier structures due to the rectangular shape for the trenches containing the MOS structure.

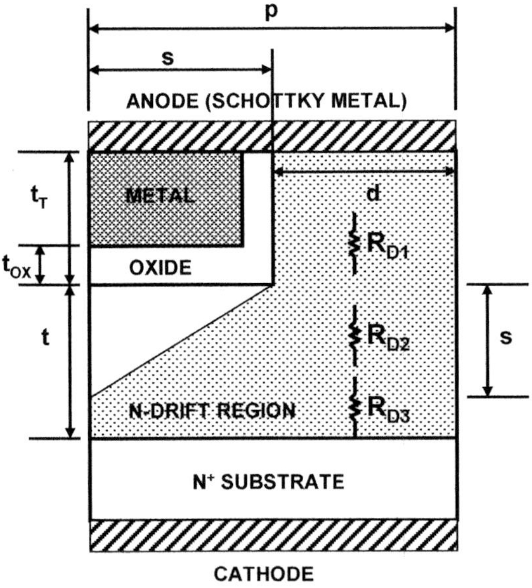


Fig. 5.4 Current Flow Pattern in the TMBS Rectifier Structure during Operation in the On-State.

The current flow pattern in the forward conduction mode for the TMBS rectifier is illustrated in Fig. 5.4 with the shaded area. The analytical model must take into account the increase in current density at the Schottky contact due to the space taken up by the trenches. The current through the Schottky contact flows through the entire space between the trenches (with dimension '2d') of the drift region at the top surface because no significant depletion layer formation occurs at the MOS interface. This provides a greater area for current transport through the Schottky contact in the TMBS structure when compared with the JBS structure. The current density at the main Schottky contact ($J_{FS}$) is related to the cell (or cathode) current density ($J_{FC}$) by:

$$J_{FS} = \left(\frac{p}{d}\right) J_{FC} \qquad [5.8]$$

where p is the cell pitch. The dimension 'd' is determined by the cell pitch (p) and the size of the window (2s) for etching the trenches:

$$d = (p - s) \qquad [5.9]$$

Depending up on the lithography used for device fabrication to minimize the space (dimension 's') taken up by the trenches, the current density at the Schottky contact can be enhanced by a factor of two or more. This must be taken into account when computing the voltage drop across the Schottky contact given by:

$$V_{FS} = \phi_B + \frac{kT}{q} \ln\left(\frac{J_{FS}}{AT^2}\right) \qquad [5.10]$$

After flowing across the Schottky contact, the current flows through the entire space between the trenches because no significant depletion region forms at the MOS interface. In the model, it is assumed that the current flows through a region with a uniform width 'd' until it reaches the bottom of the depletion region and then spreads to the entire cell pitch (p) at a 45 degree spreading angle. The current paths overlap at a distance 's' from the bottom of the trenches. The cross-sectional area for current flow then becomes uniform.

The net resistance to current flow can be calculated by adding the resistance of the three segments illustrated in the figure. The resistance of the first segment of uniform width 'd' is given by:

$$R_{D1} = \frac{\rho_D . t_T}{d.Z} \qquad [5.11]$$

The resistance of the second segment can be derived by using the same approach used for Model C for the silicon JBS rectifier:

$$R_{D2} = \frac{\rho_D}{Z} \ln\left(\frac{p}{d}\right) \qquad [5.12]$$

The resistance of the third segment with a uniform cross-section of width p is given by:

$$R_{D3} = \frac{\rho_D.(t-s)}{p.Z} \quad \textbf{[5.13]}$$

The specific resistance for the drift region can be calculated by multiplying the cell resistance ($R_{D1} + R_{D2} + R_{D3}$) by the cell-area (p.Z):

$$R_{sp,drift} = \frac{\rho_D.p.t_T}{d} + \rho_D.p.\ln\left(\frac{p}{d}\right) + \rho_D.(t-s) \quad \textbf{[5.14]}$$

In addition, it is important to include the resistance associated with the thick, highly doped $N^+$ substrate. In the case of silicon devices, the contribution from the substrate is typically 2 x $10^{-5}$ $\Omega$-cm$^2$. For 4H-SiC, the specific resistance contributed by the $N^+$ substrate is typically 4 x $10^{-4}$ $\Omega$-cm$^2$.

The on-state voltage drop for the TMBS rectifier at a forward cell current density $J_{FC}$, after including the substrate contribution, is given by:

$$V_F = \phi_B + \frac{kT}{q}\ln\left(\frac{J_{FS}}{AT^2}\right) + \left(R_{sp,drift} + R_{sp,subs}\right).J_{FC} \quad \textbf{[5.15]}$$

### 5.2.1 Silicon TMBS Rectifier: Example

In order to understand the operation of the silicon TMBS rectifier structure, it is instructive to consider a specific example of a device with breakdown voltage of 50 volts. If the edge termination limits the breakdown to 80 percent of the ideal value, the parallel-plane breakdown voltage is 62.5 volts. This voltage can be supported by a depletion region width of 2.85 microns in the drift region with doping concentration of 8 x $10^{15}$ cm$^{-3}$. For this example, it will be assumed that the barrier height for the main Schottky contact is 0.60 eV. These values are representative of using a low barrier height metal such as Chromium for the Schottky contact.

Since no significant depletion occurs at the MOS interface in the TMBS structure, it is possible to utilize a smaller cell pitch than for the JBS and TSBS rectifier structures. For the case of a typical TMBS rectifier structure with a cell pitch (p) of 0.50 microns and trenches created using a window (2s) of 0.5 microns, the dimension 'd' is found to be 0.25 microns. Consequently, the current density at the Schottky contact region where the current is transported is enhanced by a factor 2 times when compared with the cathode (or average cell) current density. This enhancement in current density is smaller than that for the silicon JBS rectifier structure because extra space is consumed in the silicon JBS rectifier structure by the lateral diffusion of the junction and the on-state depletion layer. The smaller cell pitch for the TMBS rectifier reduces the specific spreading resistance as well.

The impact of changing the cell pitch (p) on the on-state characteristics can be predicted by using the above model for the TMBS rectifier structure. The results obtained for the case of a trench depth of 0.5 microns are shown in Fig. 5.5. The *i-v* characteristic of a normal Schottky rectifier with the same drift region parameters is included for comparison. The impact on the on-state voltage drop is small as long as the pitch is more than 0.75 microns. The on-state voltage drop of the typical TMBS rectifier structure at an on-state current density of 100 A/cm$^2$ is 0.346 volts compared with 0.322 volts for the normal Schottky rectifier structure with the same barrier height as the Schottky contact. The ability to obtain a low on-state voltage drop in the silicon TMBS structure with a smaller pitch than for the JBS rectifier structure produces slightly superior on-state characteristics.

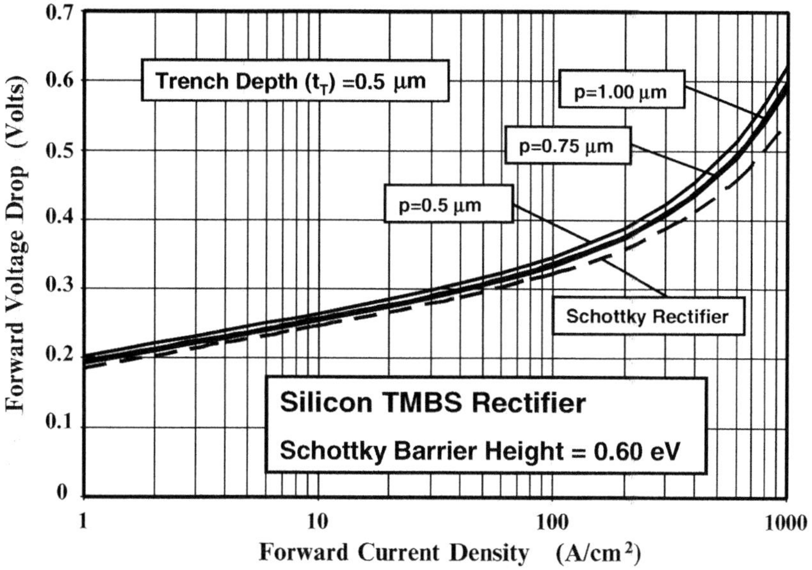


Fig. 5.5 Forward Characteristics of 50V Silicon TMBS Rectifier Structures.

The degree of shielding of the Schottky contact in the TMBS rectifier structure against the high electric fields generated in the drift region during the reverse blocking mode depends upon the depth of the trenches. A deeper trench improves the *aspect ratio* of the region between the trenches. As discussed in the previous chapter, a larger aspect ratio has been found to provide greater shielding in the case of the vertical junction field effect transistor leading to a larger blocking gain[4]. However, the resistance of the first segment in the model for the drift region resistance increases with trench depth leading to a larger on-state voltage drop. This is illustrated for the TMBS rectifier structure with a fixed cell pitch of 0.5 microns in Fig. 5.6 where the analytically calculated forward *i-v* characteristics are shown for various trench depths. Increasing the trench depth from 0.25 to 1 micron increases the on-state voltage drop at a current density of 100 A/cm$^2$ by a small amount from 0.343 to 0.352 volts. Since the specific resistance of the drift region does not change when the oxide thickness is altered, the oxide thickness has no impact on the on-state characteristics for the TMBS rectifier structure.

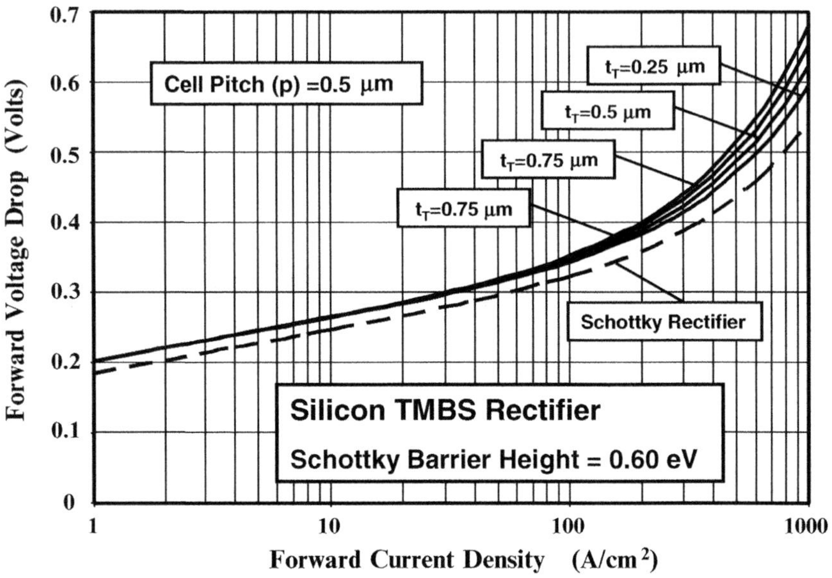


Fig. 5.6 Forward Characteristics of 50V Silicon TMBS Rectifier Structures.

**Simulation Example**

In order to validate the above model for the on-state characteristics of the silicon TMBS rectifier, the results of two-dimensional numerical simulations of a 50 V structure are described here. All the devices had a drift region thickness of 3 microns with a doping concentration of 8 x $10^{15}$ $cm^{-3}$. For all the TMBS rectifier structures, a work function of 4.80 eV was used for the Schottky contact corresponding to a barrier height of 0.60 eV. In all cases, the window for etching the trenches (twice the dimension 's' in Fig. 5.4) was kept at 0.5 microns.

The on-state *i-v* characteristic for the TMBS rectifier with a cell pitch 'p' of 0.50 microns and a trench depth of 0.50 microns is shown in Fig. 5.1E. The on-state *i-v* characteristics for a normal Schottky barrier rectifier obtained using the same drift region thickness of 3 microns with a doping concentration of 8 x $10^{15}$ $cm^{-3}$ is also shown in the figure with dashed lines. A work function of 4.80 eV was used for the Schottky contact for this device. It can be observed that the on-state voltage drop for the TMBS rectifier is increased by a small amount (0.02 volts) consistent with the increase predicted by the analytical model.

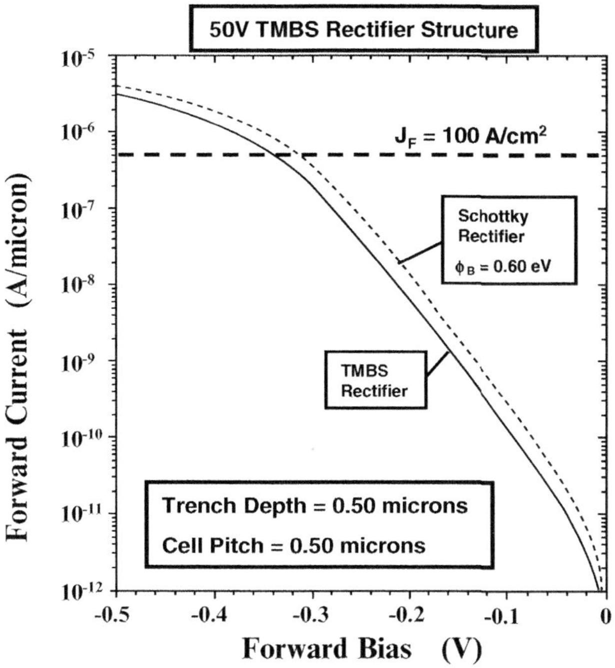


Fig. 5.1E On-State Characteristics for a 50V Silicon TMBS Rectifier.

The current flow-lines for the above TMBS rectifier structure with cell pitch (p) of 0.5 microns and trench depth of 0.5 microns are shown in Fig. 5.2E at an on-state voltage drop of 0.4 volts. The current flow pattern is consistent with pattern used to develop the analytical model with an approximately uniform cross-section between the trenches with no depletion at the MOS interface followed by a spreading of the current at an angle of about 45 degrees (see Fig. 5.4). The current then becomes uniformly distributed below a depth of 0.75 microns from the surface. This justifies the use of a three region analytical model for the series resistance of the drift region.

The on-state characteristics of 50 V TMBS rectifiers with a cell pitch of 0.50, 0.75 and 1.00 microns are compared with that of the normal Schottky rectifier with cell pitch of 0.50 microns (dashed line) in Fig. 5.3E. A barrier height of 0.60 eV was used for all the devices based upon a work function of 4.8 eV for the metal. In the TMBS rectifiers, the trench had a width (2s) of 0.5 microns and a depth of 0.5 microns. From the graph, the on-state voltage drop may appear to be reducing with increasing cell pitch. However, after accounting for the difference in the areas of the structures, the on-state voltage drop at an on-state current density of 100 A/cm$^2$ is found to be approximately equal for all the TMBS rectifier structures with cell pitch greater than 0.5 micron. These results are consistent with those predicted by the analytical model (see Fig. 5.5) confirming its utility for analysis of the TMBS rectifier structure in the on-state. However, when the pitch is reduced to 0.50 microns, the on-state voltage drop increases to 0.34 volts compared with 0.33 volts for the other TMBS rectifier structures. However, as shown later, it is necessary to use a cell pitch of 0.5 microns for the TMBS rectifier in order to suppress the electric field at the Schottky contact.

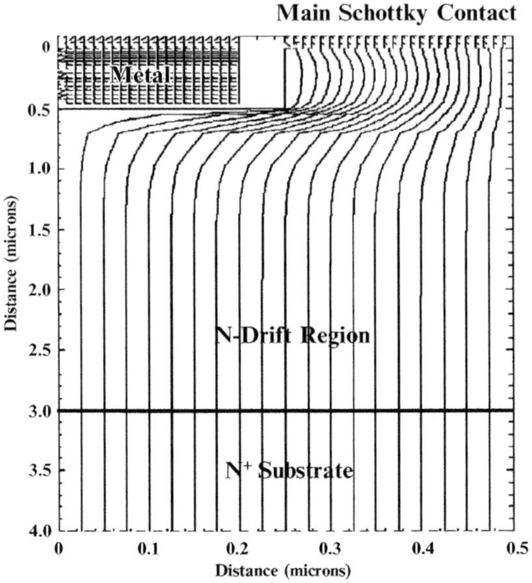


Fig. 5.2E On-State Current Distribution in a 50V Silicon TMBS Rectifier.

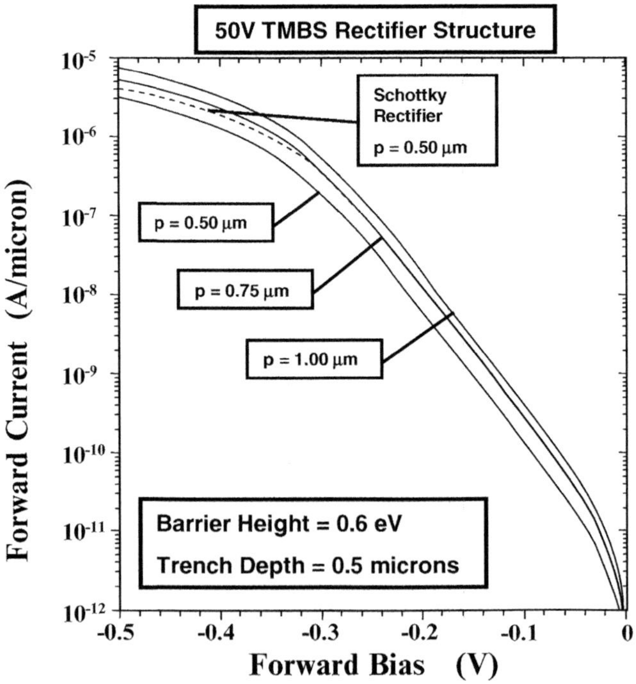


Fig. 5.3E On-State Characteristics for 50V Silicon TMBS Rectifiers.

In the TMBS rectifier structure, an increase in the trench depth improves the shielding of the Schottky contact. However, this also increases the resistance of the current path from the Schottky contact to the cathode. The impact of increasing the trench depth from 0.25 to 1.0 microns in the TMBS rectifier structure can be demonstrated using numerical simulations for the case of a cell pitch of 0.5 microns. The on-state *i-v* characteristics for these structures are compared with the normal Schottky rectifier in Fig. 5.4E. The i-v characteristics of the structures nearly coincide for the cases of a trench depth of 0.25 and 0.50 microns. The on-state voltage drop at a current density of 100 A/cm$^2$ increases slightly for the cases of deeper trench regions. These results are in good agreement with the predictions of the analytical model (see Fig. 5.6).

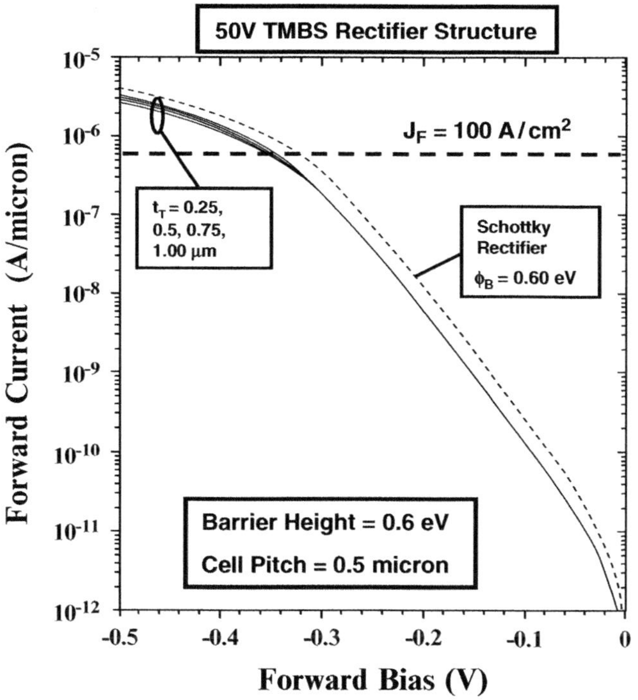


Fig. 5.4E On-State Characteristics for 50V Silicon TMBS Rectifiers.

### 5.2.2 Silicon Carbide TMBS Rectifier: Example

As mentioned earlier, the TMBS rectifier structure is not suitable for the development of silicon carbide devices because the electric field in the oxide exceeds its reliable range of operation and the oxide can undergo catastrophic failure under reverse bias operation. Despite this, in the interest of completeness, an analytical model for the on-state characteristics is provided in this section and verified using the results of two-dimensional numerical simulations.

The same on-state model can be used for the silicon carbide TMBS rectifier as described for the silicon structure because the cell structure is identical. In the case of silicon carbide TMBS rectifiers with very high breakdown voltages, the non-uniform current flow in the drift region is confined to only 2 micron under the Schottky contact with a uniform current flow over most of the drift region. The drift region resistance is therefore close to the ideal value as in the case of the silicon carbide JBS and TSBS rectifier structures.

The on-state voltage drop for the TMBS rectifier at a forward cell current density $J_{FC}$, including the substrate contribution, is given by Eq. (5.15). The Richardson's constant for 4H-SiC is 146 $A^oK^{-2}cm^{-2}$. As an example, 4H-SiC TMBS rectifiers with a blocking voltage capability of 3000 volts will be analyzed in this section. This reverse blocking capability can be obtained by using a drift region with a doping concentration of 8.5 x $10^{15}$ $cm^{-3}$ and thickness of 20 microns after accounting for loss in voltage at the edge termination. A typical 4H-SiC TMBS rectifier structure has a cell pitch of 1 micron, a trench width (2s) of 1 micron, and a trench depth of 1.0 microns.

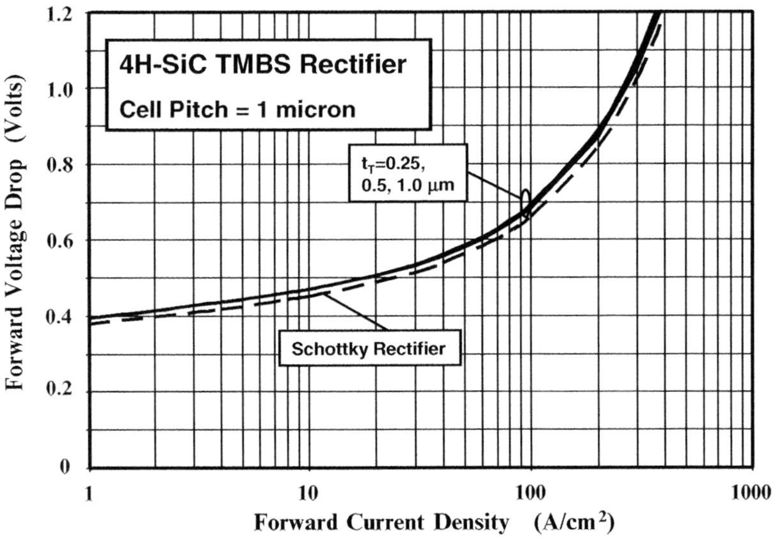


Fig. 5.7 Forward Characteristics of 3kV 4H-SiC TMBS Rectifiers.

The forward characteristics of 3kV TSBS rectifiers, calculated using the analytical model with a barrier height of 0.8 eV for the Schottky contact, are shown in Fig. 5.7 with the trench depth as a parameter. It can be observed that the on-state characteristics are not strongly affected by the trench depth. This is due to the relatively small trench depth when compared with the total thickness of 20 microns for drift region. The current spreading in these high voltage silicon carbide TSBS rectifier structures occurs within the top 2 microns making the contribution from the rest of the drift region nearly equal for all the cases. The trench depth can therefore be selected for sufficiently reducing the electric field at the Schottky contact during the reverse blocking mode.

It is worth pointing out that the area for the main Schottky contact is reduced in half by the presence of the trenches in the TMBS rectifier structures. This results in an increase in the current density at the Schottky contact which produces the observed shift in the *i-v* characteristics to larger voltages when compared with the normal Schottky rectifier. In comparison with the Schottky rectifier characteristics (shown by the dashed line in the figure), the increase in on-state voltage drop at a forward current density of 100 A/cm$^2$ is small (0.02 volts) for these TMBS rectifier structures.

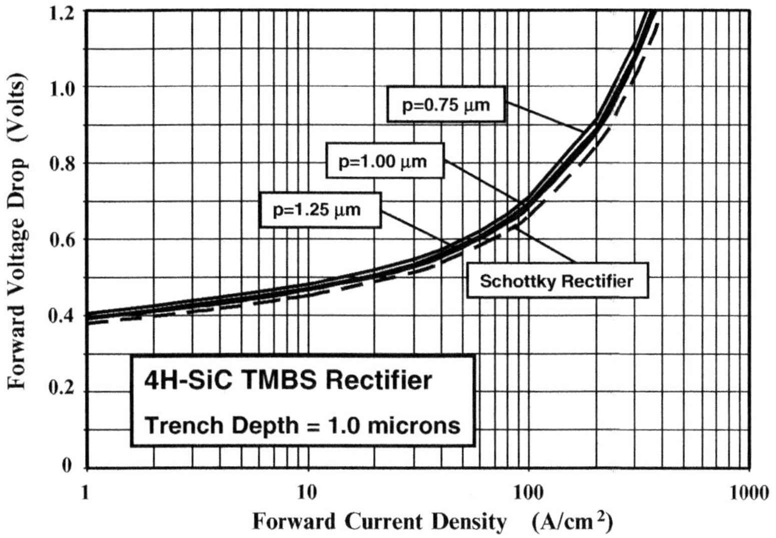


Fig. 5.8 Forward Characteristics of 3kV 4H-SiC TMBS Rectifiers.

The impact of changing the cell pitch on the on-state characteristics of the 4H-SiC TMBS rectifier is illustrated in Fig. 5.8. In this case, the trench depth was kept at the same value of 1.00 microns for the structures. It can be observed that increasing the cell pitch from 1.00 to 1.25 microns produces only a small improvement in the on-state voltage drop. In contrast, reducing the cell pitch to 0.75 microns produces a slight increase in the on-state voltage drop.

### Simulation Example

In order to validate the above model for the on-state characteristics of the silicon carbide TMBS rectifier, the results of two-dimensional numerical simulations for a 3000 V structure are described here. The structure had a drift region thickness of 20 microns with a doping concentration of 8.5 x 10$^{15}$ cm$^{-3}$. The trench region had a depth of 1.00 microns with an etching window (dimension 2s in Fig. 5.4) of

1.0 microns. A work-function of 4.5 eV was used for the Schottky contact. This corresponds to a Schottky barrier height of about 0.8 eV for the Schottky metal based upon an electron affinity of 3.7 eV for 4H-SiC. The typical device structure had an oxide thickness of 500 angstroms in the MOS-region. The oxide thickness does not have any impact on the on-state *i-v* characteristics of the TMBS rectifier structure.

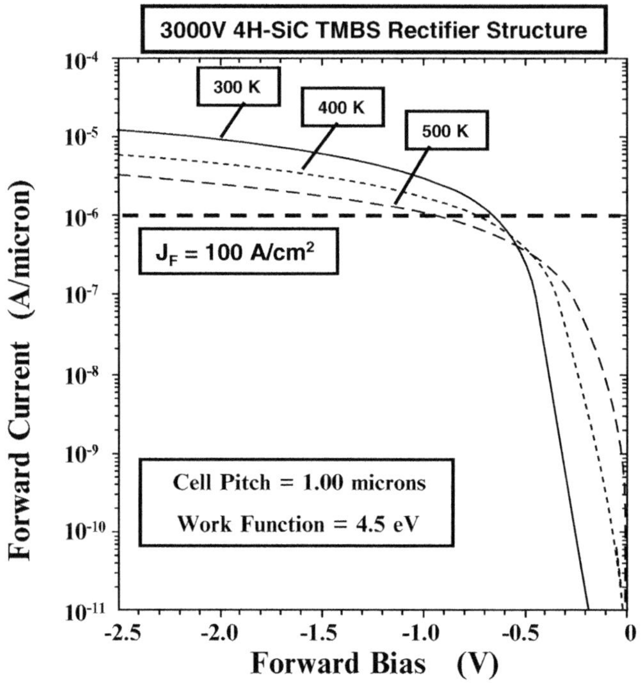


Fig. 5.5E Forward Characteristics of a Typical 3 kV 4H-SiC TMBS Rectifier.

The forward *i-v* characteristics of the typical 3000 V 4H-SiC TMBS rectifier obtained from the numerical simulations are shown in Fig. 5.5E for the case of a cell pitch of 1 micron and a trench depth of 1.00 microns. The on-state voltage drop at a current density of 100 A/cm$^2$ is only 0.7 volts at 300 $^o$K which is identical to the value obtained by using the analytical model providing validation for the model. This 4H-SiC rectifier exhibits the desirable positive temperature coefficient for the on-state voltage drop because of the substantial contribution from the drift region resistance. The drift region resistance increases with temperature due to a reduction of the mobility for electrons. A detailed analysis of the on-state characteristics for different structural parameters is not included here because of problems with the electric field in the oxide during the reverse blocking mode as discussed later in this chapter.

## 5.3 TMBS Rectifier Structure: Reverse Leakage Model

The reverse leakage current in the TMBS rectifier is reduced when compared with the Schottky rectifier due to the smaller electric field at the metal-semiconductor interface at the Schottky contact. In addition, the area of the Schottky contact is a fraction of the total cell area resulting in a smaller reverse current contribution. In the case of the silicon TMBS rectifier, the reduced electric field at the Schottky contact suppresses the barrier lowering effect. In the case of the silicon carbide TMBS rectifier, the reduced electric field not only decreases the barrier lowering but also mitigates the influence of thermionic field emission. However, the high electric field generated in the oxide precludes reliable operation of this structure.

### 5.3.1 Silicon TMBS Rectifier: Reverse Leakage Model

The leakage current model for the TMBS rectifier must take into account the smaller area for the Schottky contact within the cell and the influence of the smaller electric field generated at the Schottky contact due to the shielding effect. The leakage current contribution from the Schottky contact in the silicon TMBS rectifier can be obtained by using:

$$J_L = \left(\frac{p-s}{p}\right) AT^2 \exp\left(-\frac{q\phi_b}{kT}\right).\exp\left(\frac{q\beta\Delta\phi_{bTMBS}}{kT}\right) \qquad \textbf{[5.16]}$$

where $\beta$ is a constant to account for a smaller barrier lowering closer to the trenches as discussed previously for the JBS rectifier structure. In the previous chapter on Schottky rectifiers, it was demonstrated that the high electric field at the Schottky contact produces a reduction of the effective barrier height due to the image force lowering phenomenon. In contrast with the Schottky rectifier, the barrier lowering for the TMBS rectifier is determined by the reduced electric field $E_{TMBS}$ at the main Schottky contact:

$$\Delta\phi_{bTMBS} = \sqrt{\frac{qE_{TMBS}}{4\pi\varepsilon_S}} \qquad \textbf{[5.17]}$$

In the TMBS rectifier structure, the electric field at the Schottky contact varies with distance away from the trenches. The highest electric field is observed at the middle of the Schottky contact with a progressively smaller value closer to the trenches. In an analytical model with a worst case scenario, it is prudent to use the electric field at the middle of the Schottky contact to compute the leakage current. Until the depletion regions from the adjacent trenches produce a potential barrier under the Schottky contact, the electric field at the metal-semiconductor interface in the middle of the contact increases with the applied reverse bias voltage as in the case of the normal Schottky rectifier. A potential barrier is established by the trenches after depletion of the drift region below the Schottky contact.

The voltage at which the depletion regions from the adjacent trenches intersect under the Schottky contact is referred to as the *pinch-off voltage*. The pinch-off voltage ($V_P$) for the MOS structure is different from that for the P-N junction because some of the applied bias is supported across the oxide. The pinch-off voltage for the TMBS rectifier structure can be obtained from Eq. [5.6] as the voltage at which the MOS-depletion width becomes equal to the mesa width (dimension 'd' in Fig. 5.4):

$$V_P = \frac{qN_D}{C_{OX}} d + \frac{qN_D}{2\varepsilon_S} d^2 \quad \textbf{[5.18]}$$

where $C_{OX}$ is the specific capacitance of the gate oxide ($\varepsilon_{OX}/t_{OX}$).

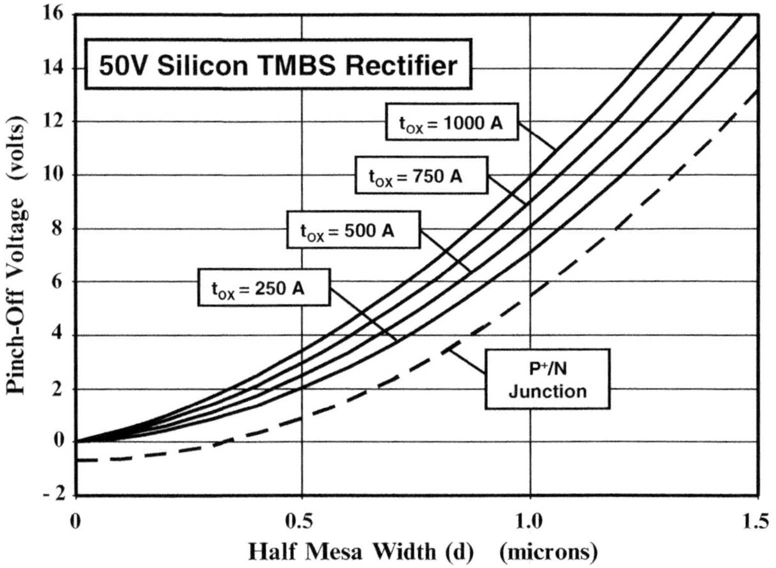


Fig. 5.9 Pinch-Off Voltage for 50V Silicon TMBS Rectifiers.

As an example, the pinch-off voltages for the case of 50 V silicon TMBS rectifier structures are shown in Fig. 5.9 for the case of various mesa widths using the oxide thickness as a parameter. For purposes of comparison, the pinch-off voltage for the case of a JBS rectifier structure has been included in the figure (see dashed line). When the mesa width is made less than 0.3 microns, the pinch of voltage for the JBS rectifier structure becomes negative because of the depletion produced by the built-in potential of the junction. This does not occur for the TMBS rectifier structure because of insignificant depletion at the MOS region at zero reverse bias. In order to produce a potential barrier under the Schottky contact at a reverse bias voltage of less than 10 percent of the breakdown voltage, it can be seen than a mesa width of nearly 2 microns can be used in the JBS rectifier structure. In contrast, a mesa width of less than 1.4 microns is required for the TMBS rectifier structure to achieve a pinch-off voltage of about 5 volts.

Although a potential barrier begins to form after the reverse bias exceeds the pinch-off voltage, the electric field continues to rise at the Schottky contact due to encroachment of the potential to the Schottky contact. In order to analyze the impact of this on the reverse leakage current, the electric field $E_{TMBS}$ can be related to the reverse bias voltage by:

$$E_{TMBS} = \sqrt{\frac{2qN_D}{\varepsilon_S}\alpha V_R + V_C} \qquad \textbf{[5.19]}$$

where $\alpha$ is a coefficient used to account for the build up in the electric field after pinch-off and $V_c$ is the contact potential of the metal-semiconductor junction.

The above analysis for the electric field at the Schottky contact is identical to that for the silicon and silicon carbide JBS rectifiers developed in the previous chapter. Consequently, the same graph relating the electric field to the reverse bias for various values of alpha shown in Fig. 3.11 is also applicable to the TMBS structure. The value for alpha for the silicon TMBS rectifier structure is different from that for the silicon JBS rectifier structure because of the rectangular shape (with an oxide layer) of the trenches in the TMBS structure.

The impact of the reduction of the electric field at the Schottky contact on the Schottky barrier lowering in the silicon TMBS rectifier structure is also the same as that shown in Fig. 3.12. Without the shielding by the metal in the trenches, a barrier lowering of 0.07 eV occurs in the normal silicon Schottky rectifier. The barrier lowering is reduced to 0.05 eV with an alpha of 0.2 in the TMBS rectifier structure. Although this may appear to be a small change, it has a large impact on the reverse leakage current as was already shown in Fig. 3.13.

## Simulation Example

In order to validate the above model for the reverse characteristics of the silicon TMBS rectifier, the results of two-dimensional numerical simulations on a 50 V structure are described here. The structure had a drift region with a doping concentration of $8.5 \times 10^{15}$ cm$^{-3}$ and a thickness of 3 microns. The typical silicon TMBS rectifier structure had a trench depth of 0.5 microns with trench etching window (dimension 2s in Fig. 5.4) of 0.5 microns. The work function of the Schottky metal was chosen to obtain a barrier height of 0.60 eV. A thickness of 500 angstroms was chosen for the oxide in the trenches.

A three dimensional view of the electric field distribution in the typical TMBS rectifier cell is shown in Fig. 5.6E. The Schottky contact is located on the lower right-hand-side in the figure with the trench region located at the top of the figure. A high electric field (~ $2 \times 10^6$ V/cm) is observed within the oxide at the bottom of the trench at a reverse bias of 50 volts. This is acceptable for reliable operation of the device structure. The electric field at the middle of the Schottky contact is greatly reduced ($1.4 \times 10^5$ V/cm) demonstrating that the field can be suppressed using an MOS-trench structure. A slightly larger electric field is generated in the silicon at the sharp corner of the trench. This can degrade the breakdown voltage within the cell. A reduction of this electric field can be achieved by rounding the bottom of the trenches.

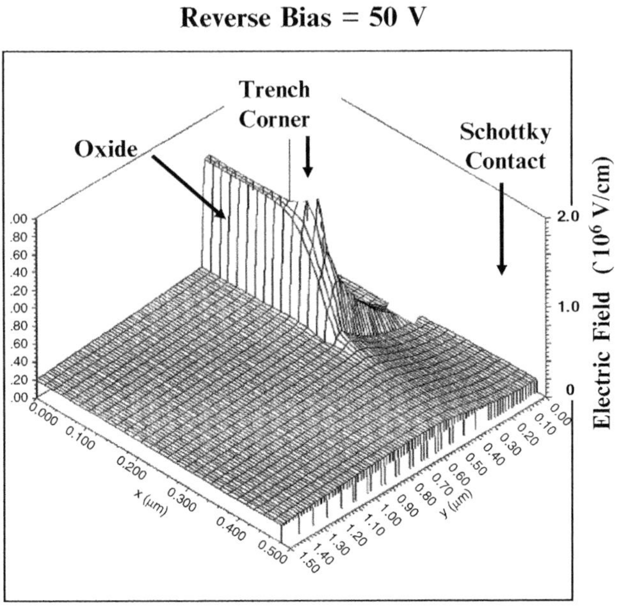


Fig. 5.6E Electric Field Distribution in a Typical 50 V Silicon TMBS Rectifier.

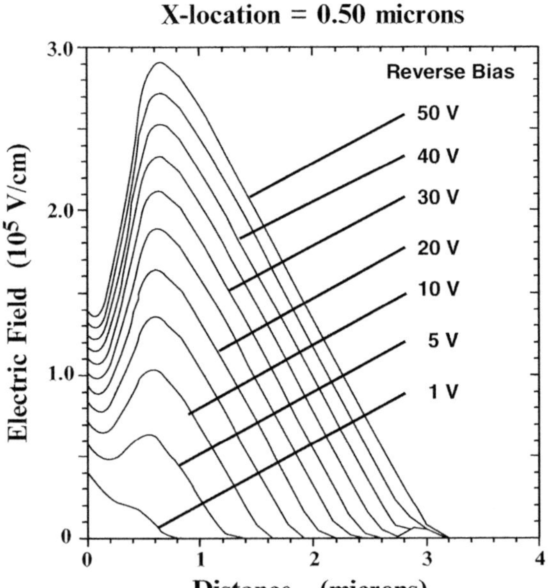


Fig. 5.7E Growth of the Electric Field at the Middle of the Schottky Contact in a Typical 50 V Silicon TMBS Rectifier.

The growth in the electric field at the middle of the Schottky contact with increasing reverse bias voltage for the typical TMBS rectifier structure with cell pitch (p in Fig. 5.6) of 0.50 microns is shown in Fig. 5.7E. In comparison with the growth of the electric field for the normal Schottky rectifier previously shown in Fig. 3.10E, it is apparent that the electric field at the Schottky contact is suppressed in the TMBS rectifier. The peak electric field occurs at a depth of about 0.6 microns in this TMBS rectifier structure. Consequently, no significant punch-through of the electric field to the $N^+$ substrate occurs at high reverse bias voltages.

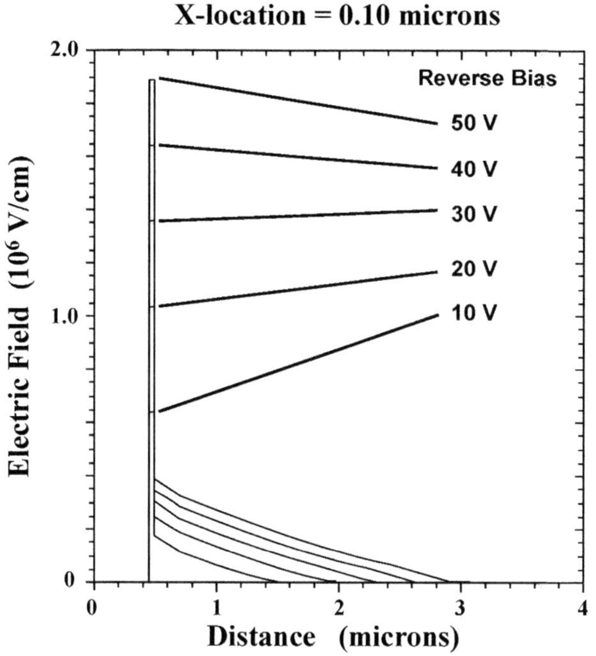


Fig. 5.8E Growth of the Electric Field at the oxide in a 50 V Silicon TMBS Rectifier.

It is instructive to examine the electric field developed in the oxide within the TMBS rectifier structure to ensure that it does not approach its rupture strength. The electric field profile taken through the trench is shown in Fig. 5.8E for various reverse bias values. The electric field in the oxide is much larger than in the silicon consistent with Eq. [5.4] and Fig. 5.2 that were used to develop the analytical model for the TMBS rectifier structure. At a reverse bias of 50 volts, the electric field in the oxide remains below 2 x $10^6$ V/cm allowing reliable operation of this silicon structure. This demonstrates that the TMBS rectifier structure is suitable for the development of silicon devices.

An even greater suppression of the electric field at the Schottky contact can be obtained by either increasing the trench depth or reducing the oxide thickness in the TMBS rectifier structure. The improvement obtained with a trench depth of 0.75 microns is shown in Fig. 5.9E for a TMBS rectifier structure with cell pitch of 0.5 microns and an oxide thickness of 500 angstroms. The electric field at the center of the Schottky contact is now reduced to 0.7 x $10^5$ V/cm at a reverse bias of 50 volts compared with 1.4 x $10^5$ V/cm for the typical TMBS rectifier structure with trench depth of 0.5 microns. The location of the peak electric field also shifts to a greater depth for the structure with the deeper trench resulting in punch-through of the electric field profile to the $N^+$ substrate at reverse bias voltages above 40 volts. These simulations are intended to illustrate the impact of changing the trench depth while maintaining the same total thickness for the drift region. In practice, it is preferable to increase the total thickness of the drift region when the trench depth is increased to avoid the punch-through of the electric field to the $N^+$ substrate.

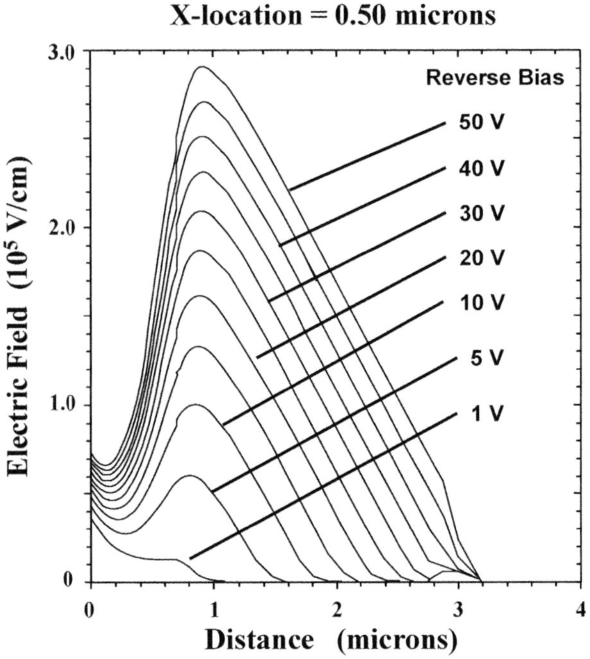


Fig. 5.9E Growth of the Electric Field at the Middle of the Schottky Contact in a 50 V Silicon TMBS Rectifier.

A similar improvement in the suppression of the electric field at the Schottky contact can be achieved by decreasing the oxide thickness while maintaining the same trench depth. This is illustrated in Fig. 5.10E for a TMBS rectifier structure with a trench depth of 0.5 microns and an oxide thickness of 200 angstroms. The electric field at the center of the Schottky contact is now reduced to 0.8 x $10^5$ V/cm compared with 1.4 x $10^5$ V/cm for the typical TMBS rectifier structure with oxide thickness of 500 angstroms. In this case, it was observed that the electric field in the oxide increases slightly but remains less than 2 x $10^6$ V/cm at a reverse bias of 50 volts. The ability to change the oxide thickness provides another degree of freedom when optimizing the TMBS rectifier structure.

The reverse *i-v* characteristic of the 50 V Silicon TMBS rectifier with a cell pitch of 0.50 microns obtained from the numerical simulations is shown in Fig. 5.11E together with that for the Schottky rectifier using a barrier height corresponding to the Schottky contact metal (0.60 eV). Both devices had a cross-sectional width (cell pitch) of 0.5 microns. The breakdown voltage for the TMBS rectifier cell is observed to be 62 volts which is larger than the 58 volts observed for the normal Schottky rectifier structure. The leakage current at small reverse bias voltages is 2x times smaller in the TMBS rectifier structure than for the Schottky rectifier with a barrier height of 0.60 eV. This is consistent with the reduction of the Schottky contact area by a factor of 2x in this TMBS rectifier structure. The leakage current for the Schottky rectifier increases by a factor of 7x when the reverse voltage increases to 50 volts. In contrast, the reverse leakage current for this TMBS rectifier increases by a factor of only 2x when the reverse voltage is increased to 50 volts. This increase is consistent with predictions of the analytical model.

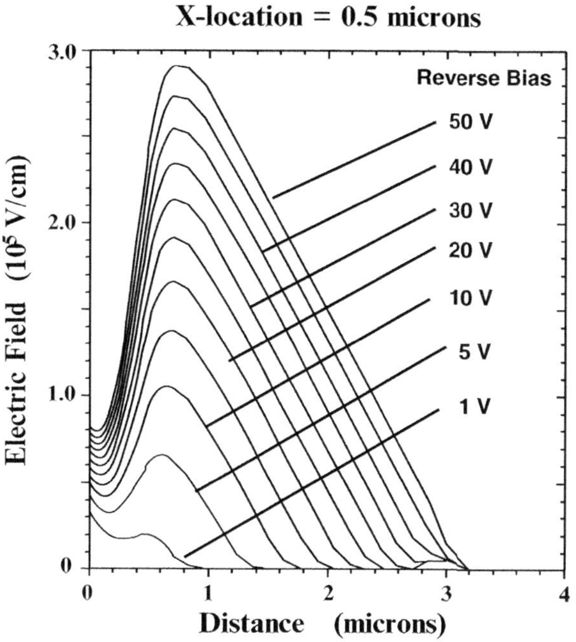


Fig. 5.10E Growth of the Electric Field at the Middle of the Schottky Contact in a 50 V Silicon TMBS Rectifier.

The coefficient alpha, that governs the rate at which the electric field increases at the middle of the Schottky contact in the analytical model for the TMBS rectifier structure, can be extracted from the results of the two-dimensional numerical simulations. The increase in the electric field at the middle of the Schottky contact, obtained from the numerical simulations, is shown in Fig. 5.12E for the TMBS rectifiers with trench depths ranging from 0.25 to 1.00 microns by the symbols. For these structures, the cell pitch was maintained constant at 0.5 microns and the oxide thickness was kept fixed at 500 angstroms. The results of calculations based upon using the analytical Eq. [5.19] are shown by the solid lines with the values for alpha adjusted to fit the results of the numerical simulations. A Schottky contact potential ($V_C$) of 0.64 volts was used based upon a metal work function of 4.8 eV.

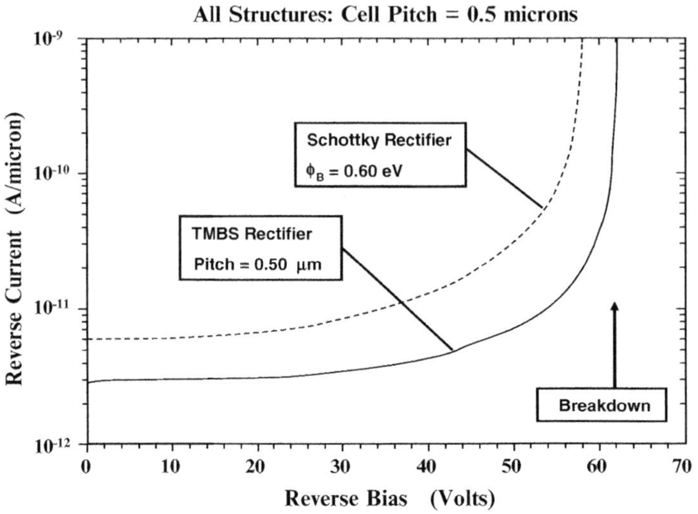


Fig. 5.11E Reverse Blocking Characteristics for a Typical Silicon 50V TMBS Rectifier compared with Schottky Rectifiers.

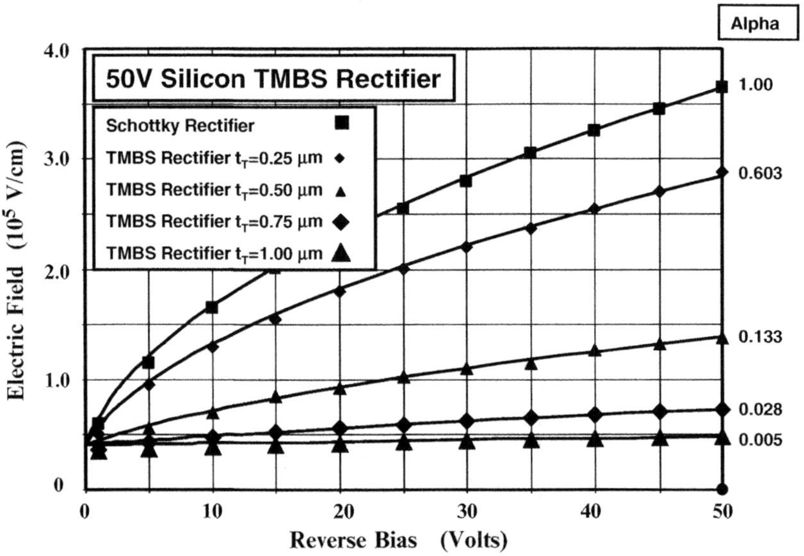


Fig. 5.12E Growth of the Electric Field at the Middle of the Schottky Contact for Silicon 50V TMBS Rectifiers.

The case with alpha of unity fits the Schottky rectifier quite well as expected. The value for alpha for the TMBS rectifier decreases with increasing cell pitch. It has a value of 0.133 for the typical TMBS rectifier structure with trench depth of 0.5 microns. When the trench depth is increased to 1 micron, the alpha value becomes very small (0.005) demonstrating that very strong suppression of the electric field is possible with the MOS-trench structure. With these values of alpha, the analytical model accurately predicts the behavior of the electric field at the middle of the Schottky contact with increasing reverse bias voltage. It can therefore be used to compute the Schottky barrier lowering and leakage current in TMBS rectifiers.

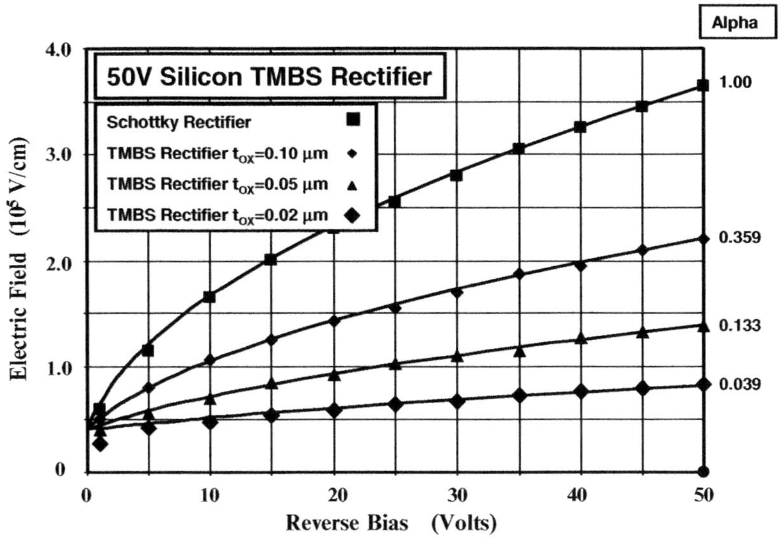


Fig. 5.13E Growth of the Electric Field at the Middle of the Schottky Contact for Silicon 50V TMBS Rectifiers.

Another degree of freedom in adjusting the suppression of the electric field is possible for the TMBS rectifier structure by varying the oxide thickness. The growth of the electric field with reverse bias voltage for the TMBS rectifier structures with the different oxide thicknesses is compared in Fig. 5.13E. For these structures, the cell pitch was maintained constant at 0.5 microns and the trench depth was kept fixed at 0.5 microns. The results of calculations based upon using the analytical Eq. [5.19] are shown by the solid lines with the values for alpha adjusted to fit the results of the numerical simulations. A Schottky contact potential ($V_C$) of 0.64 volts was used based upon a metal work function of 4.8 eV. It can be seen that the alpha is reduced from 0.359 to 0.039 when the oxide thickness is reduced from 1000 to 200 angstroms. Based upon this change in alpha by an order of magnitude, it can be concluded that the oxide thickness is a critical parameter when designing the TMBS rectifier structure.

As with the JBS and TSBS rectifier structures, the mesa width - governed by the cell pitch (p) - has a strong influence on the suppression of the electric field. The growth of the electric field with reverse bias voltage for the TMBS rectifier structures with different the cell pitches (p) is compared in Fig. 5.14E. For these structures, the oxide thickness was maintained constant at 500 angstroms and the trench depth was kept fixed at 0.5 microns. The results of calculations based upon using the analytical Eq. [5.19] are shown by the solid lines with the values for alpha adjusted to fit the results of the numerical simulations. A Schottky contact potential ($V_C$) of 0.64 volts was used based upon a metal work function of 4.8 eV. It can be seen that the alpha increases from 0.133 to 0.640 when the cell pitch is increased from 0.5 to 1.0 microns. This change is associated with a reduction of the aspect ratio.

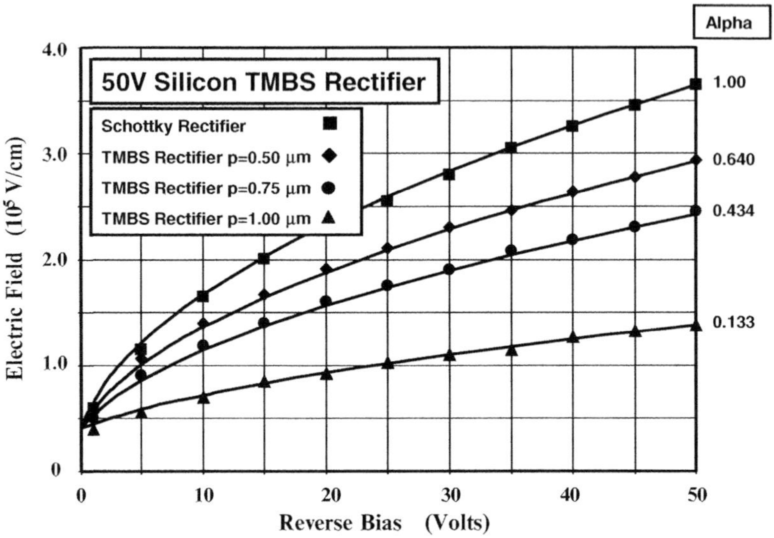


Fig. 5.14E Growth of the Electric Field at the Middle of the Schottky Contact for Silicon 50V TMBS Rectifiers.

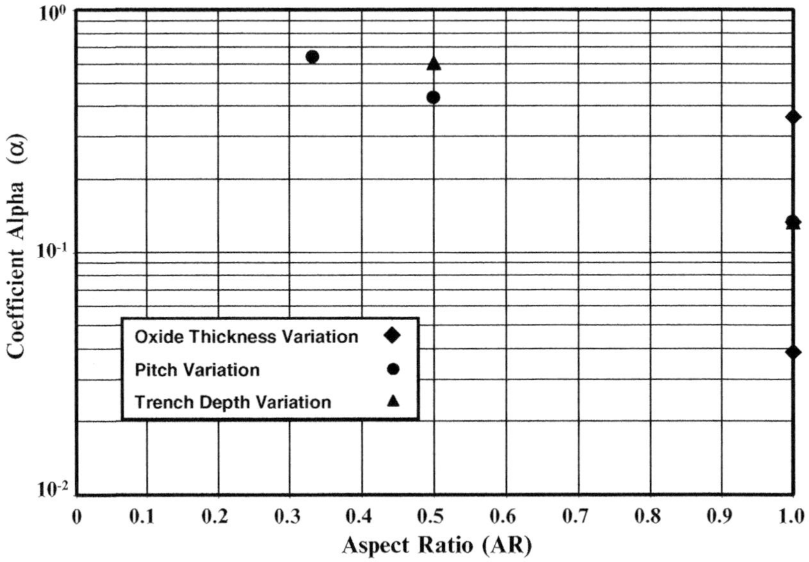


Fig. 5.15E Impact of Aspect Ratio on Alpha for TMBS Rectifiers.

As discussed previously for the JBS and TSBS rectifier structures, the aspect ratio of the current conduction region located below the Schottky contact has a strong influence on the suppression of the electric field at the contact, which is quantified by the coefficient alpha ($\alpha$) in Eq. [5.19]. The aspect ratio for the silicon JBS and TSBS rectifier structures was defined as the trench depth divided by the space between the trenches:

$$AR = \frac{t_T}{2(p-s)} \quad [5.20]$$

The variation of the coefficient alpha ($\alpha$) with aspect ratio, obtained from the numerical simulations, is shown in Fig. 5.15E for silicon TMBS rectifiers. It can be observed that, unlike in the case of the JBS and TSBS rectifier structures, the alpha ($\alpha$) values are scattered and do not fall on a single line that asymptotes to 1.00 at an aspect ratio of zero. Moreover, the alpha values for the TMBS rectifier structure are a strong function of the oxide thickness which is not factored into the aspect ratio as defined by Eq. [5.20]. An improved aspect ratio for the silicon TMBS rectifier structure can be defined by taking into account the oxide in the structure. Due to the difference in the electric field between the oxide and the semiconductor to satisfy Gausses law, the oxide can be treated as having an equivalent semiconductor thickness given by product of the oxide thickness and the ratio of the permittivity of the semiconductor and the oxide. Consequently, although the oxide thickness is small, its impact gets amplified by a factor of about 3 times.

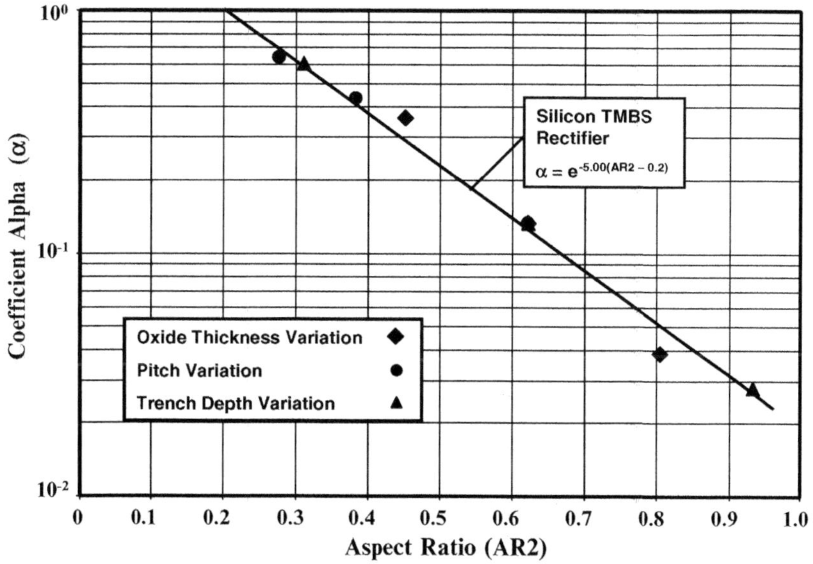


Fig. 5.16E Impact of Aspect Ratio on Alpha for TMBS Rectifiers.

One approach to defining an improved aspect ratio is by adding the equivalent oxide thickness to the mesa width (d) when computing the space between the trenches. The aspect ratio defined with this approach is given by:

$$AR2 = \frac{t_T}{2\left[d + \left(\varepsilon_S t_{OX} / \varepsilon_{OX}\right)\right]} \quad [5.21]$$

The variation of the coefficient alpha ($\alpha$) with this aspect ratio, obtained from the numerical simulations, is shown in Fig. 5.16E for silicon TMBS rectifiers with various trench depths, cell pitches, and oxide thicknesses. It can be observed that the alpha ($\alpha$) values now fall on a single line that is a remarkably good fit to all the data points. However, the line does not asymptote to an alpha of unity for an aspect ratio of zero.

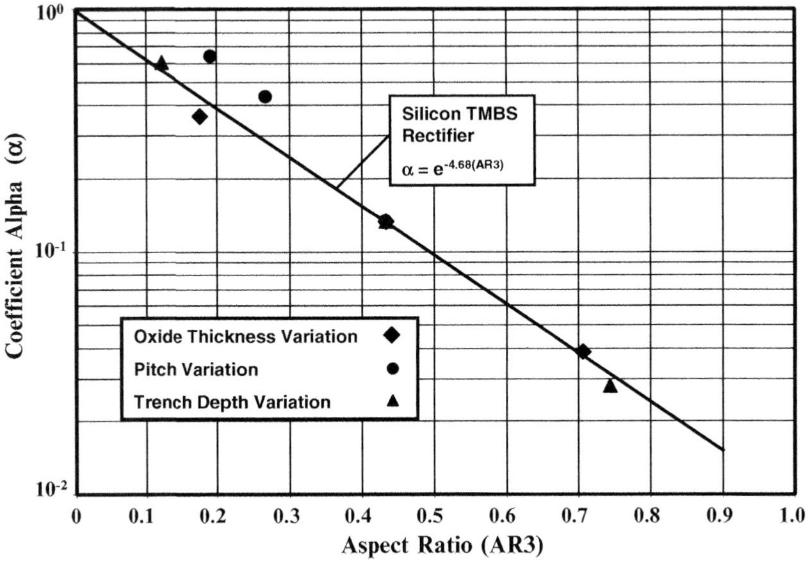


Fig. 5.17E Impact of Aspect Ratio on Alpha for TMBS Rectifiers.

This problem is overcome by accounting for the oxide thickness in the vertical dimension of the aspect ratio. The potential barrier is formed below the Schottky metal by the extension of a depletion region from the MOS interface along the vertical sidewalls of the trenches. The MOS structure extends only to the bottom of the metal in the trenches and not to the entire depth of the trenches. Consequently, it is appropriate to subtract the equivalent thickness associated with the oxide from the trench depth when determining the vertical dimension in the aspect ratio. The aspect ratio defined with this approach is given by:

$$AR3 = \frac{\left[t_T - \left(\varepsilon_S t_{OX} / \varepsilon_{OX}\right)\right]}{2\left[d + \left(\varepsilon_S t_{OX} / \varepsilon_{OX}\right)\right]} \quad [5.22]$$

The variation of the coefficient alpha ($\alpha$) with this aspect ratio, obtained from the numerical simulations, is shown in Fig. 5.17E for silicon TMBS rectifiers with various trench depths, cell pitches, and oxide thicknesses. It can be observed that the alpha ($\alpha$) values now fall on a single line that does asymptote to an alpha of unity for an aspect ratio of zero. It is therefore appropriate to use the aspect ratio as defined by Eq. [5.22] for the analytical modeling of TMBS rectifiers.

### 5.3.2 Silicon Carbide TMBS Rectifier: Reverse Leakage Model

The leakage current in the silicon carbide TMBS rectifier can be calculated using the same approach as for the silicon TMBS rectifier structure. Firstly, it is important to account for the smaller Schottky contact area in the TMBS rectifier cell. Secondly, it is necessary to include Schottky barrier lowering while accounting for the smaller electric field at the Schottky contact due to shielding by the MOS-trench structure. Third, the thermionic field emission current must be included while accounting for the smaller electric field at the Schottky contact due to shielding by the MOS-trench region. After making these adjustments, the leakage current for the silicon carbide TMBS rectifier can be calculated by using:

$$J_L = \left(\frac{p-s}{p}\right) AT^2 \exp\left(-\frac{q\phi_b}{kT}\right).\exp\left(\frac{q\Delta\phi_{bTMBS}}{kT}\right).\exp\left(C_T E_{TMBS}^2\right) \qquad [5.23]$$

where $C_T$ is a tunneling coefficient ($8 \times 10^{-13}$ $cm^2/V^2$ for 4H-SiC). In contrast to the Schottky rectifier, the barrier lowering for the TMBS rectifier is determined by the reduced electric field $E_{TMBS}$ at the contact:

$$\Delta\phi_{bTMBS} = \sqrt{\frac{qE_{TMBS}}{4\pi\varepsilon_S}} \qquad [5.24]$$

As in the case of the silicon TMBS structure, the electric field at the Schottky contact varies with distance away from the trenches. The highest electric field is observed at the middle of the Schottky contact with a progressively smaller value closer to the trenches. When developing an analytical model with a worst case scenario, it is prudent to use the electric field at the middle of the Schottky contact to compute the leakage current.

Until the depletion regions from the adjacent trenches produce a potential barrier under the Schottky contact, the electric field at the metal-semiconductor interface in the middle of the Schottky contact increases with the applied reverse bias voltage as in the case of the normal Schottky rectifier. A potential barrier is established by the MOS-trench regions after depletion of the drift region below the Schottky contact. As in the case of the silicon TMBS rectifier structure, the pinch-off voltage ($V_P$) can be obtained from the mesa width (dimension 'd' in Fig. 5.4):

$$V_P = \frac{qN_D}{C_{OX}} d + \frac{qN_D}{2\varepsilon_S} d^2 \qquad [5.25]$$

where $C_{OX}$ is the specific capacitance of the gate oxide ($\varepsilon_{OX}/t_{OX}$). Although the potential barrier begins to form after the reverse bias exceeds the pinch-off voltage, the electric field continues to rise at the Schottky contact due to encroachment of the potential to the Schottky contact. In order to analyze the impact of this on the reverse leakage current, the electric field $E_{TMBS}$ can be related to the reverse bias voltage by:

$$E_{TMBS} = \sqrt{\frac{2qN_D}{\varepsilon_S}(\alpha V_R + V_C)} \qquad \textbf{[5.26]}$$

where $\alpha$ is a coefficient used to account for the build up in the electric field after pinch-off.

In the case of the silicon carbide TMBS rectifier structure, the largest electric field generated in the semiconductor is close to the critical breakdown electric field of about 3 x $10^7$ V/cm. The maximum electric field occurs near the trenches in the TMBS rectifier structure due to the reverse bias across both the Schottky contact and the MOS structure. Based upon the application of Gausses law at the oxide interface, the electric field in the oxide then becomes 3-times larger than in the silicon carbide. Such large electric fields not only lead to long term degradation of the oxide but can produce catastrophic breakdown of the oxide when the TMBS structure is operated in its reverse bias mode. Detailed discussion of the reverse bias operation of the silicon carbide TMBS rectifier structure is therefore not provided here.

## Simulation Example

In order to demonstrate the high electric fields in the oxide for the silicon carbide TMBS rectifier, the results of two-dimensional numerical simulations on a 3000 V structure are described here. The structure had a drift region thickness of 20 microns with a doping concentration of 8.5 x $10^{15}$ $cm^{-3}$. The trench region had a depth of 0.5 microns with an etching window (dimension 2s in Fig. 5.4) of 1.0 microns. An oxide thickness of 1000 angstroms was used for this typical structure. In theses simulations, a work-function of 4.5 eV was used for the Schottky contact which corresponds to a Schottky barrier height of about 0.8 eV based upon an electron affinity of 3.7 eV for 4H-SiC.

A three-dimensional view of the electric field distribution within the silicon carbide TMBS rectifier structure is provided in Fig. 5.19E at a reverse bias of 1500 volts. Although the drift region parameters were chosen to obtain a breakdown voltage of 3000 volts, it can be observed that the electric field in the oxide has reached 1 x $10^7$ V/cm at a reverse bias of only 1500 volts. Consequently, the oxide will undergo rupture at larger voltages limiting the reverse blocking capability of the TMBS structure to well below that possible with the JBS and TSMS rectifier structures with the same drift region parameters. However, it is worth noting that the electric field at the Schottky contact is indeed suppressed in the silicon carbide TMBS structure.

The electric field profile at the center of the Schottky contact in the silicon carbide TMBS rectifier structure is shown in Fig. 5.19E for the case of a cell pitch of 1 micron. It can be seen that the electric field at the surface under the Schottky contact is significantly reduced when compared the peak electric field in the bulk. The peak of the electric field occurs at a depth of about 1 micron. At a reverse bias of 1500 volts, the electric field at the Schottky contact is only 0.7 x $10^6$ V/cm compared with nearly 2 x $10^6$ V/cm at the maxima in the bulk.

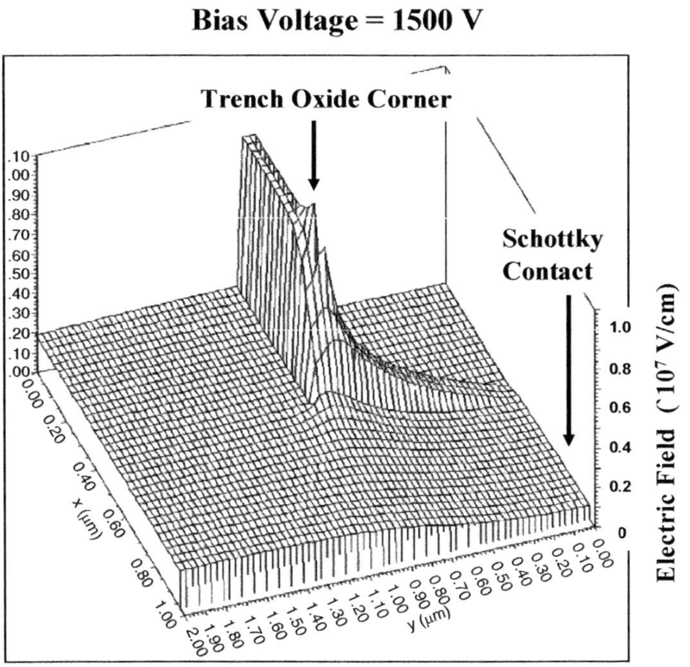


Fig. 5.18E Electric Field distribution in a 4H-SiC TMBS Rectifier.

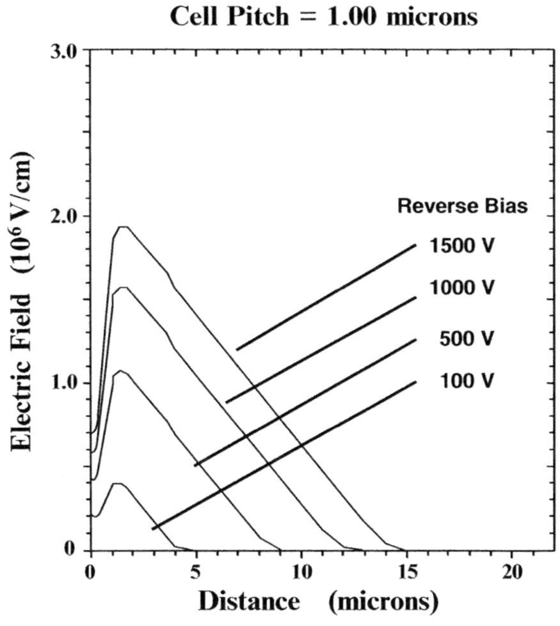


Fig. 5.19E Electric Field variation with Reverse Voltage in a 4H-SiC TMBS Rectifier.

The growth of the electric field in the oxide for the silicon carbide TMBS rectifier structure is illustrated in Fig. 5.20E. The electric field profiles in this figure were taken in the vertical direction through the trenches. It can be observed that the electric field in the oxide is much greater than in the semiconductor as expected. From this figure, it is apparent that the electric field in the oxide will exceed its rupture strength of (1 x $10^7$ V/cm) if the reverse bias is increased beyond 1500 volts. Thus, the high electric fields generated in the oxide for the silicon carbide TMBS structure limit its operation to well below the breakdown voltage capability of the drift region in the normal Schottky rectifier. This problem makes the TMBS rectifier structure unsuitable for application to silicon carbide. For this reason, a detailed analysis of the reverse blocking operation of the silicon carbide TMBS rectifier structure will not be presented in this chapter.

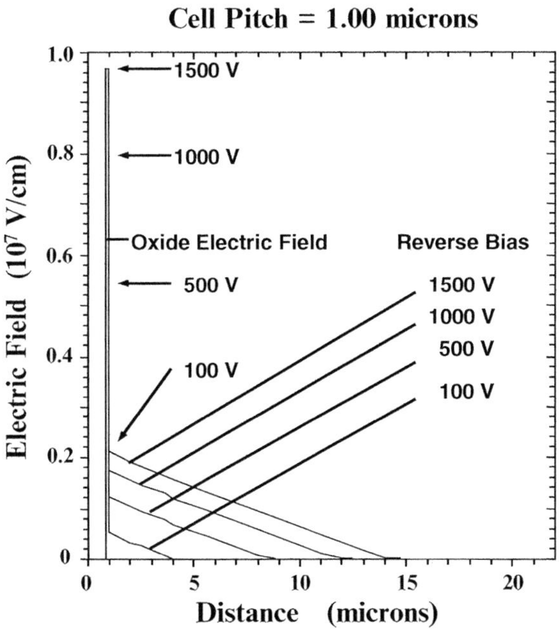


Fig. 5.20E Electric Field variation with Reverse Voltage in a 4H-SiC TMBS Rectifier.

## 5.4 Trade-Off Curve

In the textbook[3], it is demonstrated that the power dissipation can be minimized for a particular duty cycle and operating temperature by varying the Schottky barrier height during optimization of the Schottky rectifier structure. A smaller barrier height decreases the on-state voltage drop reducing conduction power losses while a large barrier height decreases the leakage current reducing reverse blocking power losses. Depending upon the duty cycle and the temperature, minimum power loss occurs at an optimum barrier height. A fundamental trade-off curve between on-state voltage drop and the leakage current was developed in the textbook based upon these considerations which does not depend upon the semiconductor material. However, the fundamental trade-off curve excludes the impact of the series resistance on the on-state voltage drop. More significantly, it excludes the strong influence of Schottky barrier lowering and pre-avalanche multiplication on the increase in leakage current for silicon Schottky rectifiers.

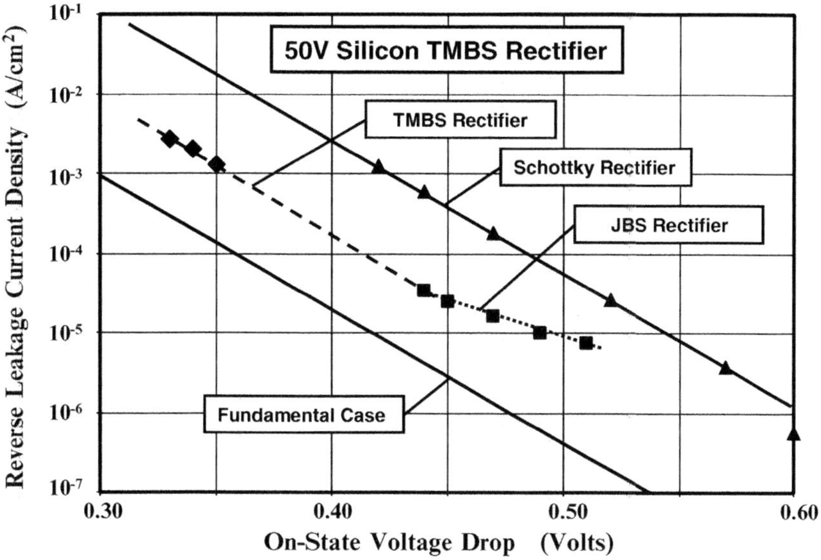


Fig. 5.10 Trade-Off Curve for the 50 V Silicon TMBS Rectifier compared with that for a Schottky Rectifier.

When the voltage drop in the drift region is included for computing the on-state voltage drop of the silicon Schottky rectifier and the influence of Schottky barrier lowering and pre-breakdown multiplication are factored into the calculation of its leakage current, the trade-off curve degrades substantially as shown in Fig. 5.10 by line corresponding to the triangular data points. In this figure, the trade-off curve for the silicon Schottky rectifier was generated by varying the Schottky barrier height. For an on-state voltage drop of 0.45 volts, the leakage current for the Schottky rectifier increases by two-orders of magnitude when compared with the fundamental trade-off curve.

In the silicon TMBS rectifier, the Schottky barrier lowering effect is ameliorated by the reduced electric field at the Schottky contact. In addition, the pre-breakdown multiplication of the current flowing through the Schottky contact is suppressed by the much lower electric field at this contact. The trade-off curve obtained using the analytical models for the TMBS rectifier structure using the alpha extracted from the numerical simulations is shown in Fig. 5.10 by the dashed line corresponding to the diamond data points. In these TMBS rectifier structures, the pitch was varied while maintaining a trench depth of 0.5 microns and oxide thickness of 500 angstroms. For an on-state voltage drop of 0.35 volts, the leakage current for the TMBS rectifier is reduced by an order of magnitude when compared with the normal Schottky rectifier. It is worth pointing out that extrapolation of the trade-off curve for the TMBS rectifier indicates that it has the same leakage current as the JBS rectifier at an on-state voltage drop of 0.45 volts demonstrating that both structures are equally effective in reducing the leakage current.

## 5.5 Summary

In this chapter, it has been shown that a significant improvement in the leakage current of Schottky rectifiers can be achieved by incorporation of a trench region with an MOS structure to shield the Schottky contact against high electric fields generated in the semiconductor. In the case of silicon devices, the reduction of the electric field at the contact suppresses the Schottky barrier lowering and pre-breakdown multiplication phenomena. This reduces the leakage current by an order of magnitude while the increase in on-state voltage due to loss of Schottky contact area is small.

In the case of silicon carbide devices, a very high electric field is generated in the oxide within the TMBS rectifier structure due to the much larger electric fields in the semiconductor. The electric field in the oxide exceeds its rupture strength when the reverse bias is only half that for the normal Schottky rectifier. Based upon this observation, it can be concluded that the TMBS rectifier structure is not suitable for the development of silicon carbide devices. Attempts to apply the TMBS structure to silicon carbide[5] have produces a low breakdown voltage of only 100 volts even when the drift region doping concentration is reduced to $6 \times 10^{15}$ cm$^{-3}$.

## References


[1] B.J. Baliga, "New Concepts in Power Rectifiers", pp. 471-481, in 'Physics of Semiconductor Devices', World Scientific Press, 1985.

[2] B.J. Baliga, 'Silicon Carbide Power Devices', World Scientific Publishing Company, 2005.

[3] B.J. Baliga, "Fundamentals of Power Semiconductor Devices", Springer Scientific, New York, 2008.

[4] B.J. Baliga, "Modern Power Devices", Chapter 4, John Wiley and Sons, 1987.

[5] V. Khemka, V. Ananthan, and T.P. Chow, "A 4H-SiC Trench MOS Barrier Schottky (TMBS) Rectifier", IEEE International Symposium on Power Semiconductor Devices, pp. 165-168, 1999.

# Chapter 6

# P-i-N Rectifiers

It was demonstrated in the previous chapters that the on-state voltage drop of silicon Schottky barrier rectifiers becomes large when the device is designed to support more than 200 volts in the reverse blocking mode. Power device applications, such as motor control, require rectifiers with blocking voltages ranging from 300 volts to 5000 volts. Silicon P-i-N rectifiers have been developed for these high voltage applications. In a P-i-N rectifier, the reverse blocking voltage is supported across a depletion region formed with a P-N junction structure. The voltage is primarily supported within the n-type drift region with the properties of the p-type region optimized for good on-state current flow. Any given reverse blocking voltage can be supported across a thinner drift region by utilizing the punch-through design[1]. Since it is beneficial to use a low doping concentration for the n-type drift region in this design, it is referred to as an i-region (implying that the drift region is intrinsic in nature). The silicon P-i-N rectifiers that are designed to support large voltages rely upon the high level injection of minority carriers into the drift region. This phenomenon greatly reduces the resistance of the thick, very lightly doped drift region necessary to support high voltages in silicon. Consequently, the on-state current flow is not constrained by the low doping concentration in the drift region. A reduction of the thickness of the drift region, by utilizing the punch-through design, is beneficial for decreasing the on-state voltage drop.

In the case of silicon carbide rectifiers, the drift region doping level is relatively large and its thickness is much smaller than for silicon devices to achieve very high breakdown voltages. This enables the design of 4H-SiC based Schottky rectifiers with reverse blocking capability of upto at least 3000 volts with low on-state voltage drop. Based upon the inherent fast switching capability of Schottky rectifiers, it is anticipated that silicon carbide based Schottky rectifiers will displace silicon P-i-N rectifiers for applications with reverse blocking capability of upto

B.J. Baliga, *Advanced Power Rectifier Concepts*, DOI: 10.1007/978-0-387-75589-2_6,

3000 volts[2]. However, this displacement will require further progress with reducing the cost of silicon carbide technology. Meanwhile, silicon P-i-N rectifiers will continue to play an important role in applications.

## 6.1 One-Dimensional Structure

The on-state current flow in the P-i-N rectifier is governed by three current transport mechanisms: (a) at very low current levels, the current transport is dominated by the recombination process within the space charge layer of the P-N junction – referred to as the *recombination current*; (b) at low current levels, the current transport is dominated by the diffusion of minority carriers injected into the drift region – referred to as the *diffusion current*; and (c) at high current levels, the current transport is dictated by the presence of a high concentration of both electrons and holes in the drift region – referred to as *high-level injection current.*

These current transport phenomena are discussed in detail in the textbook[1]. At the on-state operating current levels, current flow in the P-i-N rectifier is governed by the third process with injection of mobile carriers with concentrations far greater than the background doping concentration of the drift region. Only this process will be reviewed here for comparison with the physics of operation of the MPS rectifier that is discussed in the next chapter.

### 6.1.1 High Level Injection Current

The N-drift region in the P-i-N rectifier must be lightly doped in order to support high voltages in the reverse blocking mode. When the forward bias applied to the rectifier increases, the injected minority carrier concentration also increases in the drift region until it ultimately exceeds the background doping concentration ($N_D$) in the drift region. This is defined as *high level injection.* When the injected hole concentration in the drift region becomes much greater than the background doping concentration, charge neutrality requires that the concentrations for electrons and holes become equal:

$$n(x) = p(x) \qquad \textbf{[6.1]}$$

The large concentration of free carriers reduces the resistance of the drift region. This phenomenon is referred to as *conductivity modulation* of the drift region. Conductivity modulation of the drift region is beneficial for allowing the transport of a high current density through lightly doped drift regions with a low on-state voltage drop.

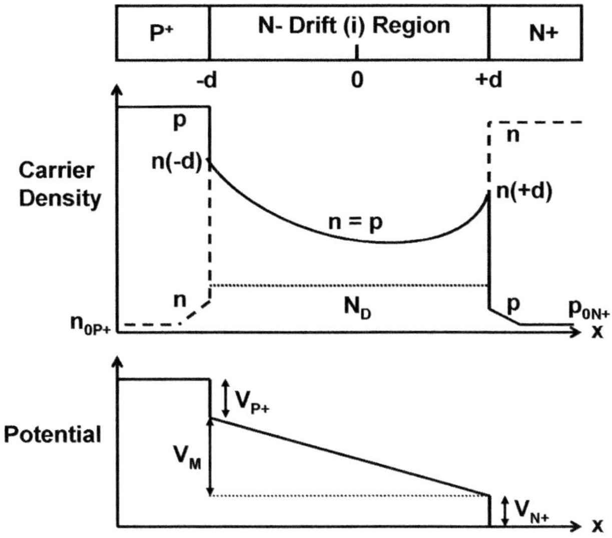


Fig. 6.1 Carrier and Potential Distribution under High-Level Injection Conditions for a P-i-N Rectifier.

The carrier distribution within the drift region n(x) can be obtained by solving the continuity equations for the N-region[3,4,5]:

$$\frac{\partial n}{\partial t} = -\frac{n}{\tau_{HL}} + D_n \frac{\partial^2 n}{\partial x^2} + \mu_n \frac{\partial}{\partial x}(nE) \quad [6.2]$$

$$\frac{\partial p}{\partial t} = -\frac{p}{\tau_{HL}} + D_p \frac{\partial^2 p}{\partial x^2} - \mu_p \frac{\partial}{\partial x}(pE) \quad [6.3]$$

where $D_n$ and $D_p$ are the diffusion coefficients for electrons and holes, respectively, and $\tau_{HL}$ is the high level lifetime in the drift region. Combining these equations after multiplying Eq. [6.2] by ($\mu_p$.p) and Eq. [6.3] by ($\mu_n$.n) gives:

$$\frac{\partial n}{\partial t} = -\frac{n}{\tau_{HL}} + \left(\frac{\mu_p p D_n + \mu_n n D_p}{\mu_p p + \mu_n n}\right)\frac{\partial^2 n}{\partial x^2} \quad [6.4]$$

In deriving this equation, it has been assumed that the transport of carriers due to the electric field can be neglected when compared with the current due to the diffusion of the carriers. The Einstein relationship between the diffusion coefficient and the mobility gives:

$$D = \frac{kT}{q}\mu \quad [6.5]$$

Since the carrier density for electrons and holes is equal in accordance with Eq. [6.1], Eq. [6.4] can be written under steady-state conditions as:

$$\frac{\partial n}{\partial t} = 0 = -\frac{n}{\tau_{HL}} + D_a \frac{\partial^2 n}{\partial x^2} \quad \textbf{[6.6]}$$

where $D_a$ is the ambipolar diffusion coefficient given by:

$$D_a = \frac{p+n}{\dfrac{p}{D_n} + \dfrac{n}{D_p}} = \frac{2D_n D_p}{D_n + D_p} \quad \textbf{[6.7]}$$

due to charge neutrality (See Eq. [6.1]). The general solution for the carrier concentration governed by Eq. [6.6] is given by:

$$n(x) = A\cosh\left(\frac{x}{L_a}\right) + B\sinh\left(\frac{x}{L_a}\right) \quad \textbf{[6.8]}$$

with the constants A and B determined by the boundary conditions for the N-drift region. The parameter $L_a$ in this equation, referred to as the *ambipolar diffusion length*, is given by:

$$L_a = \sqrt{D_a \tau_{HL}} \quad \textbf{[6.9]}$$

At the junction between the N-drift region and the $N^+$ cathode region (located at x = +d in Fig. 6.1), the total current flow occurs exclusively by electron transport:

$$J_T = J_n(+d) \quad \textbf{[6.10]}$$

and

$$J_p(+d) = 0 \quad \textbf{[6.11]}$$

Similarly, at the junction between the N-drift region and the $P^+$ anode region (located at x = –d in Fig. 6.1), the total current flow occurs exclusively by hole transport:

$$J_T = J_p(-d) \quad \textbf{[6.12]}$$

and

$$J_n(-d) = 0 \quad \textbf{[6.13]}$$

Using Eq. [6.11], the hole current due to drift and diffusion can be written as:

$$J_p(+d) = q\mu_p p(+d)E(+d) - qD_p\left(\frac{dp}{dx}\right)_{x=+d} = 0 \quad \textbf{[6.14]}$$

Combining this equation with Eq. [6.1] and the Einstein relationship:

$$E(+d) = \frac{kT}{qn(+d)}\left(\frac{dn}{dx}\right)_{x=+d} \quad \textbf{[6.15]}$$

Eq. [6.10] for the total current flow due to electron transport at this boundary can be written as:

$$J_T = q\mu_n n(+d)E(+d) + qD_n\left(\frac{dn}{dx}\right)_{x=+d} \quad \textbf{[6.16]}$$

Using Eq. [6.15] for the electric field E(+d):

$$J_T = 2qD_n\left(\frac{dn}{dx}\right)_{x=+d} \quad \textbf{[6.17]}$$

In the same manner:

$$J_T = 2qD_p\left(\frac{dn}{dx}\right)_{x=-d} \quad \textbf{[6.18]}$$

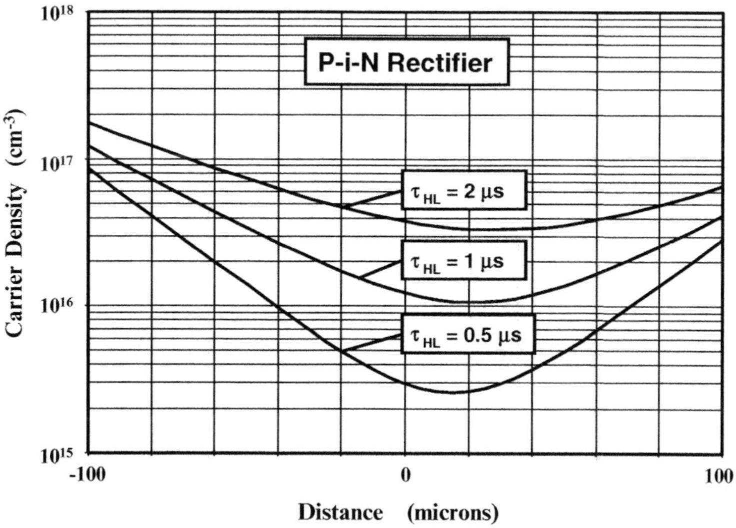


Fig. 6.2 Carrier Distribution under High-Level Injection Conditions for a P-i-N Rectifier with various High-Level Lifetime values.

The above boundary conditions can be used to obtain the constants A and B in Eq. [6.8] resulting in:

$$n(x) = p(x) = \frac{\tau_{HL} J_T}{2qL_a}\left[\frac{\cosh(x/L_a)}{\sinh(d/L_a)} - \frac{\sinh(x/L_a)}{2\cosh(d/L_a)}\right] \quad \textbf{[6.19]}$$

The catenary carrier distribution described by this equation was schematically illustrated in Fig. 6.1. As a particular example, the carrier distributions calculated by using Eq. [6.19] are shown in Fig. 6.2 for the case of three values for the high-level lifetime for a diode with drift region thickness of 200 microns. The largest concentrations for the electrons and holes in the drift region occur at its boundary with the $P^+$ and $N^+$ end-regions. The droop in the carrier density towards the center of the drift region is determined by the ambipolar diffusion length. A larger droop in concentration occurs with the smallest diffusion length, and a smaller average carrier concentration is observed when the lifetime is reduced.

The reduction of the average carrier density injected into the drift region with reduction of the lifetime can be deduced from charge control considerations. Under steady-state conditions, the current flow in the P-i-N rectifier can be related to sustaining the recombination of holes and electrons within the drift region if the recombination within the end-regions is neglected. Consequently:

$$J_T = \int_{-d}^{+d} qR\,dx \quad \textbf{[6.20]}$$

where R is the recombination rate given by:

$$R = \frac{n(x)}{\tau_{HL}} \quad \textbf{[6.21]}$$

Using an average carrier density ($n_a$) within the drift region, these equations can be combined to yield:

$$J_T = \frac{2qn_a d}{\tau_{HL}} \quad \textbf{[6.22]}$$

The average carrier density in the drift region is then given by:

$$n_a = \frac{J_T \tau_{HL}}{2qd} \quad \textbf{[6.23]}$$

From this relationship, it can be concluded that the average carrier density in the drift region will increase with the on-state current density and decrease with reduction of the lifetime. This behavior is exhibited by the carrier distribution in Fig. 6.2. For the case of an on-state current density of 100 A/cm$^2$ and a drift region thickness (2d) of 200 microns with a high-level lifetime of 1 microsecond, the average carrier concentration obtained by using the above equation is $3 \times 10^{16}$ cm$^{-3}$, which is consistent with the carrier distribution shown in Fig. 6.2.

The specific resistance of the drift region can be calculated from the average carrier density with the acknowledgement that both electrons and holes are available for current transport:

$$R_{i,SP} = \frac{2d}{q(\mu_n + \mu_p)n_a} \quad [6.24]$$

Using Eq. [6.23] for the average carrier density:

$$R_{i,SP} = \frac{4d^2}{(\mu_n + \mu_p)J_T \tau_{HL}} \quad [6.25]$$

The voltage drop across the drift region (middle-region) is then given by:

$$V_M = J_T.R_{i,SP} = \frac{4d^2}{(\mu_n + \mu_p)\tau_{HL}} \quad [6.26]$$

From this equation, it can be concluded that the voltage drop across the drift region is independent of the current density flowing through it. This unusual behavior occurs due to the presence of a high concentration of minority carriers, contrary to Ohm's law for drift regions without the conductivity modulation. Thus, the conductivity modulation phenomenon at high injection levels enables maintaining a low voltage drop across the drift region, which is extremely beneficial for obtaining a low on-state voltage drop in power P-i-N rectifiers.

A more accurate analysis of the voltage drop across the drift region can be performed by integration of the electric field. The electric field in the drift region can be obtained from the carrier distribution given by Eq. [6.19]. The hole and electron currents flowing in the drift region are given by:

$$J_p = q\mu_p\left(pE - \frac{kT}{q}\frac{dp}{dx}\right) \quad [6.27]$$

and

$$J_n = q\mu_n\left(nE + \frac{kT}{q}\frac{dn}{dx}\right) \quad [6.28]$$

The total current at any location in the drift region is constant and given by:

$$J_T = J_p + J_n \quad [6.29]$$

Combining these relationships:

$$E(x) = \frac{J_T}{q(\mu_n + \mu_p)n} - \frac{kT}{2qn}\frac{dn}{dx} \quad [6.30]$$

Here, the charge neutrality condition n(x) = p(x) was also utilized. The first term in this expression takes into account the ohmic voltage drop due to current flow through the drift region. The second term in the expression is associated with the asymmetrical carrier gradient produced by the difference in the mobility for electrons and holes.

The integration of the electric field distribution given in Eq. [6.30] yields the voltage drop across the drift (or middle) region[5,6,7]

$$\frac{V_M}{kT/q} = \left\{ \begin{array}{l} \dfrac{8b}{(b+1)^2} \dfrac{\sinh(d/L_a)}{\sqrt{1 - B^2 \tanh^2(d/L_a)}} \\ .\arctan\left[\sqrt{1 - B^2 \tanh^2(d/L_a)}\sinh(d/L_a)\right] \end{array} \right\}$$

$$+ B \ln\left[\frac{1 + B\tanh^2(d/L_a)}{1 - B\tanh^2(d/L_a)}\right] \quad [6.31]$$

where $b = (\mu_n/\mu_p)$ and $B = (\mu_n - \mu_p)/(\mu_n + \mu_p)$.

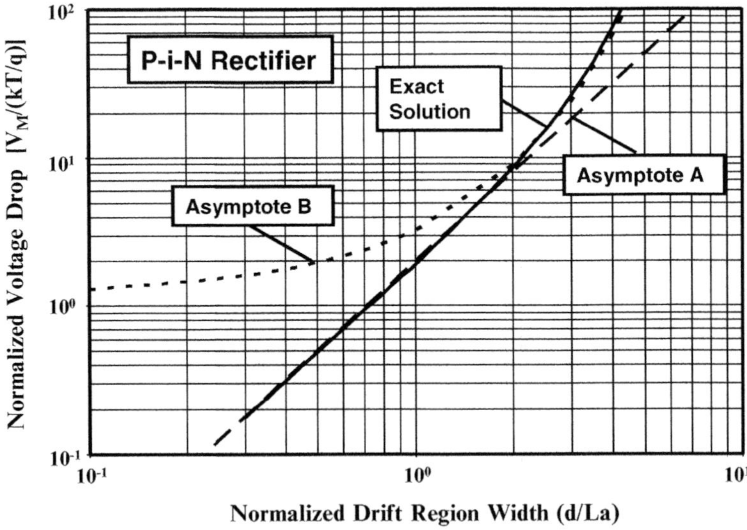


Fig. 6.3 Voltage Drop in the Drift (Middle) Region of a P-i-N Rectifier.

The above equation can be approximated by using two asymptotes as illustrated in Fig. 6.3. For $d/L_a$ ratios of up to 2, the asymptote A given by:

$$V_M = \frac{2kT}{q}\left(\frac{d}{L_a}\right)^2 \quad [6.32]$$

provides a good fit. For $d/L_a$ ratios of greater than 2, the asymptote B given by:

$$V_M = \frac{3\pi kT}{8q} e^{(d/L_a)} \quad [6.33]$$

provides a good fit. As discussed earlier in connection with Eq. [6.26], all these terms for the voltage drop across the drift region are independent of the on-state current density. The voltage drop across the drift region increases rapidly with increasing $d/L_a$ ratio. When this ratio is 0.1, the middle region voltage drop is only 0.5 mV. It increases to a voltage drop of about 50 mV for a $d/L_a$ ratio of unity and becomes 0.7 volts when the d/La ratio increases to 3. Thus, the increase in the voltage drop across the middle region degrades the on-state voltage drop when the lifetime is reduced in order to enhance the switching speed.

The on-state voltage drop in the P-i-N rectifier consists of the voltage drop across the $P^+/N$ junction, the middle region, and the $N/N^+$ interface. The voltage drop across the $P^+/N$ junction can be determined from the injected minority carrier density:

$$p(-d) = p_{0N} e^{\frac{qV_{P+}}{kT}} \quad [6.34]$$

where $p_{0N}$ is the minority carrier density in the N-type drift region in equilibrium and $V_{P+}$ is the voltage drop across the $P^+/N$ junction. Relating the minority carrier concentration in equilibrium to the doping level $N_D$ in the drift region:

$$V_{P+} = \frac{kT}{q} \ln\left[\frac{p(-d)N_D}{n_i^2}\right] \quad [6.35]$$

Similarly, applying the 'Law of the Junction' on the cathode side:

$$n(+d) = n_{0N} e^{\frac{qV_{N+}}{kT}} \quad [6.36]$$

where $n_{0N}$ is the majority carrier density in the N-type drift region in equilibrium and $V_{N+}$ is the voltage drop across the $N^+/N$ junction. Since the majority carrier concentration in equilibrium is equal to the doping level $N_D$ in the drift region:

$$V_{N+} = \frac{kT}{q} \ln\left[\frac{n(+d)}{N_D}\right] \quad [6.37]$$

The voltage drop associated with the two end regions is therefore given by:

$$V_{P+} + V_{N+} = \frac{kT}{q} \ln\left[\frac{n(+d)n(-d)}{n_i^2}\right] \quad [6.38]$$

In deriving this expression, the charge neutrality condition n(x) = p(x) under high level injection was assumed.

The voltage drop across the end regions has been combined with the voltage drop across the middle region to derive a relationship between the on-state current density ($J_T$) and the total on-state voltage drop ($V_{ON}$) for the P-i-N rectifier[3,4,5]

$$J_T = \frac{2qD_a n_i}{d} F\left(\frac{d}{L_a}\right) e^{\frac{qV_{ON}}{2kT}} \quad [6.39]$$

where

$$F\left(\frac{d}{L_a}\right) = \frac{(d/L_a)\tanh(d/L_a)}{\sqrt{1 - 0.25\tanh^4(d/L_a)}} e^{-\frac{qV_M}{2kT}} \quad [6.40]$$

From Eq. [6.39], it is apparent that, for a fixed on-state current density, the on-state voltage drop will be smaller if the function $F(d/L_a)$ is large.

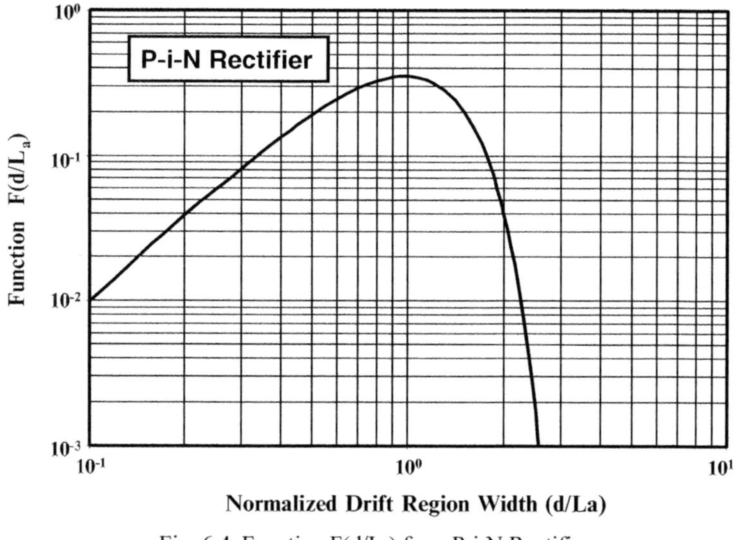


Fig. 6.4 Function $F(d/L_a)$ for a P-i-N Rectifier.

The variation of the function $F(d/L_a)$ with increasing $(d/L_a)$ ratio is plotted in Fig. 6.4. It can be seen that this function has a maximum value when $d/L_a = 1$. Thus, in order to minimize the on-state voltage drop, the lifetime should be adjusted until the diffusion length is equal to one-half of the width of the drift region. It is worth pointing out that the function $F(d/L_a)$ decrease very rapidly when the $(d/L_a)$ ratio increases beyond a value of 3. This leads to a very rapid increase in the on-state voltage drop when the diffusion length is less than one-sixth of the drift region width.

The on-state voltage drop for a P-i-N rectifier can be derived from Eq. [6.39] for a device with its half-width (d) determined by the punch-through breakdown voltage capability:

$$V_{ON} = \frac{2kT}{q} \ln\left[\frac{J_T d}{2qD_a n_i F(d/L_a)}\right] \quad \textbf{[6.41]}$$

The calculated on-state voltage drop for a silicon P-i-N rectifier with a drift region width of 200 microns is shown in Fig. 6.5 at an on-state current density of 100 A/cm$^2$. As expected, the on-state voltage drop exhibits a minimum at a $(d/L_a)$ ratio of unity and increases rapidly when the $(d/L_a)$ ratio exceeds 3.

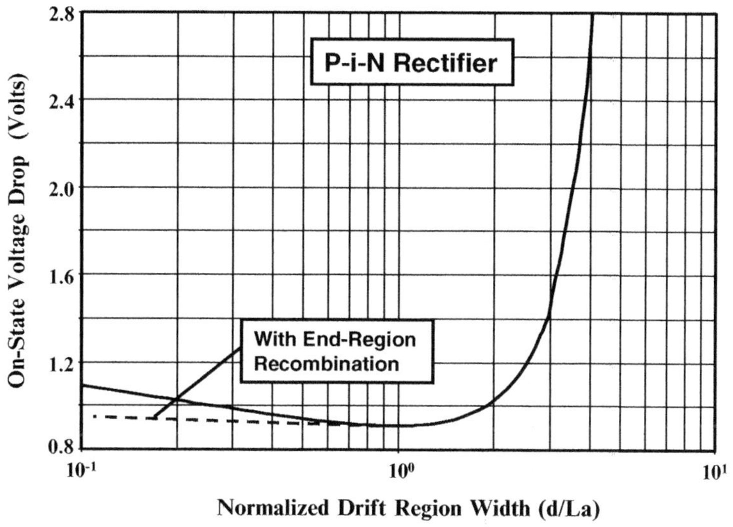


Fig. 6.5 On-state Voltage Drop for a P-i-N Rectifier.

## 6.1.2 Injection into the End-Regions

The high-level injection conditions within the drift region described above determine the on-state voltage drop at typical current densities in the range of 100 to 200 A/cm$^2$. At even greater on-state current densities, the injection of minority carriers into the end regions must be taken into account[6,7]. The total current flow must therefore accommodate not only the recombination of carriers in the drift region but also the recombination of carriers in the end regions. Thus:

$$J_T = J_{P+} + J_M + J_{N+} \qquad [6.42]$$

Consequently, the current density associated with the middle region ($J_M$) is no longer equal to the total on-state current density ($J_M$) as assumed in the previous section but has a smaller value. This reduces the injection level in the drift region corresponding to any given total current density resulting in an increase in the voltage drop across the middle region.

Due to the high doping concentrations in the end regions, the injected minority carrier density in these regions is well below the majority carrier density even during operation at very high on-state current densities. The current corresponding to the end-regions can therefore be analyzed using low-level injection theory under the assumption of a uniform doping concentration in these regions. The current in the P+ region is given by[1]:

$$J_{P+} = \frac{qD_{nP+}n_{0P+}}{L_{nP+}\tanh(W_{P+}/L_{nP+})}e^{\frac{qV_{P+}}{kT}} = J_{SP+}e^{\frac{qV_{P+}}{kT}} \qquad [6.43]$$

where $W_{P+}$ is the width of the $P^+$ region, $L_{nP+}$ is the minority carrier diffusion length in the $P^+$ region, $D_{nP+}$ is the minority carrier diffusion coefficient in the $P^+$ region, $n_{0P+}$ is the minority carrier concentration in the $P^+$ region, and $V_{P+}$ is the voltage drop at the $P^+$/N junction. In a similar manner:

$$J_{N+} = \frac{qD_{pN+}p_{0N+}}{L_{pN+}\tanh(W_{N+}/L_{pN+})}e^{\frac{qV_{N+}}{kT}} = J_{SN+}e^{\frac{qV_{N+}}{kT}} \qquad [6.44]$$

where $W_{N+}$ is the width of the $N^+$ region, $L_{pN+}$ is the minority carrier diffusion length in the $N^+$ region, $D_{pN+}$ is the minority carrier diffusion coefficient in the $N^+$ region, $p_{0N}$ is the minority carrier concentration in the $N^+$ region, and $V_{N+}$ is the voltage drop at the $N^+$/N interface. In these equations, $J_{SP+}$ and $J_{SN+}$ are referred to as the *saturation current densities* for the heavily doped $P^+$ anode and $N^+$ cathode regions, respectively. They are a measure of the quality of the end regions as determined by their doping profiles and processing conditions. Typical values for the saturation current density are in the range from $1 \times 10^{-13}$ to $4 \times 10^{-13}$ amperes per $cm^2$ for silicon devices.

The injected carrier concentrations on the two sides of the $P^+$/N junction are related under quasi-equilibrium conditions by:

$$p_{P+}(-d).n_{P+}(-d) = p(-d).n(-d) \qquad [6.45]$$

Under low-level injection conditions within the $P^+$ anode region:

$$p_{P+}(-d) = p_{0P+} \qquad [6.46]$$

and

$$n_{P+}(-d) = n_{0P+} e^{\frac{qV_{P+}}{kT}} \quad [6.47]$$

Using these relationships in Eq. [6.45]:

$$p(-d).n(-d) = p_{0P+}.n_{0P+} e^{\frac{qV_{P+}}{kT}} = n_{ieP+}^2 e^{\frac{qV_{P+}}{kT}} \quad [6.48]$$

where $n_{ieP+}$ is the effective intrinsic carrier concentration in the $P^+$ anode region including the influence of band-gap narrowing[1]. Due to charge neutrality considerations p(-d) = n(-d), leading to:

$$e^{\frac{qV_{P+}}{kT}} = \left[\frac{n(-d)}{n_{ieP+}}\right]^2 \quad [6.49]$$

Using this expression in Eq. [6.43]:

$$J_{P+} = J_{SP+}\left[\frac{n(-d)}{n_{ieP+}}\right]^2 \quad [6.50]$$

A similar derivation performed for the $N^+$ cathode side yields:

$$J_{N+} = J_{SN+}\left[\frac{n(+d)}{n_{ieN+}}\right]^2 \quad [6.51]$$

where $n_{ieN+}$ is the effective intrinsic carrier concentration in the $N^+$ cathode region including the influence of band-gap narrowing. From these equations, it can be concluded that the carrier concentration in the drift region will increase as the square root of the current density if the end region recombination becomes dominant. Under these circumstances, the middle region voltage drop is no longer independent of the current density resulting in an increase in the total on-state voltage drop. The influence of end region recombination can be mitigated by optimization of the doping profile to minimize the saturation current densities.

### 6.1.3 Forward Conduction Characteristics

The analysis of current flow in a P-i-N rectifier in the previous sections indicates that the relationship between the current density and the voltage drop across the rectifier depends upon the injection level. At very low current levels, space charge generation controls the current flow with the current proportional to ($qV_{ON}/2kT$). When the current is controlled by minority carrier injection into the drift region with the minority carrier concentration well below the background doping concentration, the current flow occurs by diffusion under low-level injection conditions. The current flow is then proportional to ($qV_{ON}/kT$). With further increase in the forward current density, the injected carrier density in the drift region exceeds the background doping concentration leading to high level injection conditions. The current flow then once again becomes proportional to $(qV_{ON}/2kT)$. In this mode of operation, the injected carrier concentration in the drift region increases in proportion to the current density resulting in a constant voltage drop across the drift region. At even larger on-state current densities, the influence of the recombination in the end-regions reduces the injected carrier density in the drift region. This produces a more rapid increase in the on-state voltage drop. This overall behavior is captured in Fig. 6.6 where a typical on-state characteristic is shown for a P-i-N rectifier under the various modes of operation.

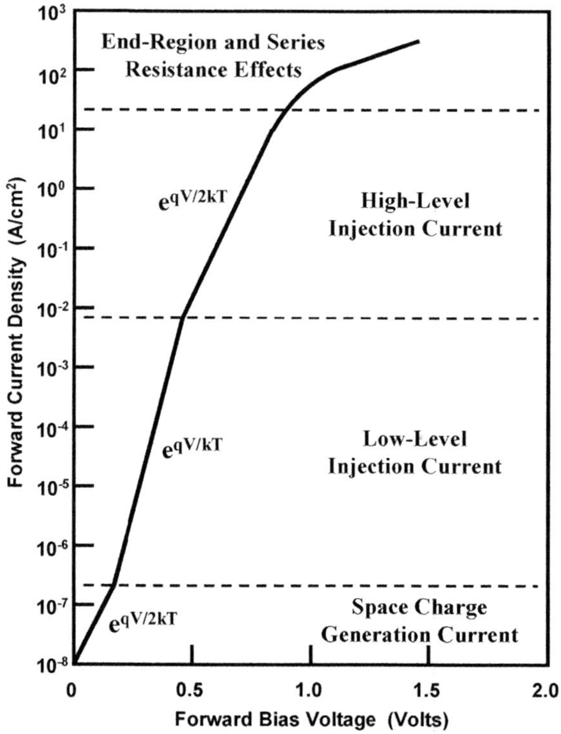


Fig. 6.6 On-state Characteristics for a P-i-N Rectifier.

**Simulation Example**

In order to gain further insight into the physics of operation for the P-i-N rectifier, the results of two-dimensional numerical simulations are provided in this section for a structure designed for supporting 3000 volts. For this case, a drift region with doping concentration of $4.6 \times 10^{13}$ cm$^{-3}$ was used with a thickness of 300 microns. The $P^+$ and $N^+$ end-regions had a surface concentration of $1 \times 10^{19}$ cm$^{-3}$ and a depth of about 5 microns. The on-state characteristics were obtained for various values for the lifetime ($\tau_{p0}$ and $\tau_{n0}$). In all cases, it was assumed that $\tau_{p0} = \tau_{n0}$. The influence of band-gap narrowing, auger recombination, and carrier-carrier scattering was included during the numerical simulations.

The on-state characteristics obtained from the numerical simulations are shown in Fig. 6.1E for the case of lifetime ($\tau_{p0}$ and $\tau_{n0}$) of 10 microseconds in the drift region. Several distinct regimes of operation are apparent in the shape of the characteristics. At current densities ranging between $10^{-7}$ and $10^{-3}$ A/cm$^2$, the device operates in the low-level injection regime. Here, the slope of the i-v characteristic exhibits the expected ($qV_{ON}/kT$) behavior with the forward voltage drop increasing at the rate of 60 mV per decade of increase in the on-state current density. At larger current densities ranging between $10^{-3}$ and $10^{1}$ A/cm$^2$, the device operates in the high-level injection regime. Here, the slope of the i-v characteristic exhibits the expected ($qV_{ON}/2kT$) behavior with the forward voltage drop increasing at the rate of 120 mV per decade of increase in the on-state current density. This behavior validates the analytical theory described in the previous sections for current conduction in the P-i-N rectifier.

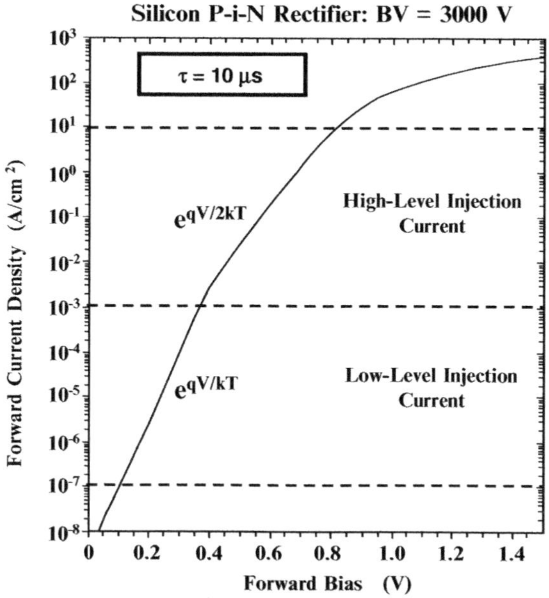


Fig. 6.1E On-state Characteristics for a 3000V Silicon P-i-N Rectifier.

The carrier distribution within the P-i-N rectifier is shown in Fig. 6.2E for the case of a lifetime ($\tau_{p0}$ and $\tau_{n0}$) of 1 microsecond in the drift region at an on-state current density of 100 A/cm$^2$. Here, the hole concentration is shown with a solid line while the electron concentration is shown by a dashed line. It can be seen that high-level injection conditions prevail in the drift region because the injected carrier concentration is far greater than the background doping concentration. The hole and electron concentrations are equal in magnitude throughout the drift region and exhibit the expected catenary shape derived in the previous sections of this chapter. The average carrier concentration ($4 \times 10^{16}$ cm$^{-3}$) calculated by using Eq. [6.23] is in good agreement with the carrier density obtained from the simulations.

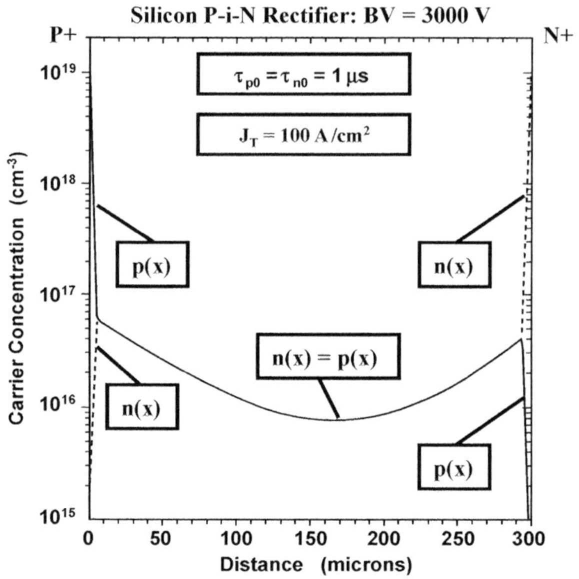


Fig. 6.2E Carrier Distribution within a 3000V P-i-N Rectifier.

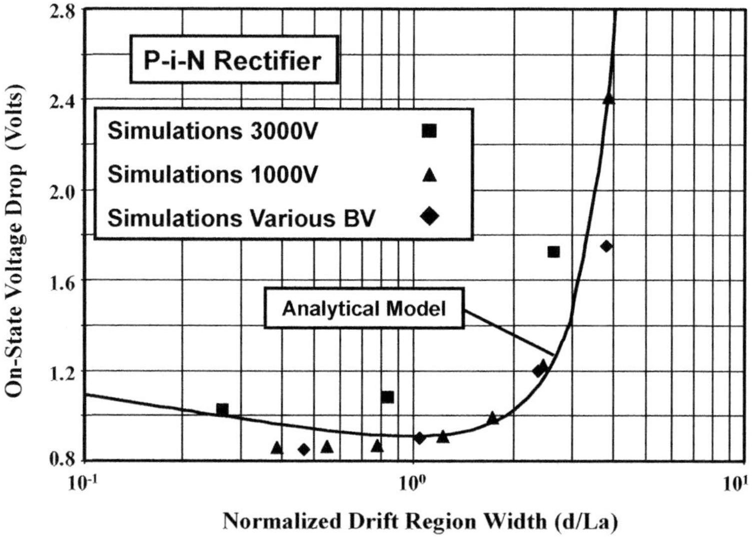


Fig. 6.3E On-State Voltage Drop for Silicon P-i-N Rectifiers.

The impact of changing the lifetime in the 3000V structure is shown in Fig. 6.3E. It can be seen that the on-state voltage drop is low for a lifetime of 100 microseconds. This is consistent with the small value of about 0.3 for the $(d/L_a)$

ratio. When the lifetime is reduced to 10 microseconds, the on-state voltage drop increases only slightly because the ($d/L_a$) ratio is still close to unity. However, the on-state voltage drop increases substantially when the lifetime is reduced to 1 microsecond because the ($d/L_a$) ratio has become significantly greater than one. The on-state voltage drop obtained with the simulations is close to that obtained using the analytical model, as shown in Fig. 6.3E by the square symbols, providing further credence to the model. In addition, the results obtained by varying the lifetime in a 1000 V P-i-N rectifier with an N-drift region width of 60 microns are shown by the triangular symbols.

During operation in power circuits, the junction temperature in the P-i-N rectifiers increases due to power dissipation. It is therefore important to evaluate the influence of the temperature upon the *i-v* characteristics in the forward conduction mode of operation. As an example, these characteristics are shown in Fig. 6.4E for the case of the 3000 V P-i-N rectifier with a lifetime ($\tau_{p0}$ and $\tau_{n0}$) of 10 microseconds in the drift region. The on-state voltage drop at a forward current density of 100 A/cm$^2$ is observed to reduce slightly with temperature. This is due to a reduction of the voltage drop at the junctions. Unfortunately, this behavior favors the development of 'hot-spots' within devices where the current density can become large. However, the positive temperature coefficient for the on-state voltage drop for current densities above 300 A/cm$^2$ indicates stable operation is possible with only moderate non-uniformities in the current distribution within silicon P-i-N rectifiers.

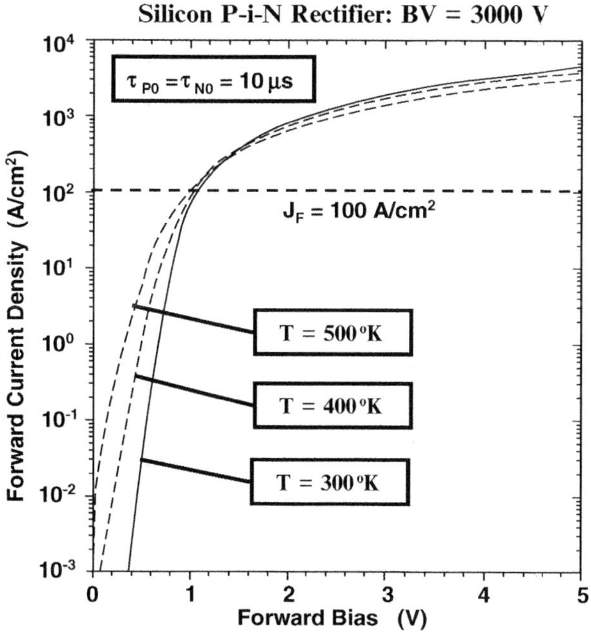


Fig. 6.4E Forward Conduction Characteristics of a Silicon P-i-N Rectifier.

## 6.2 Silicon Carbide P-i-N Rectifiers

Due to the much larger electric field that can be supported in silicon carbide, the width of the drift region is much smaller than that for the corresponding silicon device with the same breakdown voltage. This implies that the stored charge in the silicon carbide P-i-N rectifier will be much smaller than for the silicon device providing an improvement in the switching behavior. Unfortunately, the improved switching performance is accompanied by a substantial increase in the on-state voltage drop associated with the larger energy band gap for silicon carbide.

The physics of operation of the silicon carbide P-i-N rectifier is the same as that described in the previous sections. However, the parameters for the silicon carbide device defer from the silicon device. This has a strong impact on the voltage drop associated with the junctions. It was previous demonstrated that the junction voltage drop is given by:

$$V_{P+} + V_{N+} = \frac{kT}{q} \ln\left[\frac{n(+d)n(-d)}{n_i^2}\right] \quad [6.52]$$

Although the injected carrier concentrations n(+d) and n(-d) can be assumed to be similar in magnitude to those in a silicon P-i-N rectifier, the intrinsic carrier concentration for 4H-SiC is only 6.7 x $10^{-11}$ $cm^{-3}$ at 300 °K, due to its larger energy band gap, when compared with 1.4 x $10^{10}$ $cm^{-3}$ for silicon. This produces an increase in the junction voltage drop from 0.82 V for the silicon diode to 3.24 V for the 4H-SiC diode if the free carrier concentration in the drift region is assumed to be 1 x $10^{17}$ $cm^{-3}$. The power dissipation in the 4H-SiC P-i-N rectifier is therefore four times greater than in the silicon device. The expected improvement in the switching behavior is mitigated by the large on-state power loss. Consequently, it is preferable to develop silicon carbide Schottky diodes for voltage ratings of upto 5000 volts and P-i-N diodes for voltage ratings above 10,000 V.

### Simulation Example

In order to illustrate the reduction in the stored charge within silicon carbide P-i-N rectifiers, consider the case of a 4H-SiC P-i-N rectifier designed to support 10,000 volts. The thickness of the drift region for this device is only 80 microns when compared with 1200 microns required for the silicon structure. The drift region also has relatively high doping level of 2 x $10^{15}$ $cm^{-3}$. Due to the smaller thickness, good conductivity modulation of the drift region is observed even for a very small lifetime ($\tau_{p0}$ and $\tau_{n0}$) value of 100 ns as shown in Fig. 6.5E. However, poor conductivity modulation occurs for a typical lifetime ($\tau_{p0}$ and $\tau_{n0}$) value of 10 ns in the drift region observed in 4H-SiC. Methods for improving the minority carrier lifetime in 4H-SiC are required to assure good diode characteristics. The plot in Fig. 6.5E was obtained at a forward bias of 4 volts because of the larger junction potential for 4H-SiC as discussed above. Thus, the improved switching performance is obtained at the significant disadvantage of high on-state power loss.

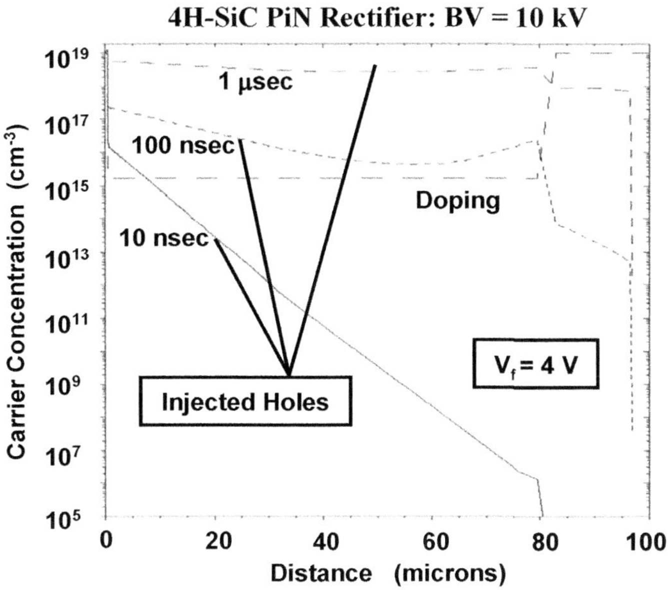


Fig. 6.5E Conductivity Modulation within 10 kV 4H-SiC PiN Rectifiers.

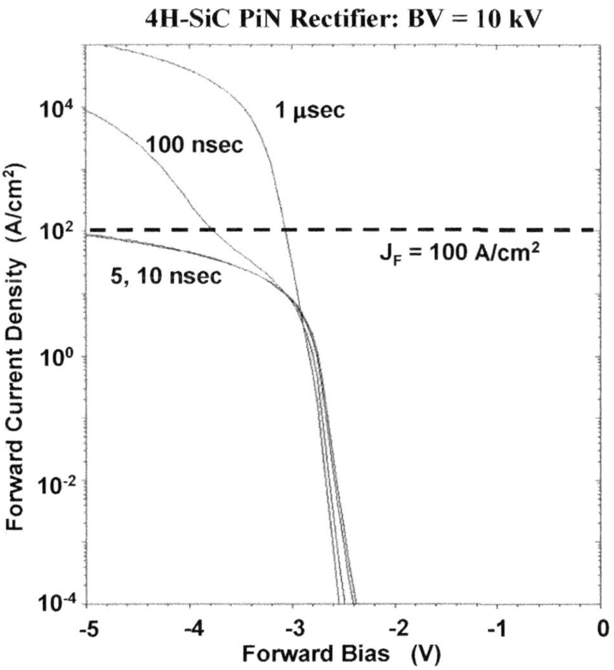


Fig. 6.6E Forward Conduction Characteristics of 10kV 4H-SiC PiN Rectifiers.

The forward i-v characteristics for the 10 kV 4H-SiC P-i-N rectifiers obtained from the numerical simulations are shown in Fig. 6.6E for various lifetime values. It can be observed that the on-state voltage drop is determined by the un-modulated resistance of the drift region when the lifetime is at 5 or 10 ns. When the lifetime is increased to 100 ns, the conductivity modulation of the drift region reduces the on-state voltage drop. With a lifetime of 1 microsecond, strong conductivity modulation occurs in the drift region as shown in Fig. 6.5E making the on-state voltage drop approximately 3 volts at an on-state current density of 100 A/cm$^2$. Lifetime values in this range have been more recently achieved and reported in silicon carbide devices[8].

## 6.3 Reverse Blocking

The reverse blocking voltage capability of the P-i-N rectifier is determined by the punch-through electric field distribution profile as described in the textbook[1]. The punch-through design enables reduction of the thickness of the drift region which is beneficial for reducing the on-state voltage drop as discussed in the previous sections of this chapter. Since the doping concentration of the drift region is small in the punch-through design, the drift region becomes completely depleted at a relatively low reverse bias voltage given by:

$$V_{PT} = \frac{qN_D(2d)^2}{2\varepsilon_S} \quad [6.53]$$

Beyond this voltage, the depletion region volume for the P-i-N rectifier remains independent of the reverse bias voltage under the assumption that the end-regions are heavily doped.

The leakage current for a reverse biased P-N junction is produced by a combination of the space charge generation current and the diffusion current. The space charge generation current, contributed by the carriers generated in the drift region, is given by:

$$J_{SC} = \frac{qW_D n_i}{\tau_{SC}} \quad [6.54]$$

where $W_D$ is the width of the depletion region. An additional component of the leakage current is associated with the generation of electron-hole pairs in the neutral regions. Any minority carriers generated in the proximity of the junction diffuse to the depletion region boundary and get swept to the opposite side of the junction by the electric field. Under reverse blocking conditions, the diffusion components of the leakage current are given by:

$$J_{LN} = \frac{qD_p p_{0N}}{L_p} = \frac{qD_p n_i^2}{L_p N_D} \quad [6.55]$$

and

$$J_{LP} = \frac{qD_n n_{0P}}{L_n} = \frac{qD_n n_i^2}{L_n N_A} \quad [6.56]$$

In the case of the P-i-N rectifier, the entire drift region is depleted at a relatively small reverse bias voltage due to its low doping concentration. The space charge generation current can therefore be assumed to arise from the entire width (2d) of the drift region while the diffusion currents are generated in the $P^+$ and $N^+$ end regions. The total leakage current for the P-i-N rectifier is then given by:

$$J_{LT} = \frac{qD_n n_i^2}{L_n N_{AP+}} + \frac{q(2d)n_i}{\tau_{SC}} + \frac{qD_p n_i^2}{L_p N_{DN+}} \quad [6.57]$$

The influence of heavy doping effects on the intrinsic carrier concentration and the diffusion lengths can enhance the leakage current arising from the end regions. However, the leakage current in the P-i-N rectifier, under reverse blocking conditions, is determined primarily by the space-charge generation current. The contributions due to the diffusion currents from the end regions become comparable to the space-charge-generation current only at elevated temperatures.

As an example, consider the case of a P-i-N rectifier with a drift region with width (2d) of 100 microns and a space-charge-generation lifetime of 10 microseconds. The leakage current components calculated using Eq. [6.57] are plotted in Fig. 6.7 between 300 and 500 °K under the assumption that the $P^+$ and $N^+$ end-regions have a doping concentration of 1 x $10^{19}$ $cm^{-3}$ and minority carrier

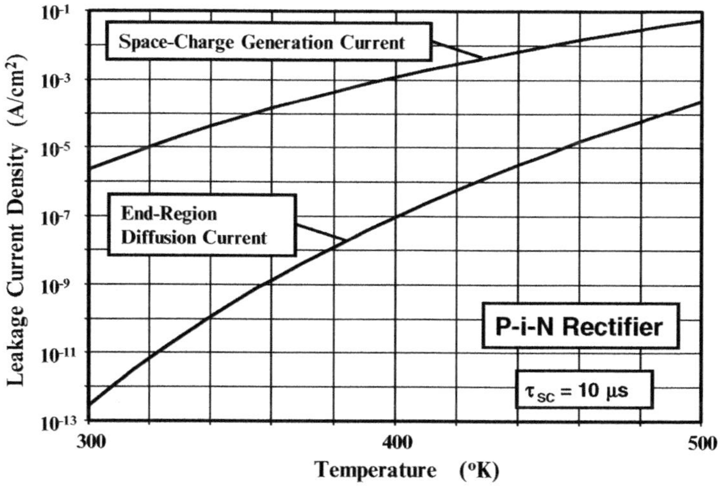


Fig. 6.7 Leakage Current Components in a Silicon P-i-N Rectifier.

lifetime of 1 ns. It is obvious that the leakage current due to the space-charge-generation process is dominant over this temperature range. It will therefore determine the total leakage current density in the silicon P-i-N rectifier.

### Simulation Example

In order to validate the above model for the leakage current in the P-i-N rectifier, the results of numerical simulations on the 3000 V silicon P-i-N rectifier structure, whose forward characteristics were discussed a previous section, are described here. A reverse bias of 1000 volts was chosen to completely deplete the drift region. The leakage current was extracted over a temperature range of 300 to 500 $^oK$. A lifetime ($\tau_{p0}$ and $\tau_{n0}$) value of 10 microseconds was used in the drift region. The end regions had a Gaussian doping profile with a surface concentration of $1 \times 10^{19}$ $cm^{-3}$ and depth of about 10 microns. A leakage current density obtained using the simulations is compared with that calculated using the model in Fig. 6.7E. The excellent agreement between the calculated values and those obtained with the numerical simulations provides confirmation that the analytical model can be utilized for the analysis of the leakage current in P-i-N rectifiers.

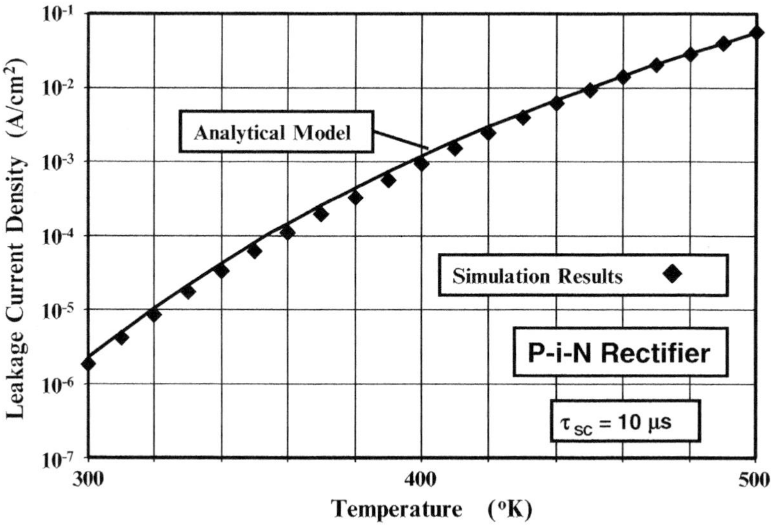


Fig. 6.7E Leakage Current in a 3kV Silicon P-i-N Rectifier.

## 6.4 Switching Performance

Power rectifiers control the direction of current flow in circuits used in various power conditioning applications. They operate for part of the time in the on-state when the bias applied to the anode is positive and for the rest of the time in the blocking state when the bias applied to the anode is negative. During each operating cycle, the diode must be rapidly switched between these states to minimize power losses. Much greater power losses are incurred when the diode switches from the on-state to the reverse blocking state than when it is turned on. The stored charge within the drift region of the power rectifier produced by the on-state current flow must be removed before it is able to support high voltages. This produces a large reverse current for a short time duration. This phenomenon is referred to as *reverse recovery*.

The presence of a large concentration of free carriers in the drift region during on-state current flow is responsible for the low on-state voltage drop of high voltage silicon P-i-N rectifiers. In order to switch the diode from its on-state mode to the reverse blocking mode, it is necessary to remove these free carriers to enable the formation of a depletion region that can support a high electric field. The process of switching the P-i-N rectifier from the on-state to the blocking state is referred to as *reverse recovery*.

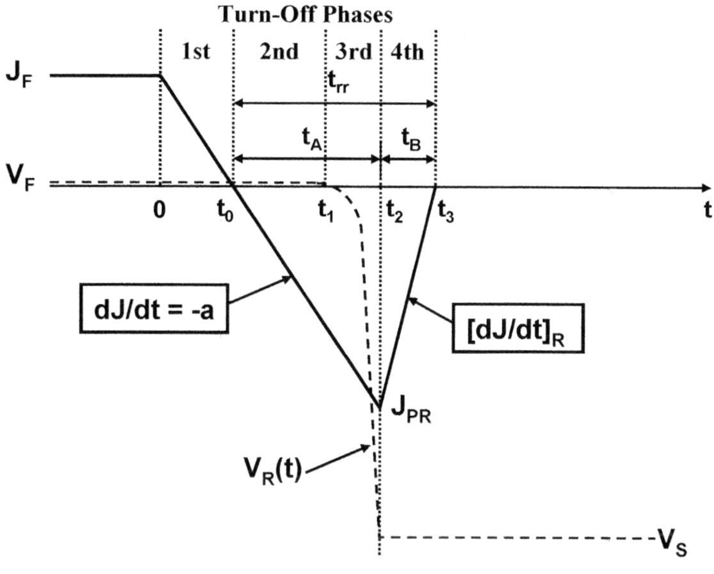


Fig. 6.8 Anode Current and Voltage Waveforms for the P-i-N Rectifier during the Reverse Recovery Process.

In power electronic circuits, it is common-place to use power rectifiers with an *inductive load*. In this case, the current reduces at a constant ramp rate ('a') as illustrated in Fig. 6.8 until the diode is able to support voltage. Consequently, a large *peak reverse recovery current* ($J_{PR}$) occurs due to the stored charge followed by the reduction of the current to zero. The power rectifier remains in its forward biased mode with a low on-state voltage drop until time $t_1$. The voltage across the diode then rapidly increases to the supply voltage with the rectifier operating in its reverse bias mode. The current flowing through the rectifier in the reverse direction reaches a maximum value ($J_{PR}$) at time $t_2$ when the reverse voltage becomes equal to the reverse bias supply voltage ($V_S$).

The simultaneous presence of a high current and voltage produces large instantaneous power dissipation in the power rectifier. The peak reverse recovery current also flows through the power switch that is controlling the switching event. This increases the power losses in the transistor. In the case of typical motor control PWM circuits that utilize IGBTs as power switches, a large reverse recovery current can trigger latch-up failure that can destroy both the transistor and the rectifier. It is therefore desirable to reduce the magnitude of the peak reverse recovery current and the time duration of the recovery transient. This time duration is referred to as the *reverse recovery time* ($t_{rr}$).

An analytical model for the reverse recovery process for the turn-off of a P-i-N rectifier under a constant rate of change of the current (*current ramp-rate*) can be created by assuming that the concentration of the free carriers in the drift region can be linearized as illustrated in Fig. 6.9[1]. As shown in the figure, the catenary carrier distribution established by the on-state current flow is approximated by an average value in the middle of the drift region and a linearly varying portion with a concentration of n(-d) at x = 0 to the average concentration of $n_a$ at a distance x = b. These carrier concentrations can be obtained from Eq. [6.19] and Eq. [6.23]:

$$n(-d) = \frac{\tau_{HL} J_F}{2qL_a}\left[\frac{\cosh(-d/L_a)}{\sinh(d/L_a)} - \frac{\sinh(-d/L_a)}{2\cosh(d/L_a)}\right] \qquad \textbf{[6.58]}$$

and

$$n_a = \frac{J_F \tau_{HL}}{2qd} \qquad \textbf{[6.59]}$$

where $J_F$ is the forward (or on-state) current density.

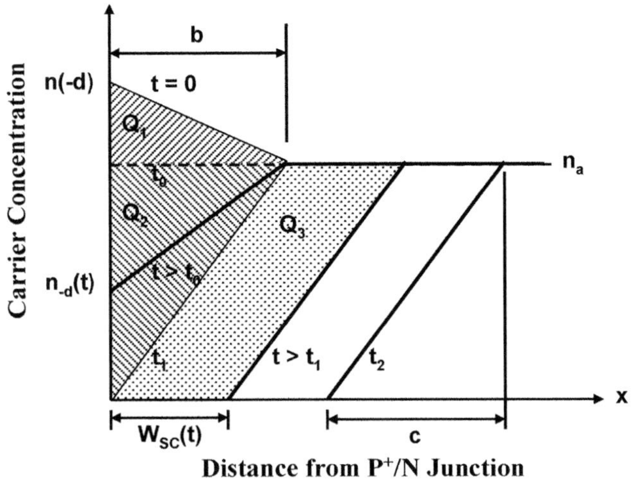


Fig. 6.9 Carrier Distribution Profiles in the P-i-N Rectifier during the Reverse Recovery Process.

The current flowing through the rectifier at any time during the turn-off transient is determined by the rate of diffusion of the carriers at the $P^+$/N junction boundary as described earlier during the discussion of on-state operation:

$$J_F = 2qD_a\left(\frac{dn}{dx}\right)_{x=-d} \quad [6.60]$$

In the first phase of the turn-off process, the current density in the P-i-N rectifier changes from the on-state current density ($J_F$) to zero at time $t_0$. The distance 'b' in Fig. 6.9 can be obtained by relating the charge $Q_1$ removed during the first phase to the current flow. At the end of the first phase, the carrier profile becomes flat at time $t_0$, as indicated by the dashed line in Fig. 6.9, because the current is zero at this time. The change in the stored charge within the drift region during the first phase can then be obtained from the cross-hatched area, indicated by $Q_1$, in the figure:

$$Q_1 = \frac{qb}{2}\left[n(-d) - n_a\right] \quad [6.61]$$

This charge can be related the current flow during the turn-off transient from t = 0 to t = $t_0$:

$$Q_1 = \int_0^{t_0} J(t)\,dt = \int_0^{t_0} (J_F - at)\,dt = J_F t_0 - \frac{at_0^2}{2} \quad [6.62]$$

The time $t_0$ at which the current crosses zero is given by:

$$t_0 = \frac{J_F}{a} \quad [6.63]$$

Combining the above relationships:

$$b = \frac{J_F^2}{qa\left[n(-d) - n_a\right]} \quad [6.64]$$

The carrier concentration n(-d) in Eq. [6.58] can be written as:

$$n(-d) = \frac{J_F \tau_{HL}}{2qL_a} K \quad [6.65]$$

where

$$K = \left[\frac{\cosh(-d/L_a)}{\sinh(d/L_a)} - \frac{\sinh(-d/L_a)}{2\cosh(d/L_a)}\right] \quad [6.66]$$

Using Eq. [6.65], in conjunction with Eq. [6.59], in Eq. [6.64]:

$$b = \frac{2dL_a J_F}{a\tau_{HL}(Kd - L_a)} \qquad [6.67]$$

The distance 'b' can therefore be calculated from the device parameters (d and $\tau_{HL}$), the on-state current density, and the ramp rate 'a'.

The second phase of the turn-off process occurs from the time $t_0$ at which the current crosses zero upto the time $t_1$ when the $P^+/N$ junction can begin to support voltage. The carrier profile at time $t_1$ is shown in Fig. 6.9 as extending from a zero concentration at the junction (located at x = 0) and the average concentration $n_a$ at a distance 'b' from the junction. After time $t_1$, a depletion region forms at the $P^+/N$ junction with the zero carrier concentration at some distance away from the junction. The time $t_1$ can be obtained by analysis of the charge removal during the turn-off transient from $t = t_0$ to $t = t_1$. In Fig. 6.9, the charge removed during this time interval is indicated by the cross-hatched area marked $Q_2$. This area is given by:

$$Q_2 = \frac{1}{2} q n_a b \qquad [6.68]$$

This charge can be related the current flow during the turn-off transient from $t = t_0$ to $t = t_1$:

$$Q_2 = \int_{t_0}^{t_1} J(t)\,dt = \int_{t_0}^{t_1} (at)\,dt = \frac{a}{2}\left(t_1^2 - t_0^2\right) \qquad [6.69]$$

Using Eq. [6.63] for $t_0$:

$$Q_2 = \frac{a}{2}\left(t_1^2 - \frac{J_F^2}{a^2}\right) \qquad [6.70]$$

Combining Eq. [6.68] and Eq. [6.70]:

$$t_1 = \sqrt{\frac{q n_a b}{a} + \frac{J_F^2}{a^2}} \qquad [6.71]$$

Making use of the Eq. [6.59] for the average carrier concentration $n_a$ and Eq. [6.67] for the distance 'b':

$$t_1 = \frac{J_F}{a}\sqrt{\frac{L_a}{(Kd - L_a)} + 1} \qquad [6.72]$$

Based upon this expression, the end of the second phase occurs earlier when the ramp rate is increased. This accelerates the point at which the rectifier can begin to support a reverse bias voltage.

During the entire time from t = 0 until time $t = t_1$, the $P^+/N$ junction within the P-i-N rectifier remains forward biased because the minority carrier density in the drift region at the junction [p(-d,t)] is above the equilibrium minority carrier concentration ($p_{0N}$). Under the assumptions of high level injection conditions in the drift region, the minority carrier density [p(-d,t)] is equal to the majority carrier density [n(-d,t)] that is illustrated in Fig. 6.9. Based upon Eq. [6.60], the current density at any point in time is given by:

$$J(t) = 2qD_a\left(\frac{dn}{dx}\right)_{x=-d} = 2qD_a\frac{[n(-d,t)-n_a]}{b} \quad \textbf{[6.73]}$$

The carrier concentration in the drift region at the junction is therefore related to the current density by:

$$p(-d,t) = n(-d,t) = n_a + \frac{J(t)b}{2qD_a} = n_a + \frac{(J_F - at)b}{2qD_a} \quad \textbf{[6.74]}$$

This expression is valid for both positive and negative values for the current density during the turn-off transient until time $t_1$. The voltage drop across the forward biased junction during this time interval can be obtained using the Boltzmann relationship:

$$V_F(t) = \frac{kT}{q}\ln\left[\frac{p(-d)}{p_{0N}}\right] = \frac{kT}{q}\ln\left[\frac{(J_F - at)b}{2qD_a p_{0N}} + \frac{n_a}{p_{0N}}\right] \quad \textbf{[6.75]}$$

This expression describes the change in the voltage drop across the P-i-N rectifier during the turn-off transient until it is able to support a reverse bias voltage.

During the third phase of the turn-off transient, the P-i-N rectifier begins to support an increasing voltage. This requires the formation of a space-charge region $W_{SC}(t)$ at the $P^+$/N junction that expands with time as illustrated in Fig. 6.9. The expansion of the space-charge region is achieved by further extraction of the stored charge in the drift region resulting in the reverse current continuing to increase after time $t_1$. The growth of the reverse bias voltage across the P-i-N rectifier can be analytically modeled under the assumption that the sweep out of the stored charge is occurring at an approximately constant current. In this case, the slope of the carrier distribution profile remains constant as shown in Fig. 6.9.

In Fig. 6.9, the charge removed at a time t, after the $P^+$/N junction is reverse biased at time $t_1$, is indicated by the shaded area marked $Q_3$. The area of this parallelogram is given by:

$$Q_3 = qn_a W_{SC}(t) \quad \textbf{[6.76]}$$

This charge can be related the current flow during the turn-off transient from time $t_1$ to time t:

$$Q_3 = \int_{t_1}^{t} J(t)dt = \int_{t_1}^{t}(J_F - at)dt = J_F(t - t_1) - \frac{a}{2}(t^2 - t_1^2) \quad \textbf{[6.77]}$$

Combining these relationships for the charge $Q_3$ provides an expression for the growth of the space-charge region as a function of time:

$$W_{SC}(t) = \frac{a}{2qn_a}\left(t^2 - t_1^2\right) - \frac{J_F}{qn_a}\left(t - t_1\right) \qquad \textbf{[6.78]}$$

The voltage supported across this space-charge region can be obtained by solving Poisson's equation:

$$\frac{d^2V}{dx^2} = -\frac{dE}{dx} = -\frac{Q(x)}{\varepsilon_S} \qquad \textbf{[6.79]}$$

where Q(x) is the charge in the space-charge region. Unlike the blocking mode of operation, where the charge in the depletion region consists of the ionized donor charge, during the turn-off process an additional charge is contributed by the large reverse current flow. This charge is due to the holes that are transiting through the space-charge region due to the removal of the stored charge. Since the electric field within the space-charge region is large, it can be assumed that these holes are moving at the saturated drift velocity ($v_{sat,p}$). The concentration of the holes within the space-charge region is then related to the current density ($J_R$) by:

$$p(t) = \frac{J_R(t)}{qv_{sat,p}} = \frac{(at - J_F)}{qv_{sat,p}} \qquad \textbf{[6.80]}$$

The voltage supported by the space-charge region is then given by:

$$V_R(t) = \frac{q\left[N_D + p(t)\right]}{2\varepsilon_S} W_{SC}(t)^2 \qquad \textbf{[6.81]}$$

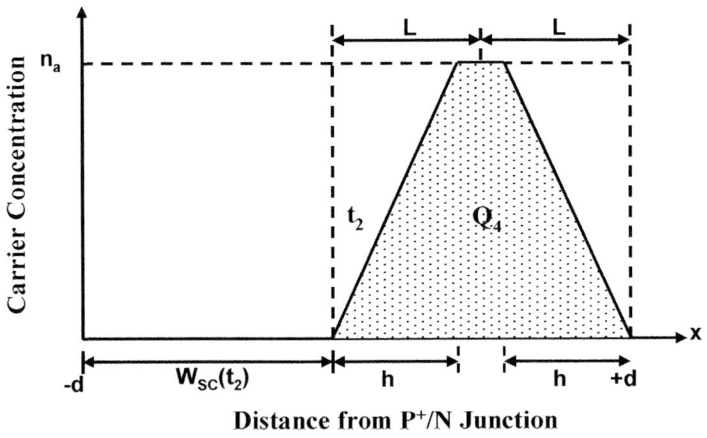


Fig. 6.10 Stored Charge within the P-i-N Rectifier after the end of the Third Phase.

This expression, in conjunction with Eq. [6.78] for the expansion of the space-charge width, indicates a rapid rise in the voltage supported by the P-i-N rectifier after time $t_1$. The end of the third phase occurs when the reverse bias across the P-i-N rectifier becomes equal to the supply voltage ($V_S$). Using this value in Eq. [6.81] together with Eq. [6.78], the time $t_2$ (and hence $J_{PR}$) can be obtained.

During the fourth phase of the turn-off process, the reverse current rapidly reduces at approximately a constant rate as illustrated in Fig. 6.8 while the voltage supported by the P-i-N rectifier remains constant at the supply voltage. The stored charge within the drift region after the end of the third phase is illustrated in Fig. 6.10 by the shaded area marked $Q_4$. At the end of the third phase of the turn-off process ($t = t_2$), the peak reverse recovery current $J_{PR}$ is flowing through the structure. This current can be related to the free carrier profile by:

$$J_{PR} = 2qD_a \frac{dn}{dx} = 2qD_a \frac{n_a}{h} \quad \textbf{[6.82]}$$

where the dimension 'h' is shown in the Fig. 6.10. Using this equation:

$$h = \frac{2qD_a n_a}{J_{PR}} \quad \textbf{[6.83]}$$

The stored charge remaining in the drift region at time $t_2$ is then given by:

$$Q_4 = qn_a\left[2d - W_{SC}(t_2) - h\right] \quad \textbf{[6.84]}$$

This charge must be removed during the fourth phase of the turn-off process. During the fourth phase, the current reduces to zero at an approximately constant rate, indicated as $[dJ/dt]_R$ in Fig. 6.8, over a time period $t_B$ extending from time $t_2$ to $t_3$. The charge removed due to the current flow during this time is given by:

$$Q_R = \frac{1}{2} J_{PR} t_B \quad \textbf{[6.85]}$$

This can be equated to the charge left in the drift region at the end of the third phase if recombination during this time is neglected due to the short duration of this time interval relative to the minority carrier lifetime. The time interval ($t_B$) for the reduction of the reverse current is then obtained:

$$t_B = (t_3 - t_2) = \frac{2qn_a}{J_{PR}}\left[2d - W_{SC}(t_2) - h\right] \quad \textbf{[6.86]}$$

The reverse ramp rate is then given by dividing the peak reverse recovery current by this time interval:

$$\left[\frac{dJ}{dt}\right]_R = \frac{J_{PR}}{t_B} = \frac{J_{PR}^2}{2qn_a\left[2d - W_{SC}(t_2) - h\right]} \qquad [6.87]$$

A smaller value for the reverse [di/dt] is desirable to reduce voltages developed across stray inductances in the circuit. These voltages cause an increase in the voltage supported by all the devices in the circuit making it necessary to enhance their breakdown voltages. This is detrimental to system performance due to an overall increase in power dissipation in the semiconductor components.

The utility of the analytical model can be illustrated by performing the analysis of the reverse recovery for a specific P-i-N rectifier structure. Consider the case of a P-i-N rectifier designed to support 1000 volts with a drift region thickness of 60 microns and doping concentration of 5 x $10^{13}$ $cm^{-3}$. The reverse recovery process in this structure is analyzed here using the analytical solutions for various ramp rates, lifetime values, and reverse supply voltages. In all cases, the reverse recovery is assumed to begin with on-state operation at a current density of 100 $A/cm^2$.

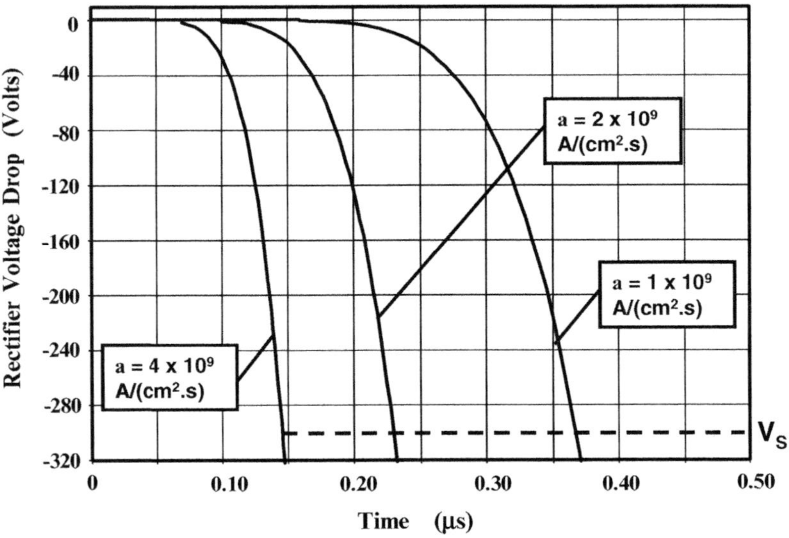


Fig. 6.11 Analytically Calculated Voltage Waveforms for a 1000 V P-i-N Rectifier during the Reverse Recovery Process using various Ramp Rates.

The voltage waveform calculated using the analytical solutions are shown in Fig. 6.11 for the case of a high level lifetime of 0.5 micro-seconds. With this lifetime, the average free carrier concentration in the drift region was found to be 5.2 x $10^{16}$ $cm^{-3}$ for this structure at the on-state current density of 100 $A/cm^2$. According to the analytical model (see Eq. [6.71], the time $t_1$ at which the junction becomes reverse biased increases from 51 nanoseconds to 80 nanoseconds to 134 nanoseconds as the ramp rate decreases from 4 x $10^9$ $A/cm^2$-s to 2 x $10^9$ $A/cm^2$-s to

$1 \times 10^9$ A/cm$^2$-s. Before this time, the voltage across the rectifier is slightly positive with a value given by Eq. [6.75]. The voltage then increases rapidly and reaches 300 volts (indicated by the dashed line in the figure) at 145, 230, and 370 nanoseconds, respectively, for the three cases. This point in the voltage waveforms defines the end of the third phase.

The peak reverse recovery current occurs at the end of the third phase. The peak reverse recovery current densities predicted by the analytical model are 480, 360, and 270 A/cm$^2$, respectively, for the three cases of the ramp rate as can be observed in Fig. 6.12 which shows the current waveforms obtained using the analytical model. After the third phase, the reverse current reduces to zero at a constant rate. The time duration ($t_B$), during which the reverse current reduces to zero, becomes smaller with a reduction in the ramp rate. The values for $t_B$ predicted by the analytical model decrease from 70 to 60 to 28 nanoseconds as the ramp rate decreases from $4 \times 10^9$ A/cm$^2$-s to $2 \times 10^9$ A/cm$^2$-s to $1 \times 10^9$ A/cm$^2$-s. These combinations of the peak reverse recovery current and period $t_B$ produce a reverse [dJ/dt] ranging from 6 to $9.5 \times 10^9$ A/cm$^2$-s.

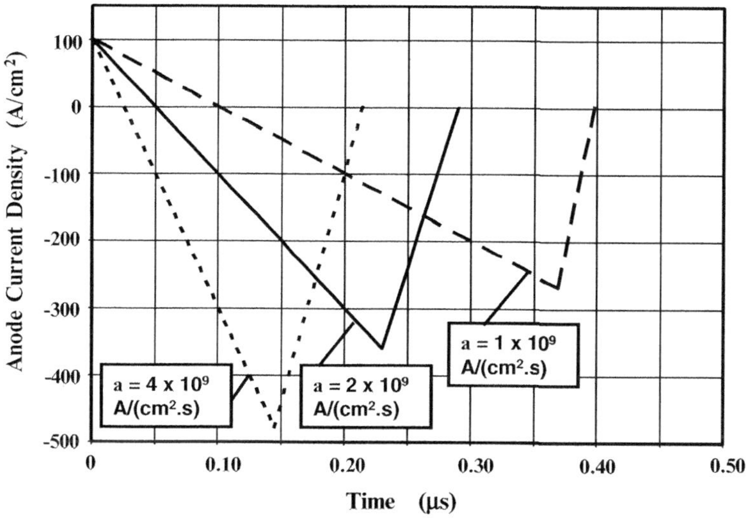


Fig. 6.12 Analytically Calculated Current Waveforms for a 1000 V P-i-N Rectifier during the Reverse Recovery Process using various Ramp Rates.

The analytical model can also be utilized to examine the influence of the minority carrier lifetime on the reverse recovery process. Consider the case of the same 1000-V P-i-N rectifier structure switched off from an on-state current density of 100 A/cm$^2$ at a ramp rate of $2 \times 10^9$ A/cm$^2$s. The voltage waveforms predicted by the analytical model for the reverse recovery process with lifetime values of 0.25, 0.5, and 1 microsecond in the drift region are shown in Fig. 6.13. The model predicts no change in the time $t_1$ for the end of the first phase and a faster rate of increase in the anode voltage during the second phase when the lifetime is reduced.

The current flow during the reverse recovery process is shown in Fig. 6.14 for the case of the three lifetime values. The peak reverse recovery current density predicted by the analytical model reduces from 480 to 360 to 270 A/cm$^2$, respectively, when the lifetime is reduced from 1 to 0.5 to 0.25 microseconds as

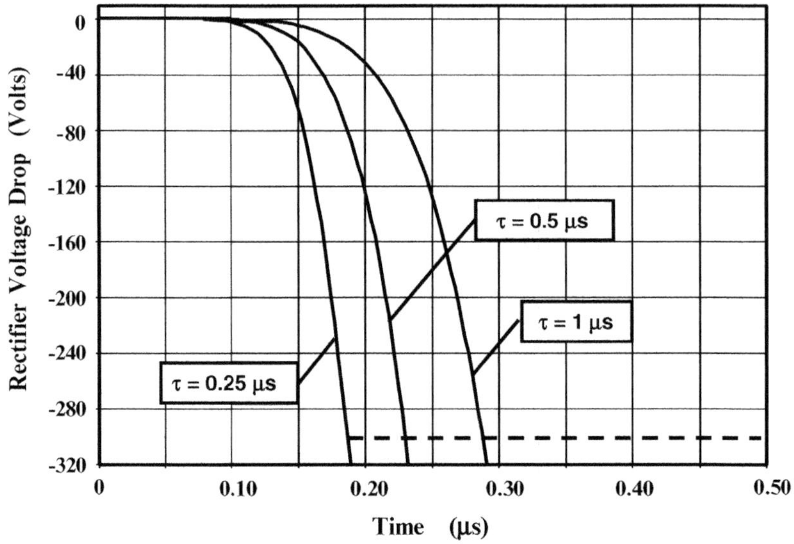


Fig. 6.13 Analytically Calculated Voltage Waveforms for a 1000 V P-i-N Rectifier during the Reverse Recovery Process for various Lifetime Values.

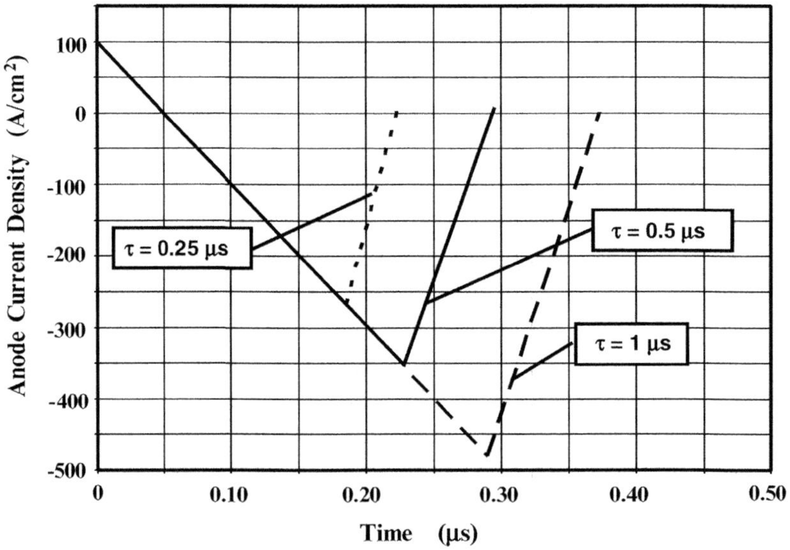


Fig. 6.14 Analytically Calculated Current Waveforms for a 1000 V P-i-N Rectifier during the Reverse Recovery Process for various Lifetime Values.

can be observed from Fig. 6.14. The time duration ($t_B$) for the fourth phase, during which the reverse current reduces to zero, also becomes smaller with a reduction in the lifetime. The values for $t_B$ predicted by the analytical model are 38 nanoseconds, 60 nanoseconds, and 84 nanoseconds for lifetime values of 0.25 microseconds, 0.5 microseconds, and 1 microsecond, respectively. These combinations of the peak reverse recovery current and period $t_B$ produce a reverse [dJ/dt] ranging from 5.7 to 7 x $10^9$ A/cm$^2$-s.

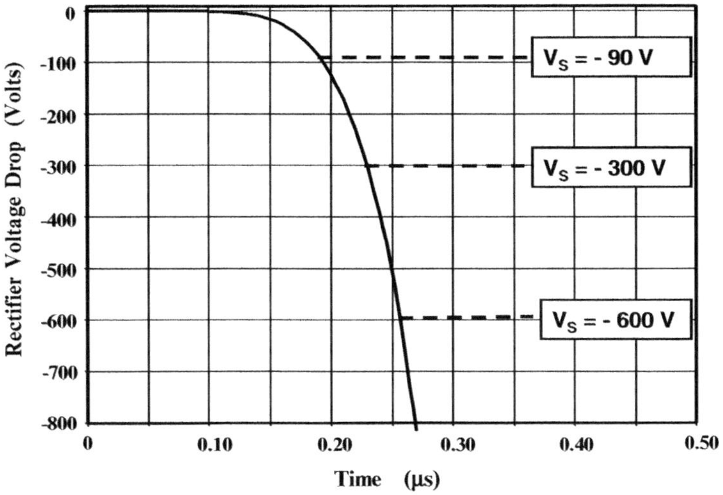


Fig. 6.15 Analytically Calculated Voltage Waveforms for a 1000 V P-i-N Rectifier during the Reverse Recovery Process for various Supply Voltages.

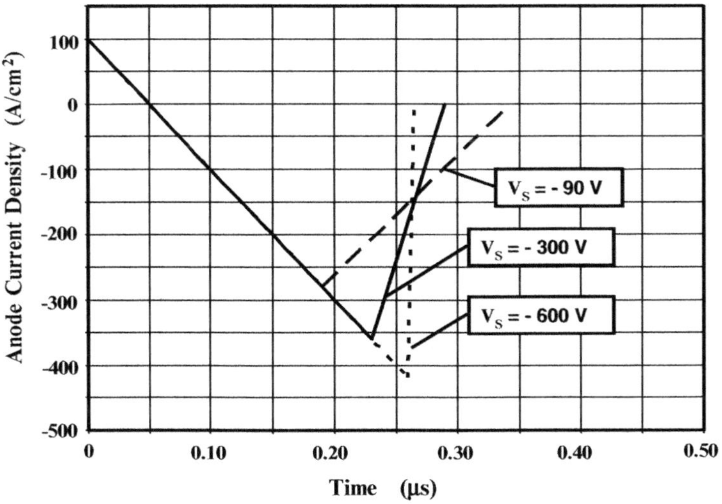


Fig. 6.16 Analytically Calculated Current Waveforms for a 1000 V P-i-N Rectifier during the Reverse Recovery Process for various Supply Voltages.

The analytical model also enables analysis of the impact of changing the reverse recovery voltage on the reverse recovery process. Consider the case of the same 1000 V P-i-N rectifier structure switched off from an on-state current density of 100 A/cm$^2$ at a ramp rate of 2 x 10$^9$ A/cm$^2$-s. The voltage waveform predicted by the analytical model is shown in Fig. 6.15 with dashed lines indicating the point at which the reverse bias voltage reaches values of 90, 300, and 600 volts during the reverse recovery transient. As the reverse voltage is increased, it takes a longer time interval to produce the wider space-charge region that is needed to support the voltage. This is accompanied by a larger value for the peak reverse recovery current as shown in Fig. 6.16. The larger space-charge region, formed at larger reverse bias voltages, removes a greater fraction of the stored charge as well. This produces a substantial reduction of the period $t_B$ resulting in very high reverse [dJ/dt] as observed in the figure.

The peak reverse recovery current density predicted by the analytical model increases from 280 to 360 to 420 A/cm$^2$, when the voltage is increased from 90 to 300 to 600 volts, respectively, as can be observed from Fig. 6.16. The time duration ($t_B$) for the fourth phase, during which the reverse current reduces to zero, also becomes smaller with an increase in the reverse voltage. The values for $t_B$ predicted by the analytical model are 153 nanoseconds, 60 nanoseconds, and only 1 nanosecond for reverse voltages of 90, 300, and 600 volts, respectively. The $t_B$ value for the 600 volt case indicates that almost all the stored charge has been extracted by the extension of the space-charge region at this large reverse voltage. These combinations of the peak reverse recovery current and period $t_B$ produce a drastic increase in the reverse [dJ/dt] ranging from 1.8 to 6 to 420 x 10$^9$ A/cm$^2$-s when the reverse voltage is increased from 90 to 300 to 600 volts, respectively.

**Simulation Example**

In order to validate the above model for the reverse recovery transient in the P-i-N rectifier, the results of numerical simulations on a 1000 V silicon P-i-N rectifier structure are described here. The structure had a drift region thickness of 60 microns with a doping concentration of 5 x 10$^{13}$ cm$^{-3}$. The cathode current was ramped from 100 A/cm$^2$ in the on-state using various values of negative ramp rates. In addition, the impact of changing the lifetime and the reverse supply voltage was examined for comparison with the analytical model.

First consider the case of varying the negative ramp rate from 1 x 10$^9$ to 2 x 10$^9$ to 4 x 10$^9$ A/cm$^2$s. For these cases, a lifetime ($\tau_{p0}$ and $\tau_{n0}$) value of 1 microsecond was used during the numerical simulations. The average carrier concentration in the drift region under steady state conditions with an on-state current density of 100 A/cm$^2$ was found to be about 5 x 10$^{16}$ cm$^{-3}$ as shown in Fig. 6.8E. This value is obtained by using a high level lifetime of 0.5 microseconds in Eq. [6.59] indicating that end-region recombination currents are significant in this structure. The carrier concentration profile exhibits a zero slope at time t = 40 ns, corresponding to time t = $t_0$ in the analytical model (see Fig. 6.8]. The slope of the carrier profile then becomes positive, as shown for the time t = 80 ns. During this time, the carrier concentration at the junction is well above the equilibrium value

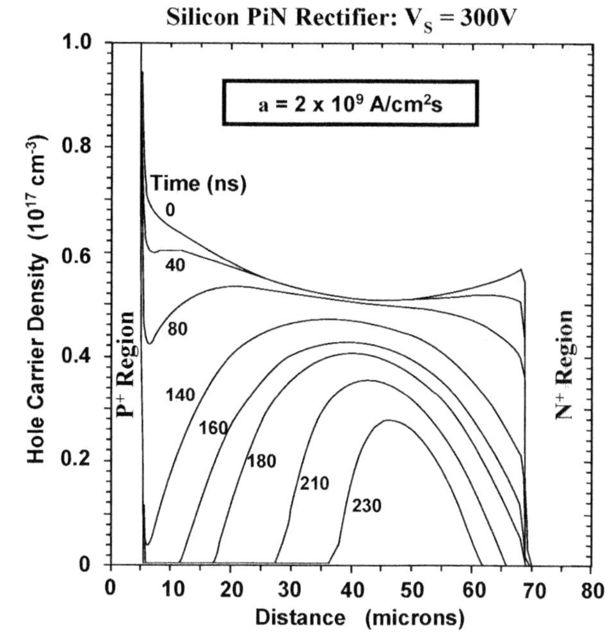


Fig. 6.8E Carrier Distribution in a 1000V Silicon P-i-N Rectifier during Phase 1, 2 and 3 of the Reverse Recovery Transient.

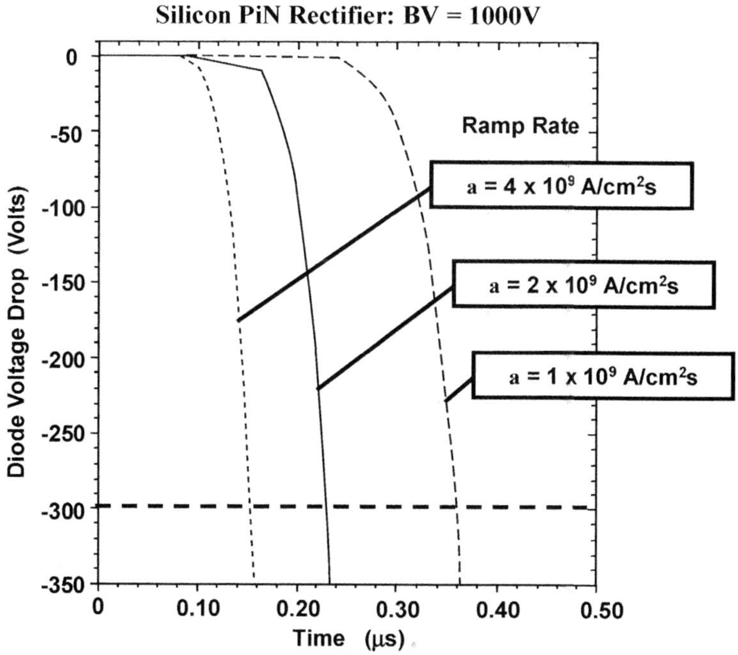


Fig. 6.9E Voltage Waveforms for a 1000V Silicon P-i-N Rectifier during the Reverse Recovery Transient with various Ramp Rates.

indicating that the $P^+/N$ junction is still forward biased. At time t = 140 ns, the carrier concentration at the junction becomes close to zero, corresponding to the time t = $t_1$ in the analytical model (see Fig. 6.8). The value for $t_1$ obtained using the analytical model is about 120 ns in good agreement with the simulations.

The carrier profiles for subsequent time instances of 160, 180, 210, and 230 ns are also shown in Fig. 6.8E. It can be observed that the depletion region expands from the $P^+/N$ junction during this time interval. The analytical model predicts a depletion region width of 38 microns, when the reverse bias voltage reaches 300 volts, in excellent agreement with the simulations. The analytical model also predicts the end of phase 3 at time $t_2$ = 230 ns in very good agreement with the simulations. Consequently, the peak reverse recovery current predicted by the model also agrees with the simulations

The diode voltage and current waveforms obtained with the aid of the numerical simulations are shown in Fig. 6.9E and 6.10E, respectively. These waveforms have the same features predicted by the analytical model (see Fig. 6.11 and Fig. 6.12). The peak reverse currents obtained using the model are in good agreement with those observed in the simulations. However, the transient observed with the numerical simulations during phase 4 occurs with a constant ramp rate followed by a more abrupt reduction of the current. The ramp rate observed with the simulations is in the range between 7 and 9 x $10^9$ $A/cm^2s$ as predicted by the analytical model.

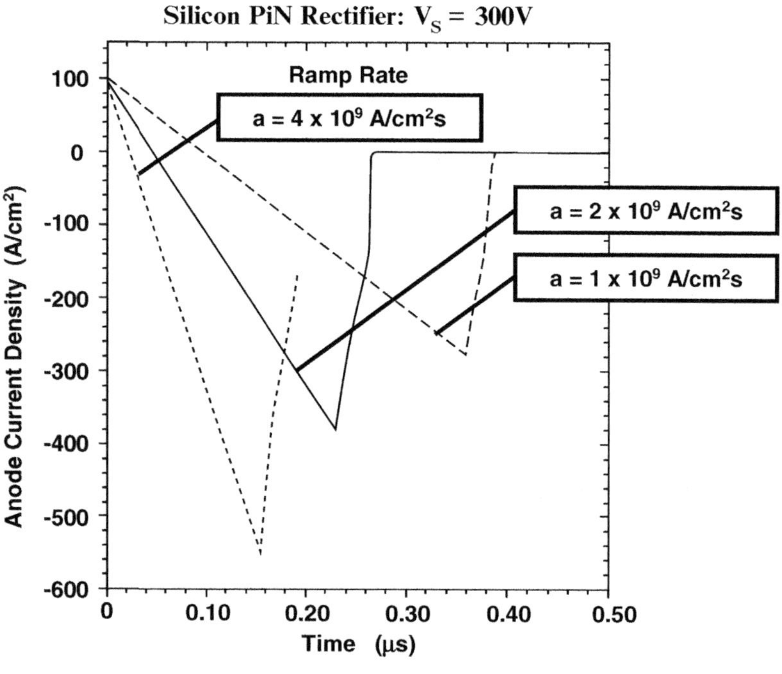


Fig. 6.10E Current Waveforms for a 1000V Silicon P-i-N Rectifier during the Reverse Recovery Transient with various Ramp Rates.

During the fourth phase of the turn-off process, the remaining free carriers in the drift region are removed by further extension of the depletion region as the current ramps down to zero. The reduction of the reverse current is accompanied by a reduction in the concentration of holes within the space-charge region as described by Eq. [6.80]. This reduces the net positive charge in the space-charge region allowing its expansion in spite of the constant reverse voltage across the diode. Since the P-i-N rectifier is designed with a punch-through architecture, the space-charge region eventually expands through the entire drift region removing all the stored charge. The removal of the stored charge, observed with the numerical simulations, is shown in Fig. 6.11E.

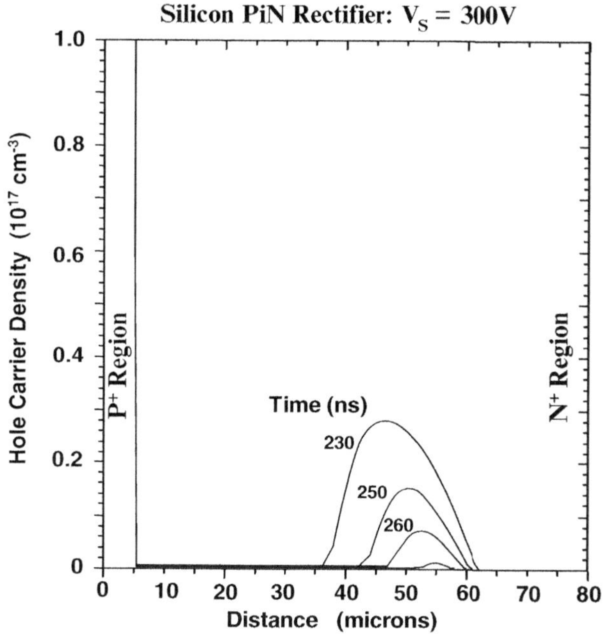


Fig. 6.11E Carrier Distribution in a 1000V Silicon P-i-N Rectifier during Phase 4 of the Reverse Recovery Transient.

The validity of the analytical model can also be examined by observation of the impact of changes in the lifetime on the reverse recovery process. In order to illustrate this, the lifetime was increased and reduced by a factor of 2x while performing the reverse recovery at ramp rate of $2 \times 10^9$ A/cm$^2$s. The diode voltage and current waveforms obtained using the numerical simulations are shown in Fig. 6.12E and 6.13E, respectively.

The waveforms shown in the above figures have the same features as the waveforms obtained using the analytical model (see Fig. 6.13 and 6.14). There is no change in the time $t_1$ for the end of the second phase while the time taken for the voltage to increase to 300 volts reduces with a reduction in the lifetime. The peak reverse recovery current decreases with a reduction of the lifetime as predicted by the model. The time duration for the fourth phase also increases with increasing lifetime as predicted by the model, resulting in a slightly smaller reverse [di/dt].

It can be concluded that the analytical model provides an accurate description of the impact of changes in the lifetime on the reverse recovery process.

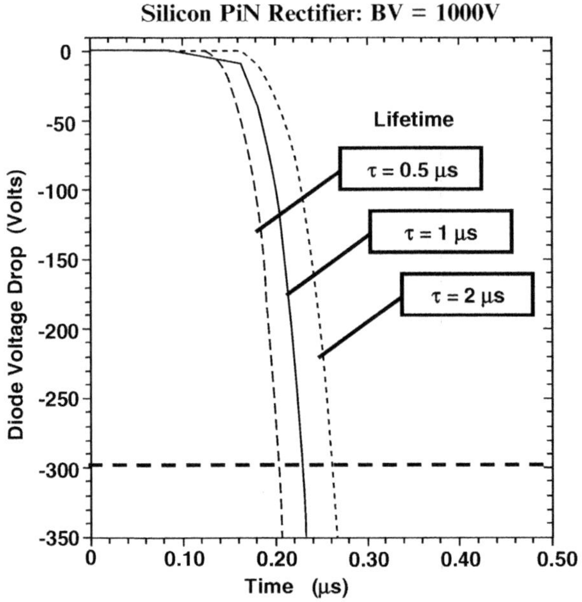


Fig. 6.12E Voltage Waveforms for a 1000V Silicon P-i-N Rectifier during the Reverse Recovery Transient with various Lifetimes.

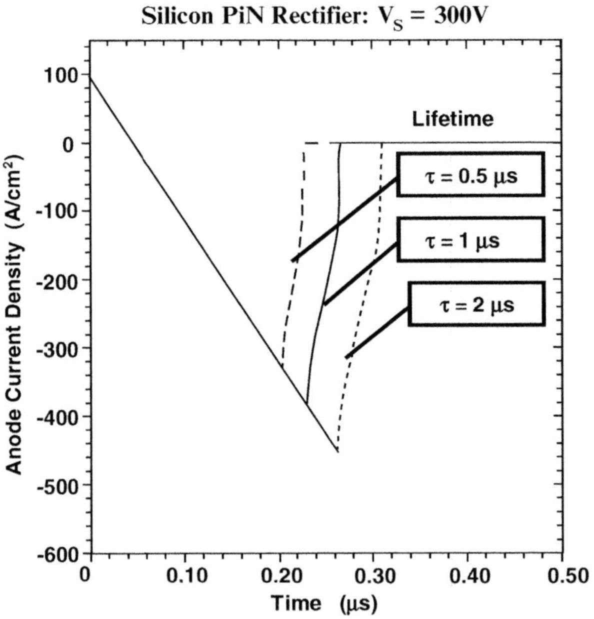


Fig. 6.13E Current Waveforms for a 1000V Silicon P-i-N Rectifier during the Reverse Recovery Transient with various Lifetimes.

As a further validation of the model, numerical simulations of the 1000 V P-i-N rectifier structure were performed with various reverse bias supply voltages. The voltage waveform is shown in Fig. 6.14E for the case of a ramp rate of 2 x $10^9$ A/cm$^2$s and lifetime of 1 microsecond. The time taken to arrive at the increasing reverse bias voltages is consistent with the prediction of the analytical model (see Fig. 6.15). Consequently, the peak reverse recovery currents predicted by the analytical model are also in agreement with those obtained from the numerical simulations. These waveforms are shown in Fig. 6.15E.

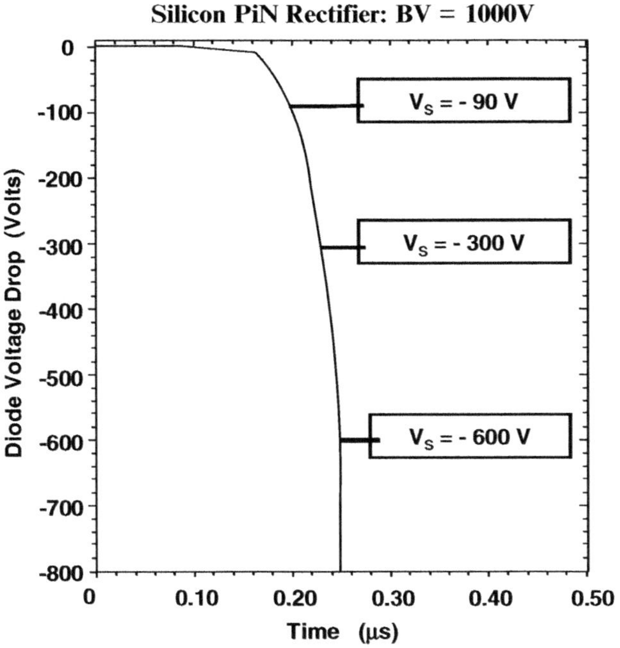


Fig. 6.14E Voltage Waveforms for a 1000V Silicon P-i-N Rectifier during the Reverse Recovery Transient with various Supply Voltages.

The current waveforms observed with the numerical simulations have the same features predicted by the analytical model (see Fig. 6.16). When the reverse voltage is increased to 600 volts, the fourth phase of the reverse recovery occurs with an abrupt reduction of the reverse current as predicted by the analytical model. The peak reverse currents obtained using the model are also in good agreement with those observed in the simulations. The extremely high reverse [di/dt] associated with an abrupt drop in the reverse recovery current is a problem in power circuits where it produces large voltage spikes across any stray inductances that are in series with the diode.

The above simulation results provide validation for the analytical model developed in this section of the chapter and give additional insight into the carrier distribution and transients. Based upon the excellent match between the predictions of the analytical model and the simulations, it can be concluded that the model is able to account for the all four phases of the reverse recovery process and predict the proper dependence of the reverse recovery current on the ramp rate, the lifetime, and the reverse supply voltage.

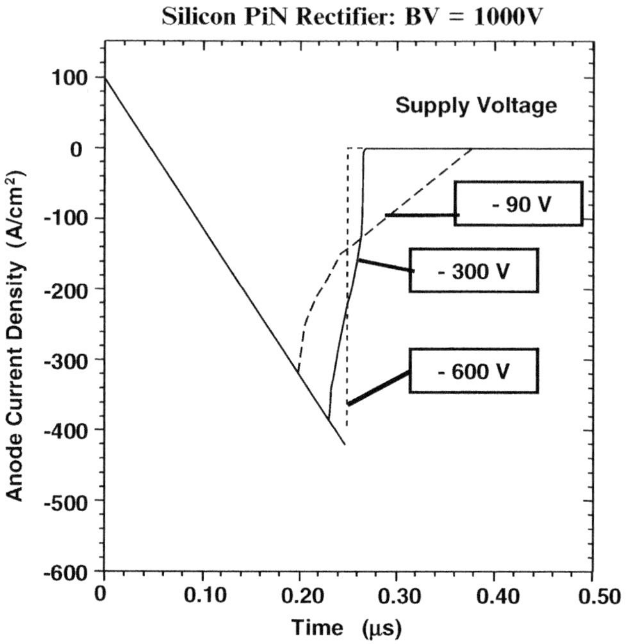


Fig. 6.15E Current Waveforms for a 1000V Silicon P-i-N Rectifier during the Reverse Recovery Transient with various Supply Voltages.

## 6.5 P-i-N Rectifier Trade-Off Curves

In the previous sections, it was demonstrated that the peak reverse recovery current and the turn-off time can be reduced by reducing the minority carrier lifetime in the drift region of the P-i-N rectifier structure. This enables reduction of the power losses during the switching transient. However, the on-state voltage drop in a P-i-N rectifier increases when the minority carrier lifetime is reduced, which produces an increase in the power dissipation during on-state current flow. For power system applications, it is desirable to reduce the total power dissipation produced in the rectifiers to maximize the power conversion efficiency. This also reduces the heat generated within the power devices maintaining a lower junction temperature which is desirable to prevent thermal runaway and reliability problems. To minimize the power dissipation, it is common-place to perform a trade-off between on-state and switching power losses for power P-i-N rectifiers by developing trade-off curves.

One type of the trade-off curve for a power P-i-N rectifier can be generated by plotting the on-state voltage drop against the reverse recovery turn-off time.

A turn-off waveform that includes a non-linear portion during the fourth phase is illustrated in Fig. 6.17. The turn-off time ($t_{rr}$) is defined as the time taken for the reverse current to reduce to 10 percent of the peak reverse recovery current ($J_{PR}$) after the current crosses zero. This time can be extracted from the measured characteristics of devices using automated test equipment. In addition, it is common practice to extract the times $t_A$ and $t_B$ as defined on the waveform in Fig. 6.17. A larger [$t_B/t_A$] ratio is considered to be desirable to mitigate the voltage spikes created in power circuits by a large [di/dt] in the reverse direction.

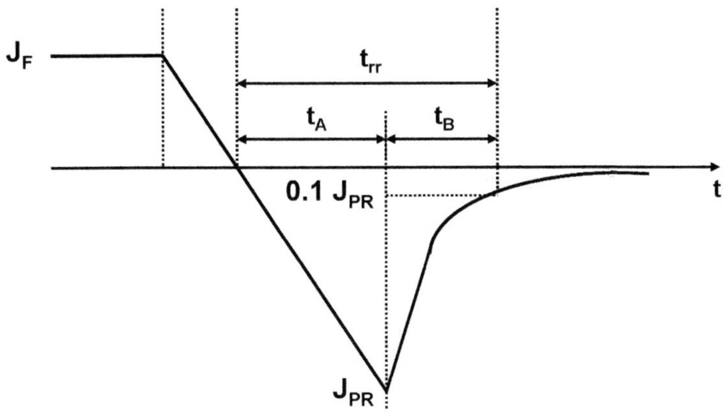


Fig. 6.17 Typical Reverse Recovery Current Waveform for the P-i-N rectifier Structure defining the Reverse Recovery Turn-Off Time.

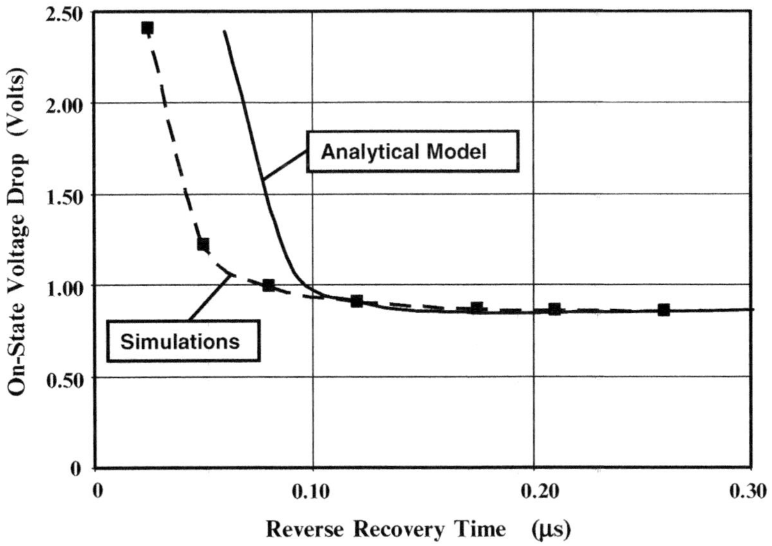


Fig. 6.18 Trade-Off Curve for the Punch-Through P-i-N rectifier Structure.

The trade-off curve between the on-state voltage drop and the reverse recovery time obtained by using the analytical models is shown in Fig. 6.18 by the solid line. The results obtained using two dimensional numerical simulations for the punch-through structure are also shown in the figure for comparison. It can be observed that there is good agreement between them until the lifetime is reduced below 0.05 microseconds to achieve a reverse recovery time of less than 0.1 microseconds. The reverse recovery time predicted by the analytical models is larger than observed with the simulations because free carrier recombination was neglected during the reverse recovery process. This assumption is not valid when the recombination lifetime is reduced to below 0.05 microseconds.

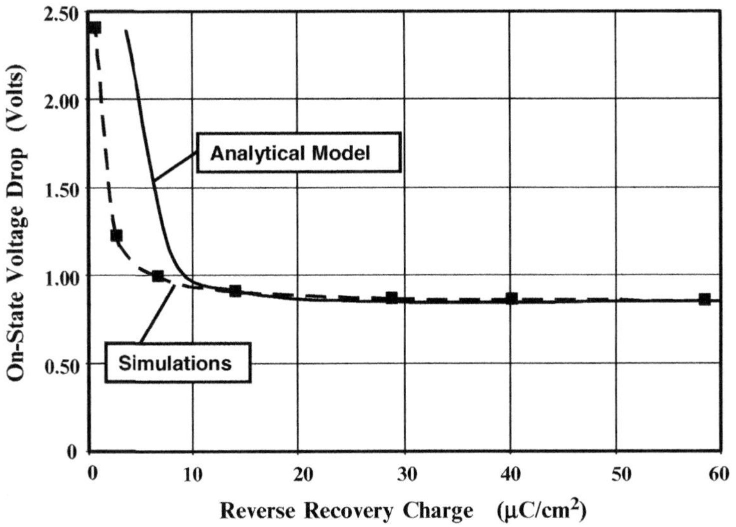


Fig. 6.19 Trade-Off Curve for the Punch-Through P-i-N rectifier Structure.

Another common practice for displaying the trade-off in power losses in power P-i-N rectifiers is by plotting the on-state voltage drop against the reverse recovery charge ($Q_{rr}$). The reverse recovery charge can be extracted by integration of the turn-off waveform for the current. This graph obtained with the analytical model and the simulations is shown in Fig. 6.19 for the punch-through structure.

## 6.6 Summary

The physics of operation of the P-i-N rectifier has been analyzed in this chapter. Analytical expressions have been derived for the on-state and blocking state, as well as the reverse recovery transients. At on-state current levels, the injected minority carrier density in the drift region exceeds the relatively low doping concentration required to achieve high breakdown voltages. This high level injection in the drift region modulates its conductivity producing a reduction in the on-state voltage drop. If recombination in the drift region is dominant, the voltage drop across the drift region becomes independent of the on-state current density. These phenomena allow operation of the silicon P-i-N rectifier with an on-state voltage drop of only 1 volt making it very attractive for power electronic applications.

The P-i-N rectifier can support a large voltage in the reverse blocking mode by appropriate choice of the doping concentration and thickness of the drift region. A punch-through design is favored because it reduces the thickness of the drift region. A narrower drift region contains a smaller amount of stored charge during on-state operation enabling faster turn-off.

The switching of the P-i-N rectifier from the on-state to the reverse blocking state is accompanied by a significant current flow in the reverse direction. This reverse current produces large power dissipation in the rectifier as well as the power switches in power converter circuits. The reverse recovery current and the reverse recovery turn-off time can be reduced by reduction of the recombination lifetime in the drift region. Since this is accompanied by an increase in the on-state voltage drop, it is customary to perform a trade-off analysis to minimize the overall power dissipation.

Silicon carbide based P-i-N rectifiers have a much narrower drift region thickness when compared with silicon devices due to the higher critical electric field for breakdown. This favors a faster switching speed with reduced reverse recovery current. However, the larger band-gap for silicon carbide produces an on-state voltage drop that is four times larger than for the silicon rectifiers. For this reason, silicon carbide P-i-N rectifiers are of interest only when the blocking voltage capability exceeds 10,000 volts.

## References

[1] B.J. Baliga, "Fundamentals of Power Semiconductor Devices", Springer Scientific, New York, 2008.

[2] B.J. Baliga, "Silicon Carbide Power Devices", World Scientific Publishing Company, 2005.

[3] S.K. Ghandhi, "Semiconductor Power Devices", pp. 112-128, John Wiley and Sons, 1977.

[4] R.N. Hall, "Power Rectifiers and Transistors", Proceedings of the IRE, Vol. 40, pp. 1512-1518, 1952.

[5] H. Benda and E. Spenke, "Reverse Recovery Processes in Silicon Power Rectifiers", Proceedings of the IEEE, Vol. 55, pp. 1331-1354, 1967.

[6] N.R. Howard and G.W. Johnson, "PIN Silicon Diodes at High Forward Current Densities", Solid State Electronics, Vol. 8, pp. 275-284, 1965.

[7] A. Herlet, "The Forward Characteristics of Silicon Power Rectifiers at High Current Densities", Solid State Electronics, Vol. 11, pp. 717-742, 1968.
[8] Q. Zhang, et al, "12-kV p-Channel IGBTs with Low On-Resistance in 4H-SiC", IEEE Electron Device Letters, Vol. EDL-29, pp. 1027-1029, 2008.

# Chapter 7

# MPS Rectifiers

Most applications for power devices, such as motor control, require rectifiers with blocking voltages ranging from 300 volts to 5000 volts. Silicon P-i-N rectifiers have been developed for these high voltage applications. The silicon P-i-N rectifiers that are designed to support large voltages rely upon the injection of a high concentration of minority carriers into the lightly doped drift region to allow conduction of the on-state current with a low on-state voltage drop[1]. The minority carriers stored within the drift region must be removed before the P-i-N rectifier is able to support a reverse bias voltage. During switching from the on-state to the reverse blocking state, a large reverse recovery current transient is observed for P-i-N rectifiers to allow the removal of the stored charge as discussed in Chapter 6. The reverse recovery transient produces significant switching power loss in the rectifier and in the transistor that is controlling the switching transient.

The power loss associated with the reverse recovery transient in the P-i-N rectifier can be reduced by decreasing the lifetime in the drift region. This traditional method for reducing the switching power loss is accompanied by an increase in the on-state voltage drop. A trade-off curve must be generated between the on-state voltage drop and the turn-off time or reverse recovery charge to optimize the total power dissipation. This trade-off curve has been found to depend upon the method used for controlling the lifetime[2]. It has been theoretically demonstrated that the trade-off curve is related to the properties of the deep levels created by each lifetime control process[3].

An alternate approach to producing a trade-off between the on-state and reverse recovery power loss was proposed in the 1980s by merging the physics of the P-i-N rectifier and the Schottky rectifier[4] to create the MPS rectifier structure illustrated in Fig. 7.1. In this structure, the drift region is designed using the same criteria as used for P-i-N rectifiers in order to support the desired reverse blocking

B.J. Baliga, *Advanced Power Rectifier Concepts*, DOI: 10.1007/978-0-387-75589-2_7,

voltage. The device structure contains a P-N junction over a portion under the metal contact and a Schottky contact for the remaining portion. It is convenient to utilize the same metal layer for making an ohmic contact to the $P^+$ region and a Schottky contact to the N- drift region.

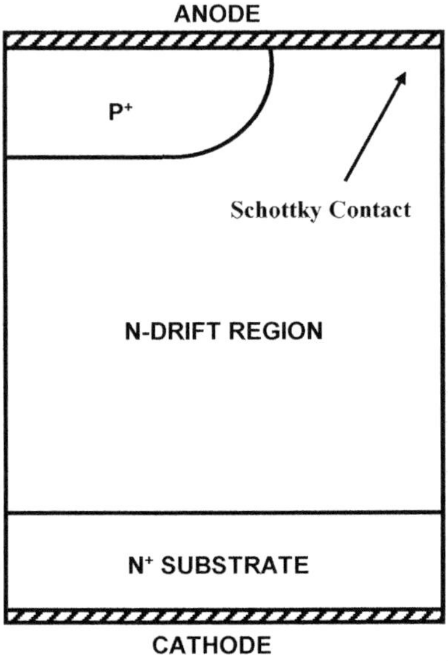


Fig. 7.1 The MPS Rectifier Structure.

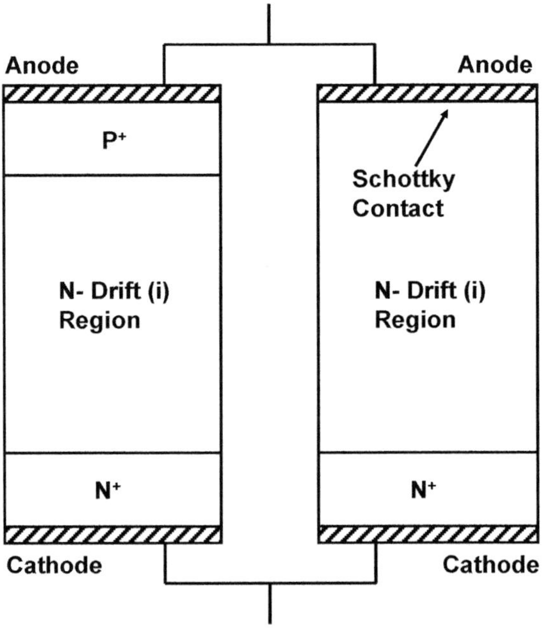


Fig. 7.2 Connection of P-i-N Rectifier and Schottky Rectifier in Parallel.

When first proposed, critics believed that the MPS rectifier would exhibit the worst attributes of both the P-i-N rectifier and the Schottky rectifier because it was considered simply as a parallel connection of the P-i-N and Schottky rectifier structures as illustrated in Fig. 7.2. In this case, the leakage current of the composite rectifier is large due to the significant leakage current observed for the Schottky rectifier resulting in poor reverse blocking characteristics. During on-state operation, most of the current flows through the P-i-N rectifier due to the large resistance of the un-modulated drift region in the Schottky rectifier. The on-state voltage drop for the composite diode is larger than for the P-i-N rectifier because the area of the Schottky rectifier must be taken into consideration when computing the average on-state current density. A large amount of stored charge is produced in the P-i-N rectifier during on-state operation of the composite diode. The stored charge produces a large reverse recovery current similar to that observed in a P-i-N rectifier resulting in large switching power losses. Consequently, the composite diode comprising a P-i-N rectifier and a Schottky rectifier connected in parallel exhibits the worst attributes of both these structures. The above conclusion is based upon the false premise that the P-i-N rectifier and the Schottky rectifier are operating in isolation within the MPS rectifier structure.

## 7.1 Device Physics

As shown in Fig. 7.1, the P-i-N rectifier and Schottky rectifier are located in close proximity in the MPS rectifier structure allowing the mingling of the physics of operation of both devices. The space between the P-N junctions in the MPS rectifier structure is designed so that it is pinched off at a relatively small reverse bias voltage. After the depletion of the space between the P-N junctions, a potential barrier is formed under the Schottky metal which screens the contact from the reverse bias applied to the cathode terminal. The electric field at the Schottky contact can be greatly reduced in the MPS rectifier when compared with the normal Schottky rectifier by judicious choice of the space between the P-N junctions. This allows suppression of the Schottky barrier lowering reducing the leakage current in silicon devices well below that for Schottky rectifiers. An even greater reduction of the leakage current can be achieved for silicon carbide devices due to suppression of the thermionic field emission current. Consequently, it is possible to achieve good reverse blocking characteristics in the MPS rectifier structure.

The on-state current flow in the MPS rectifier can be expected to occur via the P-N junction and the Schottky contact. At low on-state bias levels applied to the cathode, current transport via the Schottky contact is favored due to the larger potential required for the injection of holes into the drift region from the P-N junction. However, this current flow is limited by the large resistance of the un-modulated drift region which has a low doping concentration and large thickness as required to achieve a high reverse blocking voltage capability. As the forward bias voltage is increased, the P-N junction begins to inject a significant concentration of holes into the drift region. The drift region then operates under high-level injection conditions as in the case of the P-i-N rectifier. Since the resistance of the drift region is reduced by conductivity modulation, a large current flow can occur via the Schottky contact in the MPS rectifier structure. This allows on-state current flow with an on-state voltage drop that is smaller than that for a P-i-N rectifier.

The carrier concentration at the Schottky contact remains low in the MPS rectifier because it does not inject a significant concentration of minority carriers into the drift region. The resulting carrier profile is superior to that observed in the P-i-N rectifier in terms of charge removal during the turn-off transient. The MPS rectifier exhibits a smaller peak reverse recovery current and charge leading to reduced switching power loss. Moreover, the trade-off between the on-state voltage drop and reverse recovery power loss can be generated by changing the relative area of the P-N junction and Schottky contact in the MPS rectifier structure. A further refinement of the trade-off curve can be achieved by using the lifetime control techniques commonly used for the P-i-N rectifier.

### 7.1.1 Low Forward Bias Conditions

The N-drift region in the MPS rectifier must be lightly doped in order to support a high voltage in the reverse blocking mode. At small forward bias voltages, the voltage across the P-N junction produces low-level injection of holes into the N-type drift region. There is no conductivity modulation of the drift region at these small forward bias voltages. The resistance of the drift region is therefore determined by the doping concentration. Current transport through the MPS rectifier occurs by the thermionic emission process at the Schottky contact followed by current flow through the drift region. The high resistance of the drift region in MPS rectifiers designed to support large reverse bias voltages initially limits the current transport. The MPS rectifier characteristics resemble those of a Schottky rectifier at these small forward bias voltages.

The forward conduction *i-v* characteristics of the MPS rectifier can be analytically modeled as a metal-semiconductor (Schottky) contact with the series resistance of the drift region. In the analytical model, it is important to include the impact of the larger current density at the Schottky contact when compared with the cathode current density because the P-N junction occupies a portion of the upper surface area. The current constriction in the region between the P-N junctions and current spreading from this region into the drift region enhances the series resistance. This resistance can be modeled as previous performed for the JBS rectifier. Unlike for the JPS rectifiers, the P-N junction depth is a very small fraction of the drift region thickness in the case of the MPS rectifiers because they are designed to support large voltages. The drift region resistance in the MPS rectifier is quite close to the one-dimensional resistance of the drift region. In order to keep this resistance as small as possible, it is necessary to employ a non-punch-through design for the drift region.

Due to the large thickness of the drift region in relation to the junction depth and the width of the window used for the $P^+$ diffusions in the MPS rectifier structure, the current path in the drift region will invariably overlap before reaching the $N^+$ substrate. The current flow pattern for this case (Model C) was illustrated in Fig. 3.5 with the shaded area. In the MPS rectifier structure, the thickness of the drift region (t in Fig. 3.5) is much larger than that for the JBS rectifier.

The current density at the Schottky contact ($J_{FS}$) is enhanced in Model C due to the presence of the $P^+$ region and the depletion layer at the P-N junction. This increases the voltage drop across the Schottky contact. The current through the Schottky contact flows only within the un-depleted portion (with dimension 'd') of the drift region at the top surface. Consequently, the current density at the Schottky contact ($J_{FS}$) is related to the cell (or cathode) current density ($J_{FC}$) by:

$$J_{FS} = \left(\frac{p}{d}\right) J_{FC} \quad \textbf{[7.1]}$$

where p is the cell pitch. The dimension 'd' is determined by the cell pitch (p), the size of the $P^+$ ion-implant window (2s), the junction depth of the $P^+$ region, and the on-state depletion width ($W_{D,ON}$):

$$d = p - s - x_J - W_{D,ON} \quad \textbf{[7.2]}$$

In deriving this equation, it has been assumed that the lateral diffusion is equal to the junction depth. Depending up on the lithography used for device fabrication to minimize the size (dimension 's') of the $P^+$ region, as well as the junction depth ($x_J$) created during the diffusion process, the current density at the Schottky contact can be enhanced by a factor of two or more. This must be taken into account when computing the voltage drop across the Schottky contact given by:

$$V_{FS} = \phi_B + \frac{kT}{q} \ln\left(\frac{J_{FS}}{AT^2}\right) \quad \textbf{[7.3]}$$

After flowing across the Schottky contact, the current flows through the un-depleted portion of the drift region between the junctions. In Model C, it is assumed that the current flows through a region with a uniform width 'd' until it reaches the bottom of the depletion region and then spreads to the entire cell pitch (p) at a 45 degree spreading angle. The current paths overlap at a distance ($s + x_J + W_{D,ON}$) from the bottom of the depletion region. The current then flows through a uniform cross-sectional area.

The net resistance to current flow can be calculated by adding the resistance of the three segments shown in Fig. 3.5. The resistance of the first segment of uniform width 'd' is given by:

$$R_{D1} = \frac{\rho_D.(x_J + W_{D,ON})}{d.Z} \quad [7.4]$$

The resistance of the second segment can be derived by using the same approach used for the JBS rectifier:

$$R_{D2} = \frac{\rho_D}{Z}\ln\left(\frac{p}{d}\right) \quad [7.5]$$

The resistance of the third segment with a uniform cross-section of width p is given by:

$$R_{D3} = \frac{\rho_D.(t - s - x_J - 2W_{D,ON})}{p.Z} \quad [7.6]$$

The specific resistance for the drift region can be calculated by multiplying the cell resistance ($R_{D1} + R_{D2} + R_{D3}$) with the cell-area (p.Z):

$$R_{sp,drift} = \frac{\rho_D.p.(x_J + W_{D,ON})}{d} + \rho_D.p.\ln\left(\frac{p}{d}\right) + \rho_D.(t - s - x_J - 2W_{D,ON}) \quad [7.7]$$

The on-state voltage drop for the MBS rectifier at a small forward bias with a forward cell current density $J_{FC}$, including the substrate contribution, is then given by:

$$V_F = \phi_B + \frac{kT}{q}\ln\left(\frac{J_{FS}}{AT^2}\right) + (R_{sp,drift} + R_{sp,subs}).J_{FC} \quad [7.8]$$

When computing the on-state voltage drop using this equation, it is satisfactory to make the approximation that the depletion layer width can be computed by subtracting an on-state voltage drop of about 0.5 volts for silicon devices from the built-in potential of the P-N junction. At larger on-state bias voltages, it becomes necessary to include the impact of high-level injection in the drift region. In addition, it is important to recognize that the doping profile at the junction is linearly graded leading to half the depletion occurring on the P-side of the junction. Consequently:

$$W_{D,ON} = 0.5.\sqrt{\frac{2\varepsilon_S(V_{bi} - 0.5)}{qN_D}} \quad [7.9]$$

where $V_{bi}$ is the built-in potential of the P/N junction. Since the on-state voltage drop of MPS rectifiers is close to the built-in potential at the P-N junction, it is possible to neglect the depletion width during device analysis in the forward bias state.

### 7.1.2 High Level Injection Conditions

When the forward bias applied to the MPS rectifier increases, the injected minority carrier concentration from the P-N junction also increases in the drift region until it ultimately exceeds the background doping concentration ($N_D$) in the drift region resulting in *high level injection*. When the injected hole concentration in the drift region becomes much greater than the background doping concentration, charge neutrality requires that the concentrations for electrons and holes become equal:

$$n(x) = p(x) \tag{7.10}$$

The large concentration of free carriers reduces the resistance of the drift region resulting in *conductivity modulation* of the drift region. As in the case of the P-i-N rectifier, conductivity modulation of the drift region is beneficial for allowing the transport of a high current density through lightly doped drift regions with a low on-state voltage drop.

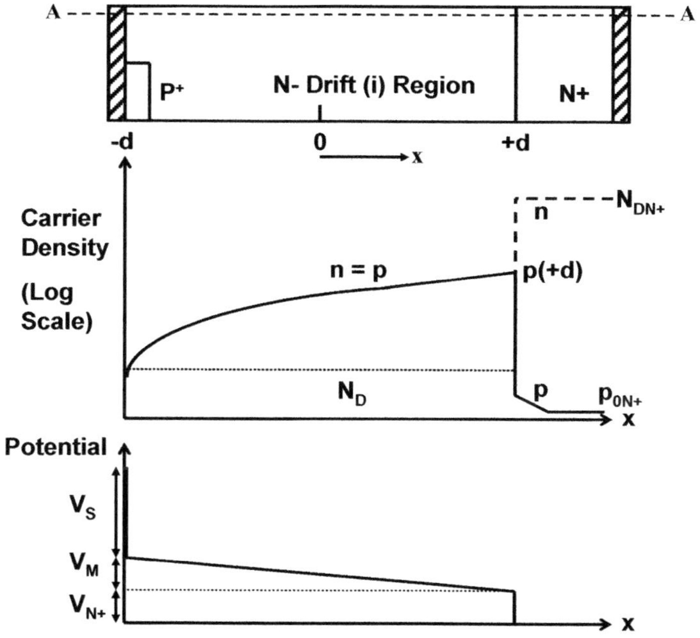


Fig. 7.3 Carrier and Potential Distribution under High-Level Injection Conditions for the MPS Rectifier.

The carrier distribution within the drift region for the MPS rectifier is different from that observed for the P-i-N rectifier because of the presence of the Schottky contact. The carrier distribution p(x) can be obtained by solving the continuity equation for holes in the N-region[1]:

$$\frac{d^2p}{dx^2} - \frac{p}{L_a^2} = 0 \quad [7.11]$$

where $L_a$ is the ambipolar diffusion length given by:

$$L_a = \sqrt{D_a \tau_{HL}} \quad [7.12]$$

The general solution for the carrier concentration governed by Eq. [7.12] is given by:

$$p(x) = A\cosh\left(\frac{x}{L_a}\right) + B\sinh\left(\frac{x}{L_a}\right) \quad [7.13]$$

with the constants A and B determined by the boundary conditions for the N-drift region.

For the MPS rectifier, it is appropriate to solve for the carrier profile along the path indicated by the dashed line marked 'A-A' in Fig. 7.3 which is located through the Schottky contact. At the interface between the N-drift region and the $N^+$ cathode region (located at x = +d in Fig. 7.3), the total current flow occurs exclusively by electron transport:

$$J_{FC} = J_n(+d) \quad [7.14]$$

and

$$J_p(+d) = 0 \quad [7.15]$$

Using these equations:

$$J_{FC} = 2qD_n\left(\frac{dp}{dx}\right)_{x=+d} \quad [7.16]$$

The second boundary condition occurs at the junction between the N-drift region and the Schottky contact (located at x = –d in Fig. 7.3). Here, the hole concentration becomes zero due to negligible injection at the Schottky contact:

$$p(-d) = 0 \quad [7.17]$$

The above boundary conditions can be used to obtain the constants A and B in Eq. [7.13]:

$$A = -\frac{L_a J_{FC}}{2qD_n}\left[\frac{\sinh(-d/L_a)}{\cosh(-d/L_a)\cosh(d/L_a) - \sinh(-d/L_a)\sinh(d/L_a)}\right] \quad [7.18]$$

$$B = -\frac{L_a J_{FC}}{2qD_n}\left[\frac{\cosh(-d/L_a)}{\cosh(-d/L_a)\cosh(d/L_a) - \sinh(-d/L_a)\sinh(d/L_a)}\right] \quad [7.19]$$

Using these constants in Eq. [7.13] and simplifying the expression yields:

$$p(x) = n(x) = \frac{L_a J_{FC}}{2qD_n} \frac{\sinh[(x+d)/L_a]}{\cosh[2d/L_a]} \qquad [7.20]$$

The carrier distribution described by this equation was schematically illustrated in Fig. 7.3. It has a maximum value at the interface between the drift region and the $N^+$ substrate with a magnitude of:

$$p(+d) = n(+d) = \frac{L_a J_{FC}}{2qD_n} \frac{\sinh[2d/L_a]}{\cosh[2d/L_a]} = \frac{L_a J_{FC}}{2qD_n} \tanh\left(\frac{2d}{L_a}\right) \qquad [7.21]$$

and reduces monotonically when proceeding towards the Schottky contact in the negative x direction. The concentration becomes equal to zero at the Schottky contact as required to satisfy the boundary condition used to derive the expression.

As a particular example, the carrier distributions calculated at an on-state current density of 100 A/cm$^2$ by using Eq. [7.20] are shown in Fig. 7.4 for the case of three values for the high-level lifetime for a silicon MPS rectifier with drift region thickness of 70 microns corresponding to a breakdown voltage of 500 volts. The largest concentrations for the electrons and holes in the drift region occur at its boundary with the $N^+$ end-region. The carrier concentration at this boundary decreases when the lifetime is reduced. It has a magnitude of 6.2 x $10^{16}$ cm$^{-3}$ for a high lifetime of 100 microseconds, 5.8 x $10^{16}$ cm$^{-3}$ for a moderate lifetime of 10 microseconds, and 3.7 x $10^{16}$ cm$^{-3}$ for a low lifetime of 1 microsecond.

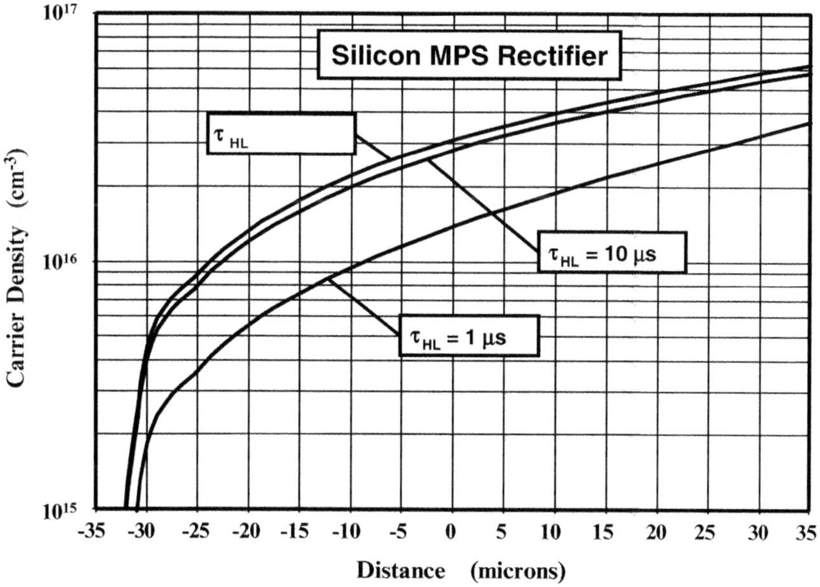


Fig. 7.4 Carrier Distribution under High-Level Injection Conditions for the Silicon MPS Rectifier with various High-Level Lifetime values.

When the lifetime in the drift region is large, it is possible to derive the carrier distribution while neglecting recombination in the drift region. This is equivalent to the case of a very large value for the ambipolar diffusion length ($L_a$) in Eq. [7.11] leading to the expression:

$$\frac{d^2p}{dx^2} = 0 \qquad [7.22]$$

The solution for the carrier concentration governed by Eq. [7.22] is a linear carrier distribution given by:

$$p(x) = Cx + D \qquad [7.23]$$

with the constants C and D determined by the boundary conditions for the N-drift region. At the junction between the N-drift region and the $N^+$ cathode region (located at x = +d in Fig. 7.3), the total current flow occurs exclusively by electron transport:

$$J_{FC} = J_n(+d) \qquad [7.24]$$

and

$$J_p(+d) = 0 \qquad [7.25]$$

Using these equations:

$$J_{FC} = 2qD_n\left(\frac{dp}{dx}\right)_{x=+d} \qquad [7.26]$$

The second boundary condition occurs at the junction between the N-drift region and the Schottky contact (located at x = –d in Fig. 7.3). Here, the hole concentration becomes zero due to negligible injection at the Schottky contact:

$$p(-d) = 0 \qquad [7.27]$$

The above boundary conditions can be used to obtain the constants C and D in Eq. [7.23]:

$$C = \frac{J_{FC}}{2qD_n} \qquad [7.28]$$

$$D = \frac{J_{FC}d}{2qD_n} \qquad [7.29]$$

Using these constants in Eq. [7.23] and simplifying the expression yields:

$$p(x) = n(x) = \frac{J_{FC}}{2qD_n}(x + d) \qquad [7.30]$$

where x extends from –d to +d across the drift region. The carrier distribution described by this equation has a maximum value at the interface between the drift region and the $N^+$ substrate with a magnitude of:

$$p(+d) = n(+d) = \frac{J_{FC}d}{qD_n} \qquad [7.31]$$

and reduces monotonically when proceeding towards the Schottky contact in the negative x direction. The concentration becomes equal to zero at the Schottky contact as required to satisfy the boundary condition used to derive the expression.

As a particular example, the carrier distribution calculated by using Eq. [7.31] is shown in Fig. 7.5 by the dashed line for a silicon MPS rectifier with drift region thickness of 70 microns corresponding to a breakdown voltage of 500 volts. It is worth pointing out that a linear scale has been used for the carrier concentration in this figure in contrast to the logarithmic scale used in Fig. 7.4. The largest concentrations for the electrons and holes in the drift region occur at its boundary with the $N^+$ end-region with a value of 6.3 x $10^{16}$ $cm^{-3}$. The carrier concentration profiles computed by using Eq. [7.20] with recombination in the drift region are also shown in Fig. 7.5 with the solid lines. As expected, the carrier profiles using the two models coincide when the lifetime value is large, and there is agreement between the values computed by using Eq. [7.21] derived using recombination in the drift region with a high value for the lifetime and that computed by using Eq. [7.31] with no recombination. Even with lower lifetime values, the carrier profile can be observed to be nearly linear in shape although the value of the concentration at the interface with the $N^+$ substrate becomes smaller.

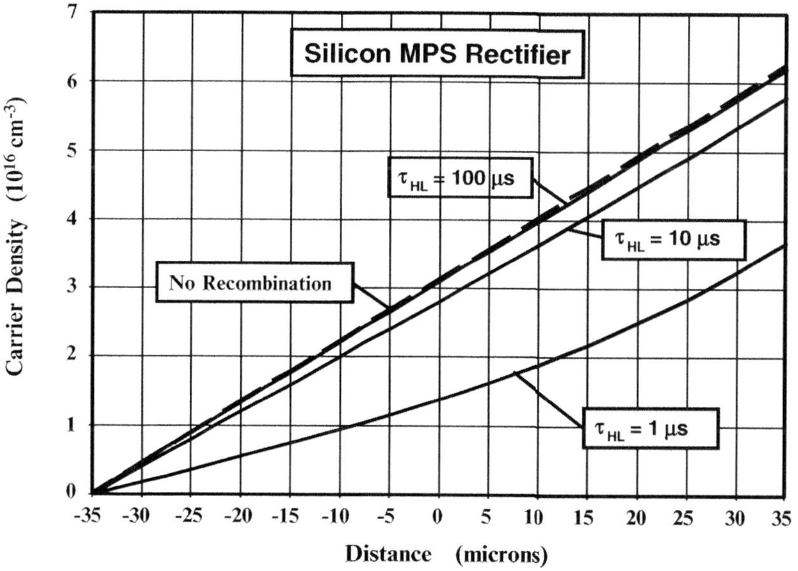


Fig. 7.5 Carrier Distribution under High-Level Injection Conditions for the Silicon MPS Rectifier with various High-Level Lifetime values.

### 7.1.3 On-State Voltage Drop

The on-state voltage drop for the MPS rectifier can be obtained by summing the voltage drops along the path marked 'A-A' in Fig. 7.3 through the Schottky contact. The total voltage drop along this path consists of the voltage drop across the Schottky contact ($V_{FS}$), the voltage drop across the drift region (middle region voltage $V_M$), and the voltage drop at the interface with the $N^+$ substrate ($V_{N+}$):

$$V_{ON} = V_{FS} + V_M + V_{N+} \quad [7.32]$$

The voltage drop across the Schottky contact is given by:

$$V_{FS} = \phi_{BN} + \frac{kT}{q} \ln\left(\frac{J_{FS}}{AT^2}\right) \quad [7.33]$$

where the current density at the Schottky contact ($J_{FS}$) is related to the cell or cathode current density ($J_{FC}$) by Eq. [7.1]. In the case of JBS rectifiers, it is customary to utilize low barrier heights in order to reduce the on-state voltage drop despite the accompanied increase in reverse leakage current. In the case of MPS rectifiers, it is preferable to utilize a large barrier height to reduce the reverse leakage current because the impact on the total on-state voltage drop is small. Using Eq. [7.33] with a barrier height of 0.9 eV, the voltage drop across the Schottky contact at an on-state cathode current density of 100 A/cm$^2$ is found to be 0.62 volts for the silicon MPS rectifier if the Schottky contact occupies half the cell area. This is smaller than the voltage drop across the P-N junction in the P-i-N rectifier.

Analysis of the voltage drop across the drift (middle) region can be performed by integration of the electric field. The electric field in the drift region can be obtained from the carrier distribution. The hole and electron currents flowing in the drift region are given by:

$$J_p = q\mu_p\left(pE - \frac{kT}{q}\frac{dp}{dx}\right) \quad [7.34]$$

and

$$J_n = q\mu_n\left(nE + \frac{kT}{q}\frac{dn}{dx}\right) \quad [7.35]$$

The total current at any location in the drift region is constant and given by:

$$J_{FC} = J_p + J_n \quad [7.36]$$

Combining these relationships:

$$E(x) = \frac{J_{FC}}{q(\mu_n + \mu_p)n} - \frac{kT}{2qn}\frac{dn}{dx} \quad [7.37]$$

The charge neutrality condition n(x) = p(x) was utilized when deriving this expression.

In the case of the MPS rectifier, it was found in the previous section that the carrier distribution has a linear shape. The case with no recombination in the drift region is treated here with the carrier profile given by Eq. [7.30]. Substituting this profile in Eq. [7.37] yields:

$$E(x) = \frac{2D_n}{(\mu_n + \mu_p)(x+d)} - \frac{kT}{2q(x+d)} \quad [7.38]$$

The voltage drop in the drift (middle) region can be obtained by integration of the electric field across the drift region:

$$V_M = \left[\frac{2D_n}{(\mu_n + \mu_p)} - \frac{kT}{2q}\right]\ln\left(\frac{2d}{x_J}\right) \quad [7.39]$$

In deriving this expression, the integration was terminated just below the Schottky contact at the junction depth of the P-N junction to avoid an indeterminate solution. It is worth pointing out that the voltage drop across the middle region is independent of the current density as in the case of the P-i-N rectifier because of conductivity modulation by the injected carriers. For the case of a MPS rectifier with a drift region thickness (2d) of 70 microns and P-N junction depth of 1 micron, the voltage drop across the middle region at room temperature is found to be 0.091 volts. This voltage drop is much smaller than the voltage drop of 8.47 volts for an un-modulated drift region with doping concentration of 3.8 x $10^{14}$ $cm^{-3}$ and thickness of 70 microns. This demonstrates the advantage of the MPS rectifier structure over the Schottky rectifier.

The voltage drop across the interface between the drift region and the $N^+$ substrate can be determined from the carrier density at x = +d:

$$V_{N+} = \frac{kT}{q}\ln\left[\frac{n(+d)}{N_D}\right] \quad [7.40]$$

The carrier concentration in the drift region at this interface is given by Eq. [7.31] without recombination in the drift region. Using this equation in Eq. [7.40]:

$$V_{N+} = \frac{kT}{q}\ln\left[\frac{J_{FC}d}{qD_nN_D}\right] \quad [7.41]$$

The voltage drop across the interface between the drift region and the $N^+$ substrate at an on-state cathode current density of 100 A/$cm^2$ is found to be 0.132 volts for the silicon MPS rectifier when the drift region doping concentration is $3.8 \times 10^{14}$ cm$^{-3}$. The on-state voltage drop for the silicon MPS rectifier with a drift region width of 70 microns at an on-state current density of 100 A/cm$^2$ is then found to be 0.86 volts by summing the voltage drops across the Schottky contact, the middle region, and the N/N$^+$ interface.

The impact of reducing lifetime on the on-state voltage drop for the MPS rectifier can be modeled by taking into account the change in the carrier concentration in the drift region as described by Eq. [7.20] derived with recombination in the drift region. In order to simplify the analysis, it will be assumed that a linear approximation for the carrier concentration in the drift region can be used. Under these assumptions, based upon Eq. [7.21], the carrier profile is given by:

$$p(x) = n(x) = \frac{J_{FC} L_a \tanh(2d/L_a)}{4qD_n d}(x+d) \qquad \textbf{[7.42]}$$

Using this carrier profile in Eq. [7.37] for the electric field:

$$E(x) = \frac{4D_n d}{(\mu_n + \mu_p) L_a \tanh(2d/L_a)(x+d)} - \frac{kT}{2q(x+d)} \qquad \textbf{[7.43]}$$

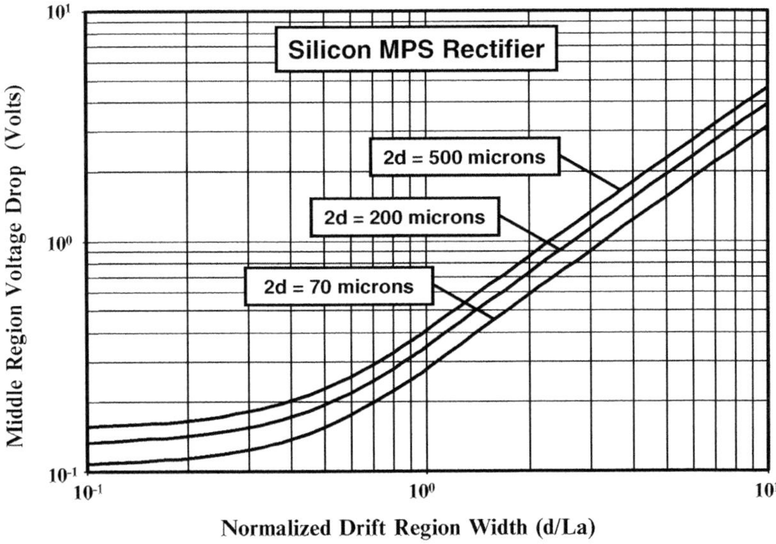


Fig. 7.6 Middle Region Voltage Drop for a MPS Rectifier.

The voltage drop in the drift (middle) region can be obtained by integration of the electric field across the drift region:

$$V_M = \left\{ \frac{4D_n}{(\mu_n + \mu_p)} \left[ \frac{(d/L_a)}{\tanh(2d/L_a)} \right] - \frac{kT}{2q} \right\} \ln\left(\frac{2d}{x_J}\right) \qquad \textbf{[7.44]}$$

According to this expression, the voltage drop across the drift (middle) region is now dependent on the lifetime via the ambipolar diffusion length. This expression has been written using the ratio ($d/L_a$) in a similar manner to the equation previously derived for the P-i-N rectifier. The values for the middle region voltage drop for silicon MPS rectifiers with various drift region thicknesses are shown in Fig. 7.6 as a function of the ($d/L_a$) ratio. It can be observed from this figure that the voltage drop across the drift region begins to increase rapidly when the ($d/L_a$) ratio becomes larger than 0.5 for all cases of the drift region thickness. This corresponds to a high-level lifetime of 10 microseconds for a MPS rectifier with drift region thickness of 70 microns.

The voltage drop across the interface between the drift region and the $N^+$ substrate is also a function of lifetime for the MPS rectifier because the carrier concentration at this interface depends upon the lifetime. Substituting Eq. [7.21] into Eq. [7.40] yields:

$$V_{N+} = \frac{kT}{q} \ln\left[\frac{J_{FC} L_a \tanh(2d/L_a)}{2qD_n N_D}\right] \quad [7.45]$$

As an example, the voltage drop across the interface between the drift region and the $N^+$ substrate for the silicon MPS rectifier at an on-state cathode current density of 100 A/cm$^2$ is provided in Fig. 7.7 for the case of a drift region doping concentration of $3.8 \times 10^{14}$ cm$^{-3}$. This voltage drop increases with increasing values for $L_a$ because the carrier concentration at the interface increases with increasing lifetime (see Fig. 7.5).

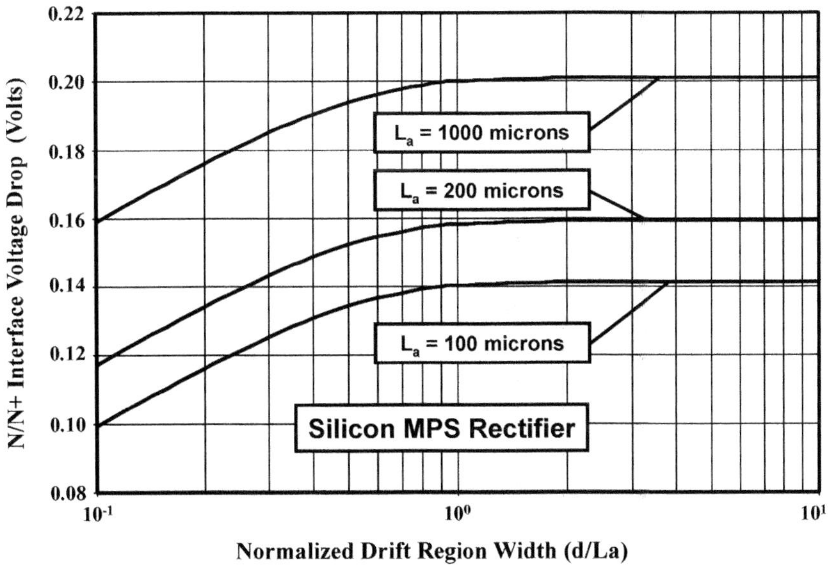


Fig. 7.7 Voltage Drop at N/N+ Interface for a Silicon MPS Rectifier.

The on-state voltage drop for the MPS rectifier can be computed by utilizing the three components discussed above. Using Eq. [7.32] with Eq. [7.33], Eq. [7.44], and Eq. [7.45] yields:

$$V_{ON} = \phi_{BN} + \frac{kT}{q}\ln\left(\frac{J_{FC}p}{AT^2 d}\right) + \left\{\frac{4D_n}{(\mu_n + \mu_p)}\left[\frac{(d/L_a)}{\tanh(2d/L_a)}\right] - \frac{kT}{2q}\right\}\ln\left(\frac{2d}{x_J}\right) + \frac{kT}{q}\ln\left[\frac{J_{FC}L_a \tanh(2d/L_a)}{2qD_n N_D}\right] \quad [7.46]$$

The on-state voltage drop for the MPS rectifier is a function of the lifetime in the drift region because the middle region and $N/N^+$ interface voltage drops change with lifetime. As an example, consider the case of a silicon MPS rectifier with a drift region doping concentration of 3.8 x $10^{14}$ $cm^{-3}$ and thickness of 70 microns capable of supporting 500 volts in the reverse blocking mode. The variation of the on-state voltage drop with high-level lifetime in the drift region for this case as predicted by the analytical model is provided in Fig. 7.8. It can be observed that the on-state voltage drop begins to increase when the lifetime is reduced below 1 microsecond. The three components of the on-state voltage drop are also shown in the figure. The voltage drop at the Schottky contact is independent of the lifetime. The voltage drop at the $N/N^+$ interface increases slightly when the lifetime becomes larger than 1 microsecond. The most significant increase in the voltage drop occurs for the middle region when the lifetime is reduced below 1 microsecond.

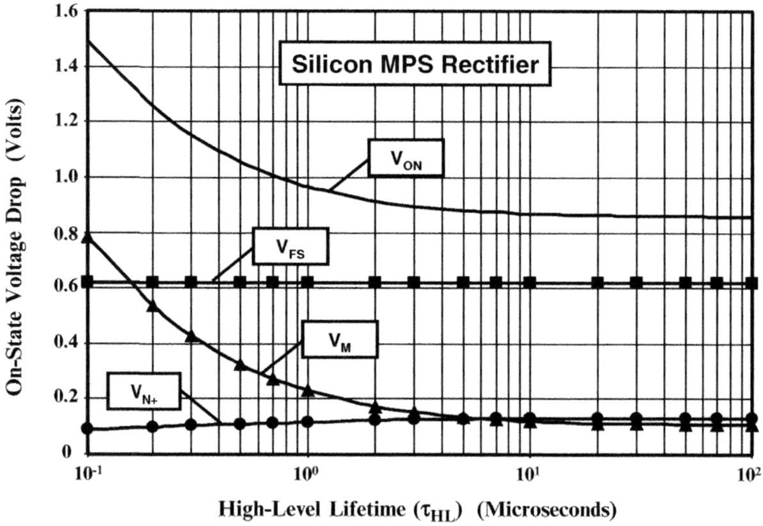


Fig. 7.8 Voltage Drops in the 500V Silicon MPS Rectifier.

### 7.1.4 Forward Conduction Characteristics

The analysis of current flow in the MPS rectifier in the previous sections indicates that the relationship between the on-state current density and the voltage drop across the rectifier depends upon the injection level. At low current levels, where the current is controlled by current flow via the Schottky contact with low-level injection across the P-N junction, the characteristics are like those for a Schottky rectifier. In this regime of operation, the on-state voltage drop of the MPS rectifier is smaller than that for the P-i-N rectifier.

At larger forward current densities, the injected carrier density in the drift region exceeds the background doping concentration leading to high level injection conditions. In this mode of operation, the injected carrier concentration in the drift region increases in proportion to the current density resulting in a constant voltage drop across the drift region. From Eq. [7.46], an expression for the on-state current density can be derived:

$$J_{FC} = \sqrt{\frac{2qAT^2D_nN_D}{p}\frac{(d/L_a)}{\tanh(2d/L_a)}e^{-\frac{q(\phi_{BN}+V_M)}{2kT}}}\; e^{\frac{qV_{ON}}{2kT}} \quad [7.47]$$

The current flow is observed to become proportional to ($qV_{ON}/2kT$). This behavior is similar to that observed for the P-i-N rectifier under high-level injection conditions.

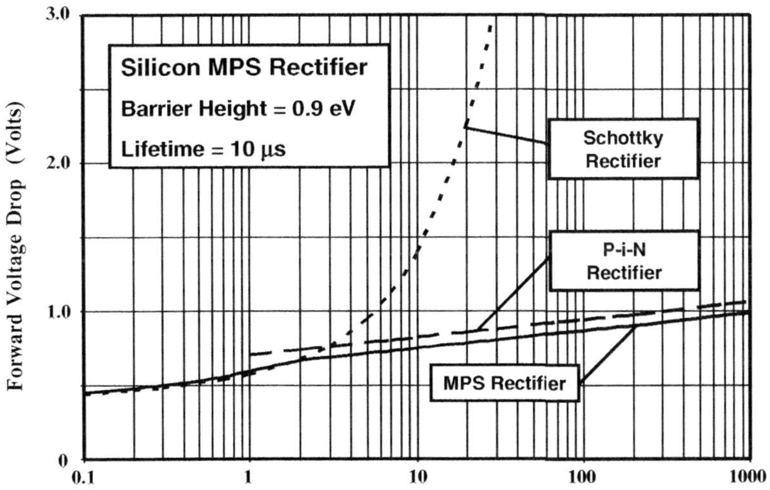


Fig. 7.9 On-state Characteristics for the Silicon MPS Rectifier with Lifetime of 10 microseconds in the Drift Region.

The on-state characteristics for the silicon MPS rectifier are shown in Fig. 7.9 for the case of a lifetime of 10 microseconds in the drift region. For comparison, the on-state characteristic for a silicon P-i-N rectifier with a lifetime of 10 microseconds in the drift region is also shown in the figure by the dashed line. In addition, the on-state characteristic for the Schottky rectifier with no modulation of the drift region is included in the figure for discussion. It can be observed that at forward bias voltages below 0.65 volts, the MPS rectifier behaves like the Schottky rectifier with a slightly larger on-state voltage drop due to the enhanced current density at the Schottky contact. When the forward bias voltage exceeds 0.65 volts, the on-state characteristics of the MPS rectifier resemble those for the P-i-N rectifier but the on-state voltage drop is less than that of the P-i-N rectifier. This demonstrates that it is possible to obtain a smaller on-state voltage drop in the MPS rectifier while also obtaining a smaller stored charge in the drift region.

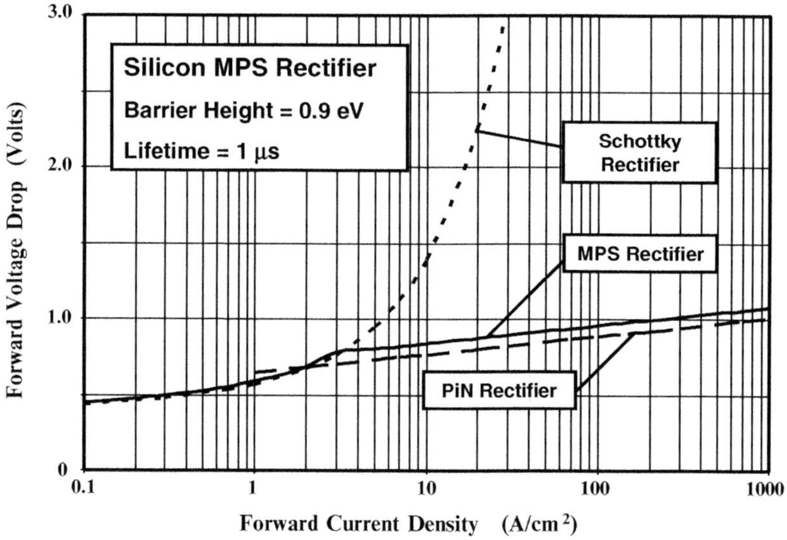


Fig. 7.10 On-state Characteristics for the Silicon MPS Rectifier with Lifetime of 1 microsecond in the Drift Region.

The impact of reducing the lifetime in the drift region on the on-state characteristics for the silicon MPS rectifier is shown in Fig. 7.10. The on-state voltage drop for the MPS rectifier increases when the lifetime is reduced to 1 microsecond in the drift region. In contrast, the on-state voltage drop for the P-i-N rectifier decreases slightly due to an increase in the ($d/L_a$) ratio making it closer to unity. Although the on-state voltage drop for the MPS rectifier is larger under these conditions, its turn-off switching power loss is much smaller than that for the P-i-N rectifier as shown later in the chapter.

The on-state voltage drop for the silicon MPS rectifier is also determined by the barrier height of the Schottky metal. A reduction in the on-state voltage drop of the MPS rectifier can be achieved by reducing the barrier height as illustrated in Fig. 7.11 where the characteristics of devices with two barrier heights can be compared. The analytical model predicts a decrease in on-state voltage drop that is equal to the reduction of the barrier height. This behavior is similar to that observed in Schottky rectifiers.

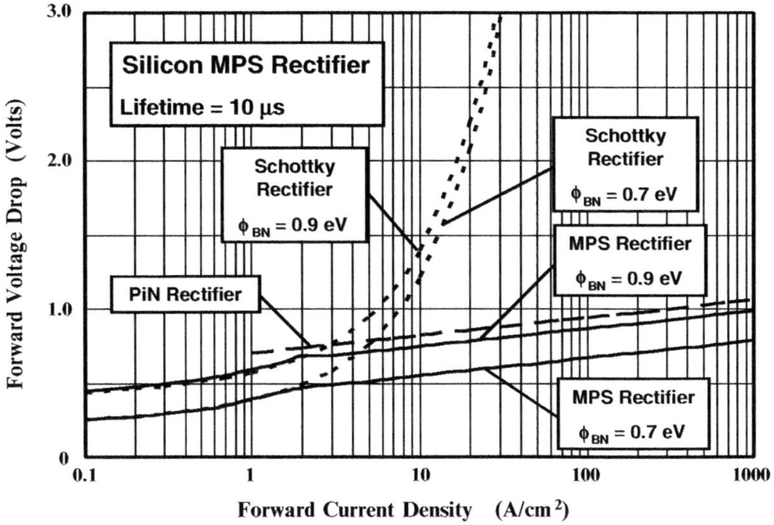


Fig. 7.11 On-state Characteristics for Silicon MPS Rectifiers with Different Schottky Barrier Heights.

### 7.1.5 Injection into the $N^+$ End-Region

In the case of P-i-N rectifiers, recombination in the end regions ($P^+$ and $N^+$ regions) has been shown to have a strong impact on the on-state characteristics especially when the lifetime in the drift region is large. In the presence of recombination in the end regions, the carrier concentration in the drift region no longer increases in proportion to the on-state current density. The voltage drop in the middle region is then no longer constant but increases as the current density is increased. This produces a significant increase in the on-state voltage drop. A similar effect can be expected to occur in the MPS rectifier. An analytical model that takes the recombination in the end region ($N^+$ region for the MPS rectifier) into account is derived in this section.

With the presence of recombination in the end region for the MPS rectifier, the total current flow must accommodate not only the recombination of carriers in the drift (middle) region but also the recombination of carriers in the end regions. Thus:

$$J_{FC} = J_M + J_{N+} \qquad [7.48]$$

Consequently, the current density associated with the middle region ($J_M$) is no longer equal to the total (cathode) on-state current density ($J_{FC}$) as assumed in the previous section but has a smaller value. This reduces the injection level in the drift region corresponding to any given total current density resulting in an increase in the voltage drop across the middle region.

Due to the high doping concentration in the $N^+$ end region, the injected minority carrier density in this region is well below the majority carrier density even during operation at very high on-state current densities. The current corresponding to the end-region can therefore be analyzed using low-level injection theory under the assumption of a uniform doping concentration in the $N^+$ region. Under low-level injection conditions:

$$J_{N+} = \frac{qD_{pN+}p_{0N+}}{L_{pN+}\tanh(W_{N+}/L_{pN+})} e^{\frac{qV_{N+}}{kT}} = J_{SN+} e^{\frac{qV_{N+}}{kT}} \qquad \textbf{[7.49]}$$

where $W_{N+}$ is the width of the $N^+$ region, $L_{pN+}$ is the minority carrier diffusion length in the $N^+$ region, $D_{pN+}$ is the minority carrier diffusion coefficient in the $N^+$ region, $p_{ON}$ is the minority carrier concentration in the $N^+$ region, and $V_{N+}$ is the voltage drop at the $N^+/N$ interface. The tanh term becomes equal to unity when the width ($W_{N+}$) of the $N^+$ region is large relative to the diffusion length for holes ($L_{pN+}$). In this equation, $J_{SN+}$ are referred to as the *saturation current density* for the heavily doped $N^+$ cathode region.

The injected carrier concentrations on the two sides of the $N^+/N$ junction are related under quasi-equilibrium conditions by:

$$p_{N+}(+d).n_{N+}(+d) = p(+d).n(+d) \qquad \textbf{[7.50]}$$

Under low-level injection conditions within the $N^+$ cathode region:

$$n_{N+}(+d) = n_{0N+} \qquad \textbf{[7.51]}$$

and

$$p_{N+}(+d) = p_{0N+} e^{\frac{qV_{N+}}{kT}} \qquad \textbf{[7.52]}$$

Using these relationships in Eq. [7.50]:

$$p(+d).n(+d) = p_{0N+}.n_{0N+} e^{\frac{qV_{N+}}{kT}} = n_{ieN+}^2 e^{\frac{qV_{N+}}{kT}} \qquad \textbf{[7.53]}$$

where $n_{ieN+}$ is the effective intrinsic carrier concentration in the $N^+$ cathode region including the influence of band-gap narrowing. Due to high-level injection conditions in the drift region $p(+d) = n(+d)$, leading to:

$$e^{\frac{qV_{N+}}{kT}} = \left[\frac{n(+d)}{n_{ieN+}}\right]^2 \qquad \textbf{[7.54]}$$

Using this expression in Eq. [7.49]:

$$J_{N+} = J_{SN+}\left[\frac{n(+d)}{n_{ieN+}}\right]^2 \quad [7.55]$$

Recombination in the drift region can be neglected when the lifetime in the drift region is large, allowing equating the total current density to the recombination current given by Eq. [7.55] for the $N^+$ end region. The carrier concentration at the interface between the drift region and the $N^+$ region is then given by:

$$n(+d) = n_{ieN+}\sqrt{\frac{J_{FC}}{J_{SN+}}} \quad [7.56]$$

From this equation, it can be concluded that the carrier concentration in the drift region will increase as the square root of the total current density if the end region recombination becomes dominant. Under these circumstances, the middle region voltage drop is no longer independent of the current density resulting in an increase in the total on-state voltage drop.

If recombination in the drift region is neglected, the carrier concentration profile determined by solving Eq. [7.22] is given by:

$$p(x) = n(x) = \frac{n(+d)}{2d}(x+d) \quad [7.57]$$

where x extends from –d to +d across the drift region. The voltage drop across the middle region can be computed by first obtaining the electric field. Substituting Eq. [7.57] into Eq. [7.37] derived earlier for the electric field yields:

$$E(x) = \frac{2dJ_{FC}}{q(\mu_n+\mu_p)n(+d)(x+d)} - \frac{kT}{2q(x+d)} \quad [7.58]$$

The voltage drop in the drift (middle) region can be obtained by integration of the electric field across the drift region:

$$V_M = \left[\frac{2J_{FC}d}{q(\mu_n+\mu_p)n(+d)} - \frac{kT}{2q}\right]\ln\left(\frac{2d}{x_J}\right) \quad [7.59]$$

Using Eq. [7.56] for the carrier concentration at the interface between the drift region and the $N^+$ substrate in this expression yields:

$$V_M = \left[\frac{2d\sqrt{J_{FC}.J_{SN+}}}{q(\mu_n+\mu_p)n_{ieN+}} - \frac{kT}{2q}\right]\ln\left(\frac{2d}{x_J}\right) \quad [7.60]$$

This expression indicates that the voltage drop across the middle region is no longer independent of the current density when the recombination in the end region becomes dominant. Using Eq. [7.49] and replacing the minority carrier density in the $N^+$ end region in terms of the doping concentration, the effective intrinsic concentration in the $N^+$ region can be eliminated from the equation leading to:

$$V_M = \left[\frac{2d}{(\mu_n + \mu_p)}\sqrt{\frac{D_{pN+}J_{FC}}{qL_{pN+}N_{DN+}\tanh(W_{N+}/L_{pN+})}} - \frac{kT}{2q}\right]\ln\left(\frac{2d}{x_J}\right) \quad \textbf{[7.61]}$$

The voltage drop across the interface between the drift region and the $N^+$ substrate also gets altered when recombination in the end region becomes dominant. This voltage drop can be determined from the carrier density at x = +d:

$$V_{N+} = \frac{kT}{q}\ln\left[\frac{n(+d)}{N_D}\right] \quad \textbf{[7.62]}$$

Since the carrier concentration in the drift region at this interface is now given by Eq. [7.56]:

$$V_{N+} = \frac{kT}{q}\ln\left[\frac{n_{ieN+}}{N_D}\sqrt{\frac{J_{FC}}{J_{SN+}}}\right] = \frac{kT}{2q}\ln\left[\frac{J_{FC}L_{pN+}\tanh(W_{N+}/L_{pN+})}{qN_{DN+}D_{pN+}}\right] \quad \textbf{[7.63]}$$

In the presence of recombination in the end region, the on-state voltage drop for the MPS rectifier can be computed by utilizing the three components of the on-state voltage drop. Using Eq. [7.32] with Eq. [7.33], Eq. [7.61], and Eq. [7.63] yields:

$$\begin{aligned} V_{ON} = {} & \phi_{BN} + \frac{kT}{q}\ln\left(\frac{J_{FC}p}{AT^2d}\right) \\ & + \left\{\frac{2d}{(\mu_n + \mu_p)}\sqrt{\frac{D_{pN+}J_{FC}}{qL_{pN+}N_{DN+}\tanh(W_{N+}/L_{pN+})}} - \frac{kT}{2q}\right\}\ln\left(\frac{2d}{x_J}\right) \\ & + \frac{kT}{2q}\ln\left[\frac{J_{FC}L_{pN+}\tanh(W_{N+}/L_{pN+})}{qN_{DN+}D_{pN+}}\right] \end{aligned} \quad \textbf{[7.64]}$$

Since the middle region voltage drop now increases with increasing current density, the on-state voltage drop increases at a faster rate than for the analytical model without the end region recombination.

The on-state characteristic for the silicon MPS rectifier obtained by using the above analytical model with end region recombination is shown in Fig. 7.12 for the case of a lifetime of 10 microseconds in the drift region. For comparison, the on-state characteristic for a P-i-N rectifier with a lifetime of 10 microseconds in the drift region is also shown in the figure by the dashed line. In addition, the on-state characteristic for the Schottky rectifier with no modulation of the drift region is included in the figure for discussion. In order to compare the models with and without recombination in the end region, the characteristics obtained without end region recombination is also displayed in this figure. It can be observed that at forward bias voltages below 0.55 volts, the MPS rectifier behaves like the Schottky rectifier with a slightly larger on-state voltage drop due to the enhanced current density at the Schottky contact. When the forward current density exceeds 100 A/cm$^2$, the on-state voltage drop of the MPS rectifier with end region recombination is larger than without end region recombination. It is worth pointing out that the slope of the characteristics with end region recombination is not as steep as without end region recombination. It is also important to note that the characteristic shown in the figure for the P-i-N rectifier does not include the impact of end region recombination. When end region recombination is accounted for in the P-i-N rectifier, its on-state voltage drop also increases more rapidly than shown in this figure[1].

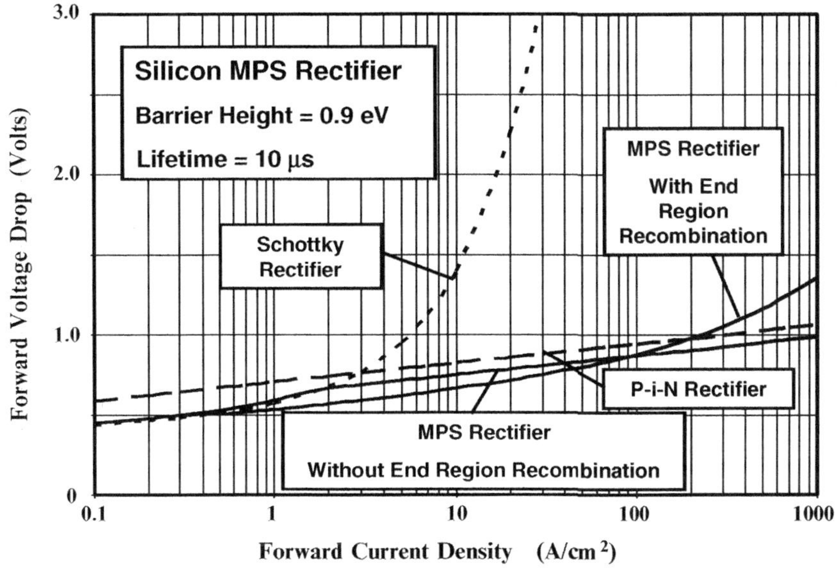


Fig. 7.12 On-State Characteristic for the Silicon MPS Rectifier after accounting for End Region Recombination.

## Simulation Example

In order to gain further insight into the physics of operation for the MPS rectifier, the results of two-dimensional numerical simulations are provided in this section for a structure designed for supporting 500 volts. For this case, a drift region with doping concentration of 3.8 x 10$^{14}$ cm$^{-3}$ was used with a thickness of 70 microns. The P$^+$ and N$^+$ end-regions had a surface concentration of 1 x 10$^{19}$ cm$^{-3}$ and a depth of about 1 micron. In all cases, it was assumed that $\tau_{p0}$ = $\tau_{n0}$. The influence of band-gap narrowing, auger recombination, and carrier-carrier scattering was included during the numerical simulations. The on-state characteristics were obtained for various values for the lifetime ($\tau_{p0}$ and $\tau_{n0}$). The impact of changes in the Schottky barrier height on the on-state characteristics was also analyzed to check the validity of the analytical models.

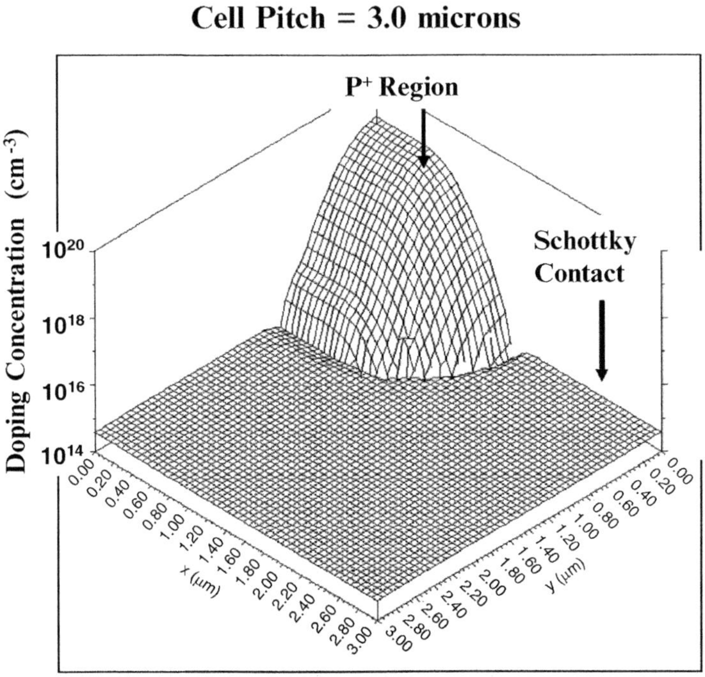


Fig. 7.1E Doping Distribution in the Baseline Silicon MPS Rectifier Structure.

A three-dimensional view of the doping concentration in the upper portion of the baseline silicon MPS device structure with cell pitch of 3 microns is provided in Fig. 7.1E. The $P^+$ region is located on the upper left-hand-side. It is formed by using a 1 micron deep diffusion from a 0.5 micron wide window in the cell. The Schottky contact has a width of 1.5 microns because of the lateral diffusion of the $P^+$ region. Consequently, the $P^+$ region and the Schottky contact have an equal area in the baseline silicon MPS rectifier structure used for the simulations.

The on-state characteristic for the baseline silicon MPS rectifier structure obtained from the numerical simulations is shown in Fig. 7.2E for the case of lifetime ($\tau_{p0}$ and $\tau_{n0}$) of 10 microseconds in the drift region. Several distinct regimes of operation are apparent in the shape of the characteristics. In order to relate these to the behavior of the inherent Schottky and P-i-N rectifiers within the MPS rectifier structure, the current flow through the contact to the $P^+$ region and the Schottky contact are also shown in the figure. It can be observed that current flow through the Schottky contact is dominant throughout the characteristics justifying the development of the analytical model at the Schottky contact. The on-state voltage drop at a current density of 100 A/cm$^2$ obtained using the numerical simulation is 0.87 volts. The value of 0.86 volts predicted by the analytical model is in excellent agreement providing validation of the analytical model.

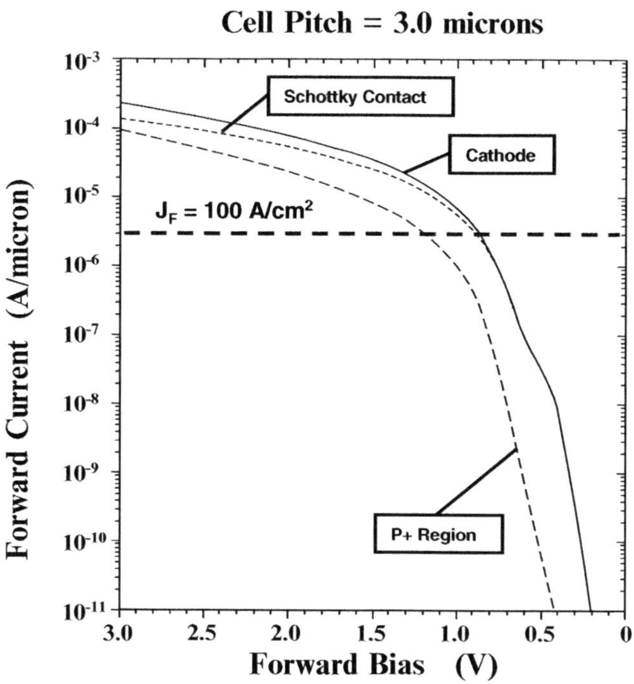


Fig. 7.2E On-State Characteristic of the 500V Baseline Silicon MPS Rectifier.

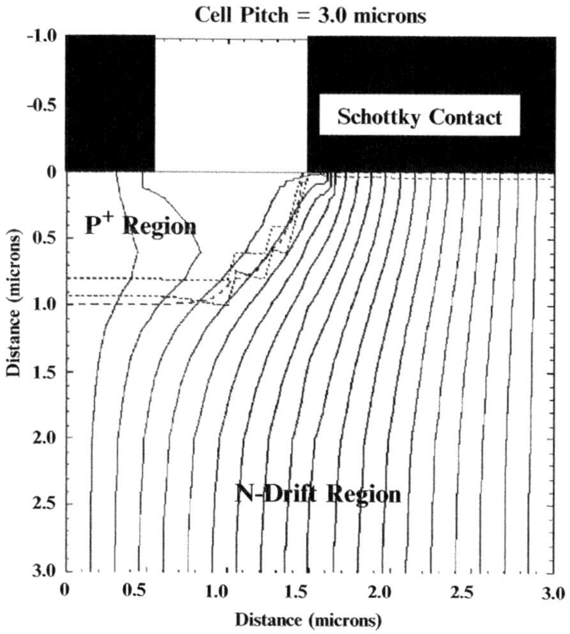


Fig. 7.3E Current Distribution within the 500V Baseline Silicon MPS Rectifier.

The current distribution within the silicon MPS rectifier can also be confirmed by examination of the current flow lines in the on-state. This is shown in Fig. 7.3E at a forward bias of 1 volt. It can be observed that most of the current flow occurs into the Schottky contact. This validates the basic premise of the MPS rectifier concept which is based upon a current flow via the Schottky contact with sufficient injection from the P-N junction to provide conductivity modulation of the drift region. In the above figure, the depletion region at the P-N junction has also been delineated. It can be observed that the depletion width at the P-N junction is negligible in width due to the forward bias across the junction. This justifies the assumptions used in determination of the area of the Schottky contact for current flow in the analytical model.

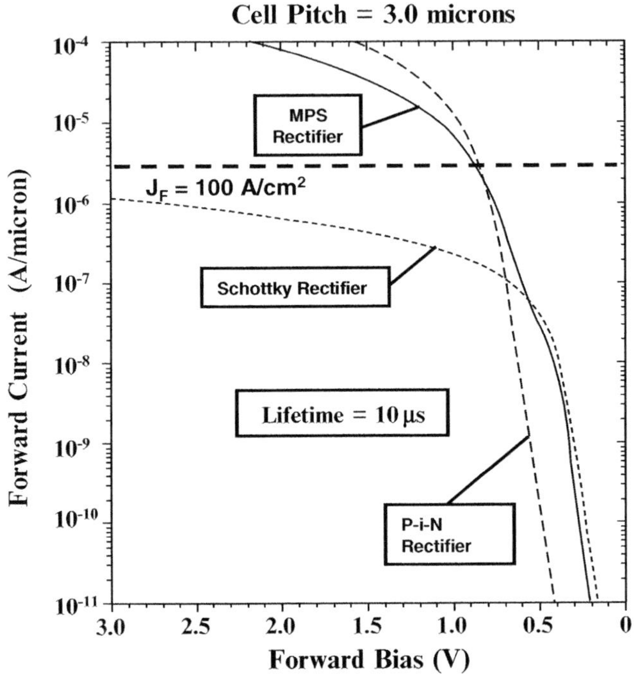


Fig. 7.4E On-state Characteristic of the 500V Baseline Silicon MPS Rectifier.

It is instructive to compare the characteristics of the silicon MPS rectifier with those of the P-i-N and Schottky rectifiers made using the same drift region parameters. This comparison is provided in Fig. 7.4E for the case of a lifetime of 10 microseconds. The P-i-N rectifier characteristics in the figure were obtained by simulations performed with a $P^+$ region across the entire cell width of 3 microns using the same surface concentration and junction depth as used in the MPS rectifier. In the case of the Schottky rectifier, no P-N junction was formed across the entire 3 micron cell and the same barrier height of 0.9 eV was used for the Schottky contact as in the case of the MPS rectifier. It can be observed that, at on-state current densities below 100 A/cm$^2$, the MPS rectifier exhibits a lower on-state voltage drop than the P-i-N rectifier. This demonstrates the premise that the MPS rectifier can exhibit superior on-state characteristics to those of the P-i-N rectifier.

The on-state voltage drop for the MPS rectifier resembles that for the Schottky rectifier at on-state voltage drops below 0.5 volts. It can be observed that the slope of the *i-v* characteristic for the MPS rectifier is not as steep as that for the P-i-N rectifier at the cross-over point. This behavior is properly predicted by the analytical model for the MPS rectifier after accounting for recombination in the $N^+$ end region (see Fig. 7.12) because the lifetime in the drift region is large.

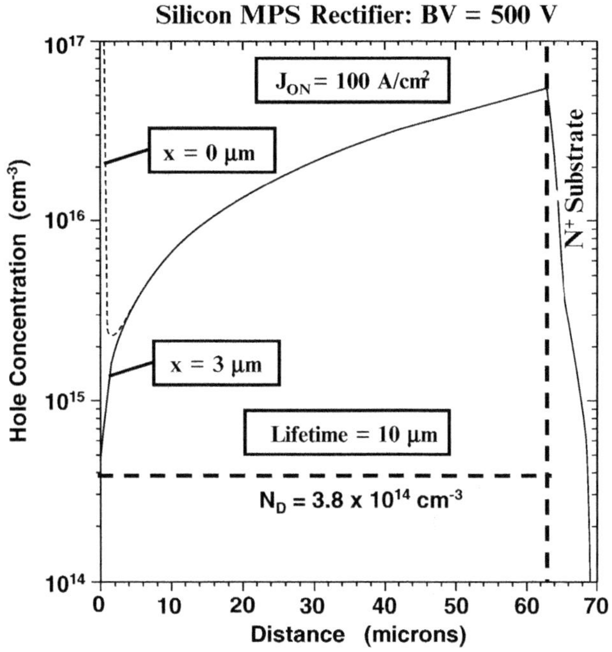


Fig. 7.5E Carrier Distribution within the Silicon MPS Rectifier.

The carrier distribution within the silicon MPS rectifier is shown in Fig. 7.5E for the case of a lifetime ($\tau_{p0}$ and $\tau_{n0}$) of 10 microsecond in the drift region at an on-state current density of 100 A/cm$^2$. Here, the hole concentration is shown with a solid line at the Schottky contact (x = 3 microns) and at the P-N junction (x = 0 microns) by the dotted line. It can be seen that high-level injection conditions prevail in the drift region because the injected carrier concentration is much greater than the background doping concentration except in close proximity of the Schottky contact. The hole concentration profile obtained from the simulations is similar to that illustrated in Fig. 7.3 at the Schottky contact. In spite of injection of holes from the P-N junction, it can be observed from the figure that the hole distribution below the junction is very similar to that at the Schottky contact. This justifies utilization of a one-dimensional model for deriving the carrier distribution in the MPS rectifier structure. Due to the high-level injection conditions, the hole and electron concentrations are equal in magnitude throughout the drift region. From this figure, significant injection of holes can also be observed in the $N^+$ substrate region. It is therefore appropriate to include end region recombination when analyzing the current flow in the MPS rectifier.

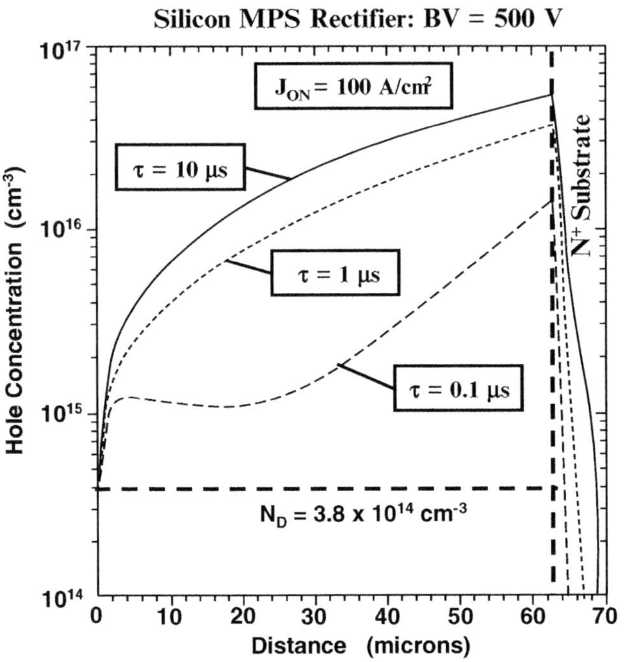


Fig. 7.6E Carrier Distribution within the Silicon MPS Rectifier.

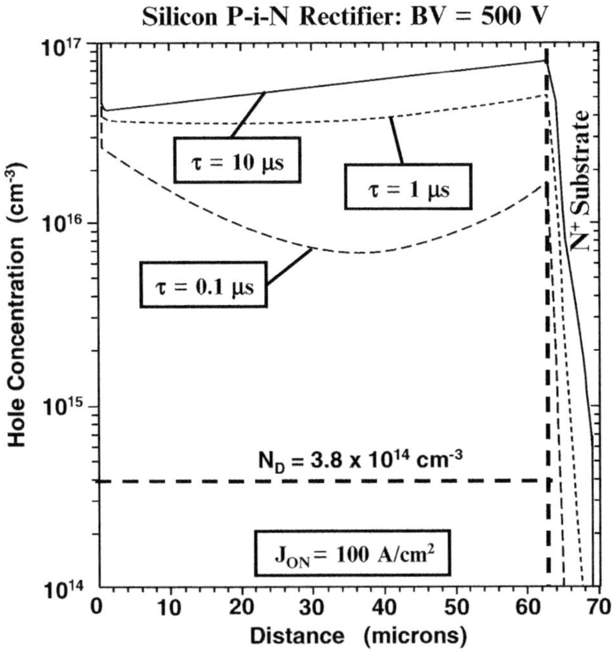


Fig. 7.7E Carrier Distribution within the Silicon P-i-N Rectifier.

The injection carrier concentration in the drift region reduces when the lifetime is reduced in the silicon MPS rectifier structure. This is illustrated in Fig. 7.6E where the hole concentration profile obtained by using numerical simulations is shown for three lifetime values. The maximum hole concentration located at the interface between the drift region and the $N^+$ substrate reduces as the lifetime is reduced resulting in less modulation of the drift region. This behavior is modeled quite well by the analytical model with recombination in the drift region (see Fig. 7.4). A similar behavior, namely reduced modulation of the drift region with smaller lifetime in the drift region, is also observed in the P-i-N rectifier as illustrated in Fig. 7.7E. A comparison of the hole concentration profiles for the MPS and P-i-N rectifier at the same lifetime can be performed using these figures. As an example, for a lifetime of 1 microsecond, the stored charge in the MPS rectifier can be seen to be about one-half the stored charge in the P-i-N rectifier. This favors superior turn-off characteristics with lower switching power losses for the MPS rectifier.

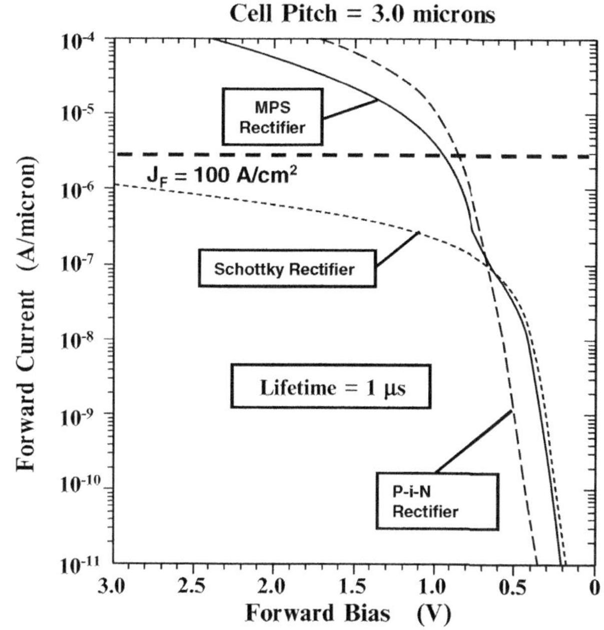


Fig. 7.8E On-State Characteristic of a 500V Silicon MPS Rectifier.

A reduction in the carrier concentration in the drift region with smaller lifetime in the MPS rectifier leads to an increase in the voltage drop across the middle region. This can be observed in Fig. 7.8E where the on-state characteristic for the MPS rectifier obtained using the numerical simulations is shown for the case of a lifetime of 1 microsecond. The characteristic for the P-i-N rectifier with the same lifetime is also provided in this figure for comparison. It can be observed that the slope of the characteristics for the two devices is similar in the range of current densities from 10 to 100 A/cm$^2$. This indicates that the recombination in the drift region is becoming dominant for the reduced lifetime of 1 microsecond making the slopes the same as predicted by the analytical model.

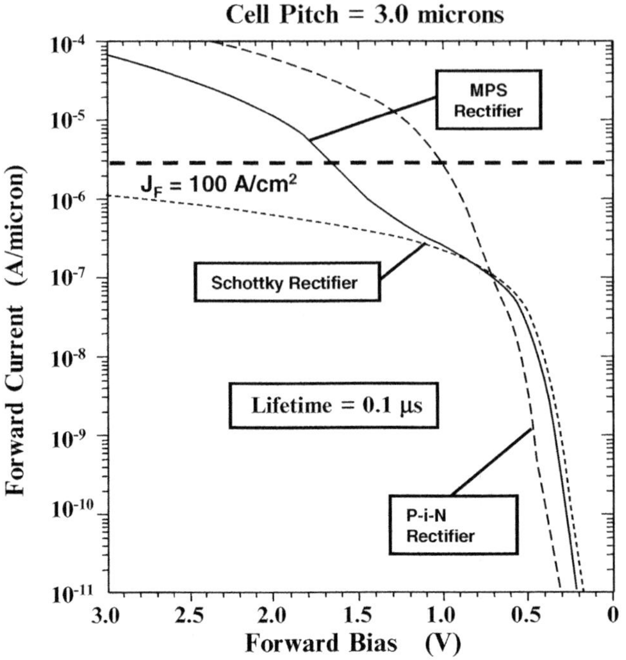


Fig. 7.9E On-State Characteristic of a 500V Silicon MPS Rectifier.

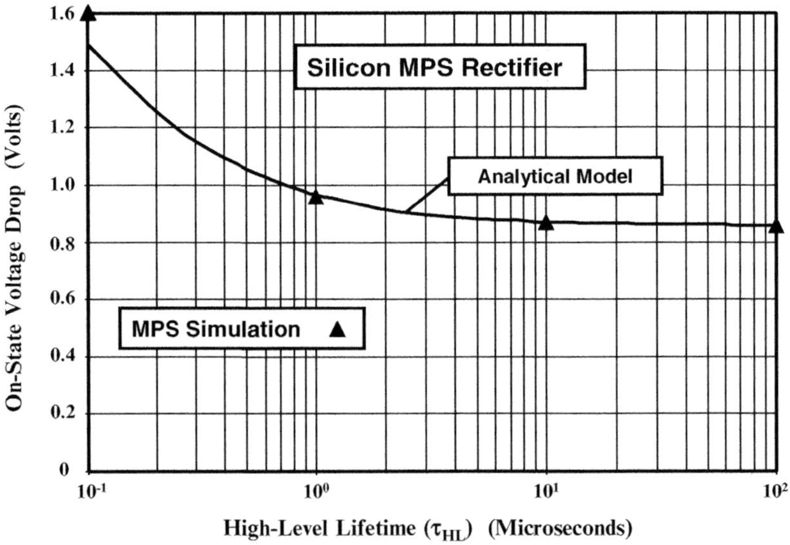


Fig. 7.10E On-State Voltage Drop for Silicon MPS Rectifiers.

The impact of reducing the lifetime to 0.1 microseconds on the on-state characteristic of the silicon MPS rectifier is shown in Fig. 7.9E. The characteristic obtained by numerical simulations of the P-i-N rectifier with the same lifetime is also included in the figure. It can be observed that the on-state voltage drop for the MPS rectifier has increased to 1.64 volts. This increase is due to the poor modulation of the drift region as previously shown in Fig. 7.6E. Since the stored charge in the MPS rectifier is much smaller than for the P-i-N rectifier, it is not necessary to reduce the lifetime in the MPS rectifier to same degree as required for the P-i-N rectifier to achieve low turn-off switching power losses. The increase in the on-state voltage drop in the MPS rectifier with reduced lifetime observed with the simulations is quite well predicted by the analytical model that takes into account recombination in the drift region. This can be observed in Fig. 7.10E where the on-state voltage drop predicted by the analytical model with recombination in the drift region is compared with the values obtained from the numerical solutions.

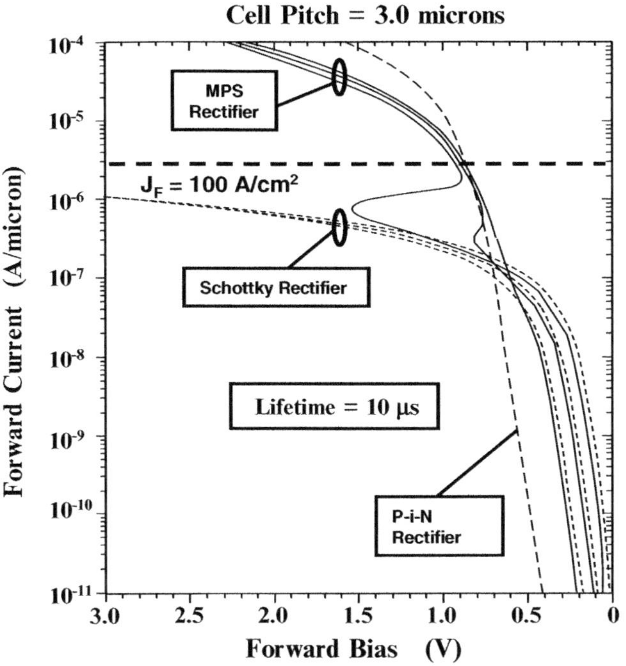


Fig. 7.11E On-State Characteristics of 500V Silicon MPS Rectifiers.

The analytical model predicts a reduction in the on-state voltage drop for the silicon MPS rectifier when the barrier height of the Schottky contact is reduced (see Fig.7.11). In order to verify this, numerical simulations were performed for the MPS rectifier structure using various work functions for the Schottky metal contact. For comparison purposes, the characteristics for the Schottky rectifier were also obtained using numerical simulations with various work functions for the Schottky metal contact. The characteristics of all of these structures can be compared in Fig. 7.11E. When the barrier height for the Schottky contact is reduced, the on-state characteristics for the MPS rectifier shift towards lower voltages as predicted by the analytical models for these devices when the on-state current density is below 3 A/cm$^2$. This is consistent with the operation of the MPS rectifier as a Schottky rectifier at these current levels. When the current density increases beyond 10 A/cm$^2$, the characteristic for the MPS rectifier follows the analytical model for the largest barrier height case (0.9 eV). However, when the barrier height is reduced, a snap back in the characteristics is observed and the on-state voltage drop for the MPS rectifier remains larger than that for the P-i-N rectifier even after snap-back. This behavior is related to the inability to generate enough voltage across the P-N junction for the injection of holes when the barrier height is made less than 0.8 eV. As stated earlier in the chapter, it is appropriate to use a large barrier height for the Schottky contact in the MPS rectifier to reduce the leakage current. The utilization of a large barrier height also prevents the snap-back in the on-state characteristics of the silicon MPS rectifier.

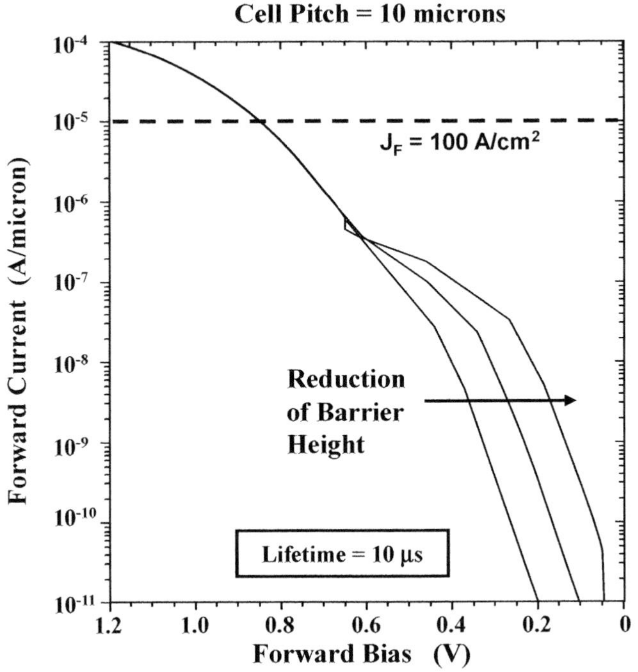


Fig. 7.12E On-State Characteristics of 500V Silicon MPS Rectifiers.

The snap-back in the characteristics of the MPS rectifier at lower barrier heights for the Schottky contact can be suppressed by enlarging the relative area occupied by the P$^+$ region. This is demonstrated in Fig. 7.12E using results of numerical simulations for a silicon MPS rectifier structure with cell pitch of 10 microns and Schottky contact width of 1.5 microns (same as for the baseline device). At on-state voltage drops below 0.6 volts, the devices operate like Schottky rectifiers and the voltage drop becomes smaller when the barrier height is reduced. However, at larger forward bias voltages, the characteristic for the MPS rectifier becomes independent of the barrier height. No snap-back of the characteristics is observed with the larger area for the P$^+$ region in this MPS rectifier structure.

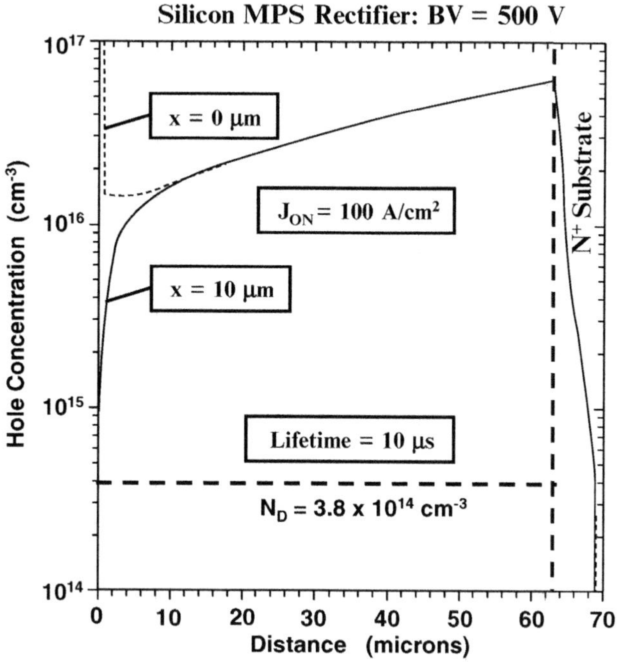


Fig. 7.13E Carrier Distribution within the MPS Rectifier with Cell Pitch of 10 microns.

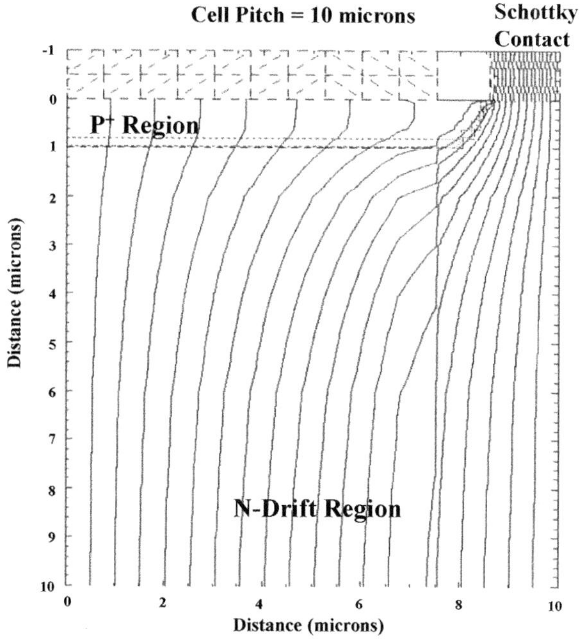


Fig. 7.14E Current Distribution within the Silicon MPS Rectifier.

The carrier profile within the silicon MPS rectifier with the larger cell pitch of 10 microns is shown in Fig. 7.13E. The profile shown by the solid line at the Schottky contact (x = 10 microns) is similar to that for the baseline MPS rectifier structure (see Fig. 7.5E). However, the hole concentration at the middle of the $P^+$ region (at x = 0 microns) is larger than for the baseline MPS rectifier structure (see Fig. 7.5E). The enhanced injection of holes at the middle of the $P^+$ region suppresses the snap-back effect. The current flow through the $P^+$ region is also much larger as shown by the current flow-lines obtained from the numerical simulations as shown in Fig. 7.14E. With the enlarged $P^+$ region, about half the current flows through the $P^+$ region as compared with only 10 percent for the baseline structure (see Fig. 7.3E).

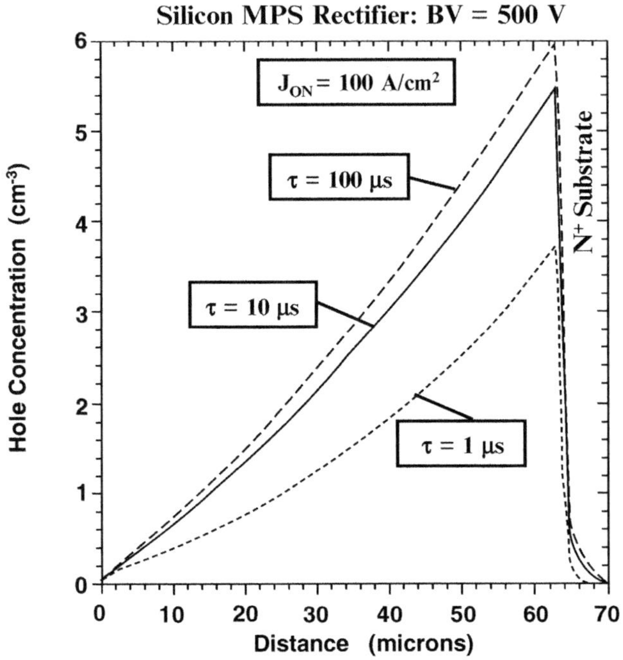


Fig. 7.15E Carrier Distribution within the Silicon MPS Rectifier.

The analytical model developed in this section indicates that the carrier distribution profile in the drift region is linear in shape (see Eq. [7.30]) if recombination in the drift region can be neglected. It is worth verifying this conclusion from the results of numerical simulations of the baseline MPS rectifier structure. The hole concentration profiles for this structure are shown in Fig. 7.15E for the case of three values for the lifetime using a linear scale for the vertical axis. It can be observed that the profile is close to linear in shape when the lifetime is large (10 microseconds). Although the hole concentration reduces at the interface between the drift region and the $N^+$ substrate when the lifetime is reduced, the profile can still be approximated by a straight line. A linear carrier profile can therefore be used as a good approximation when computing the stored charge in the drift region, voltage drop across the drift region, and for development of turn-off switching model for the MPS rectifier structure.

## 7.2 Silicon Carbide MPS Rectifiers

In chapter 6, it was demonstrated that for reverse blocking voltages above 5000 volts, the silicon carbide P-i-N rectifier is preferable due to larger on-state voltage drop of the silicon carbide Schottky rectifier. Due to the much larger electric field that can be supported in silicon carbide, the width of the drift region of the silicon carbide P-i-N rectifier is much smaller than that for the corresponding silicon device with the same breakdown voltage. This implies that the stored charge in the silicon carbide P-i-N rectifier will be much smaller than for the silicon device providing an improvement in the switching behavior. A further improvement in the switching performance can be achieved by using the MPS structure while reducing the on-state voltage drop.

The analytical theory for operation of the silicon carbide MPS rectifier structure is identical to that already developed for the silicon device. However, the large band-gap for silicon carbide produces a larger built-in potential for the P-N junction that must be overcome to obtain substantial injection of minority carriers into the drift region. In the case of the silicon MPS rectifier, it was shown in the previous section that the injection is suppressed when the Schottky barrier height is reduced. This phenomenon can also be expected to occur for the silicon carbide MPS rectifier.

In the presence of recombination in the drift region, the carrier distribution was found to be given by:

$$p(x) = n(x) = \frac{L_a J_{FC}}{2qD_n} \frac{\sinh\left[(x+d)/L_a\right]}{\cosh\left[2d/L_a\right]} \quad [7.65]$$

The carrier distribution described by this equation was schematically illustrated in Fig. 7.3. It has a maximum value at the interface between the drift region and the $N^+$ substrate with a magnitude of:

$$p(+d) = n(+d) = \frac{L_a J_{FC}}{2qD_n} \frac{\sinh\left[2d/L_a\right]}{\cosh\left[2d/L_a\right]} = \frac{L_a J_{FC}}{2qD_n} \tanh\left(\frac{2d}{L_a}\right) \quad [7.66]$$

and reduces monotonically when proceeding towards the Schottky contact in the negative x direction. The concentration becomes equal to zero at the Schottky contact as required to satisfy the boundary condition used to derive the expression.

As a particular example for the case of the silicon carbide 10 kV MPS rectifier, the carrier distributions calculated at an on-state current density of 100 $A/cm^2$ by using Eq. [7.65] are shown in Fig. 7.13 for the case of three values for the high-level lifetime for a MPS rectifier with drift region thickness of 80 microns and doping concentration of 2 x $10^{15}$ $cm^{-3}$. The largest concentrations for the electrons and holes in the drift region occur at its boundary with the $N^+$ end-region. The carrier concentration at this boundary decreases when the lifetime is reduced. It has a magnitude of 8.2 x $10^{16}$ $cm^{-3}$ for a high lifetime of 100 microseconds, 6.3 x $10^{16}$ $cm^{-3}$ for a moderate lifetime of 10 microseconds, and 2.6 x $10^{16}$ $cm^{-3}$ for a low lifetime of 1 microsecond. When the lifetime is reduced to 1 microsecond, the analytical model predicts that the entire drift region is not conductivity modulated. As previously discussed for the silicon MPS rectifier, the carrier profiles for the higher lifetime cases in the silicon carbide MPS rectifier can also be assumed to close to linear in shape.

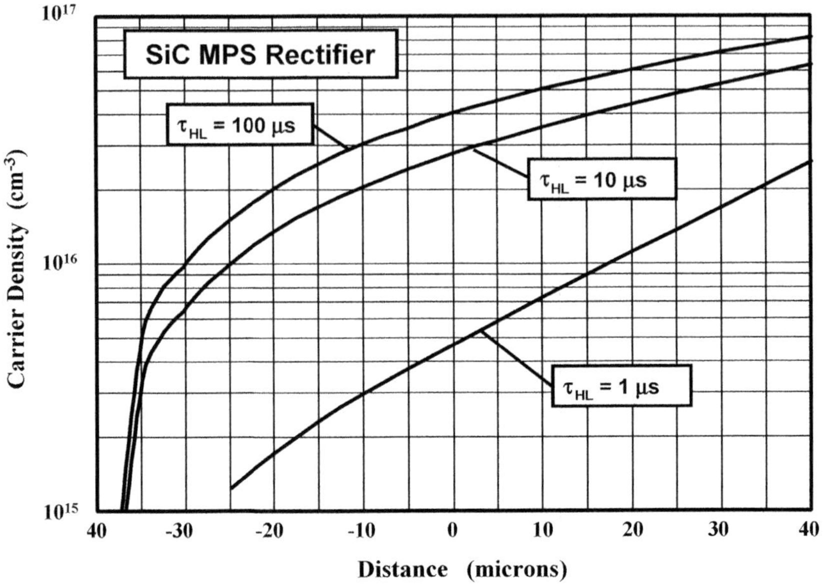


Fig. 7.13 Carrier Distribution under High-Level Injection Conditions for the 10 kV SiC MPS Rectifier with various High-Level Lifetime values.

At low on-state current density levels, the on-state characteristics for the silicon carbide MPS rectifier will resemble those for the Schottky rectifier with an enhanced current density at the Schottky contact. At larger current levels with high-level injection in the drift region, the on-state voltage drop for the silicon carbide MPS rectifier can be obtained by summing the voltage drops along the path marked 'A-A' in Fig. 7.3 through the Schottky contact. The total voltage drop along this path consists of the voltage drop across the Schottky contact ($V_{FS}$), the voltage drop across the drift region (middle region voltage $V_M$), and the voltage drop at the interface with the $N^+$ substrate ($V_{N+}$):

$$V_{ON} = V_{FS} + V_M + V_{N+} \quad [7.67]$$

The voltage drop across the Schottky contact is given by:

$$V_{FS} = \phi_{BN} + \frac{kT}{q}\ln\left(\frac{J_{FS}}{AT^2}\right) \quad [7.68]$$

where the current density at the Schottky contact ($J_{FS}$) is related to the cell or cathode current density by Eq. [7.1]. The voltage drop in the drift (middle) region with inclusion of recombination is given by:

$$V_M = \left\{ \frac{4D_n}{(\mu_n + \mu_p)} \left[ \frac{(d/L_a)}{\tanh(2d/L_a)} \right] - \frac{kT}{2q} \right\} \ln\left( \frac{2d}{x_J} \right) \quad [7.69]$$

The voltage drop across the interface between the drift region and the $N^+$ substrate is given by:

$$V_{N+} = \frac{kT}{q} \ln\left[ \frac{J_{FC} L_a \tanh(2d/L_a)}{2qD_n N_D} \right] \quad [7.70]$$

The on-state voltage drop for the MPS rectifier can be computed by utilizing the three components discussed above. Using these equations yields:

$$V_{ON} = \phi_{BN} + \frac{kT}{q} \ln\left( \frac{J_{FC} p}{AT^2 d} \right) + \left\{ \frac{4D_n}{(\mu_n + \mu_p)} \left[ \frac{(d/L_a)}{\tanh(2d/L_a)} \right] - \frac{kT}{2q} \right\} \ln\left( \frac{2d}{x_J} \right) + \frac{kT}{q} \ln\left[ \frac{J_{FC} L_a \tanh(2d/L_a)}{2qD_n N_D} \right] \quad [7.71]$$

The on-state voltage drop for the MPS rectifier is a function of the lifetime in the drift region because the middle region and $N/N^+$ interface voltage drops change with lifetime.

As an example, consider the case of a silicon carbide MPS rectifier with a drift region doping concentration of $2 \times 10^{15}$ $cm^{-3}$ and thickness of 80 microns capable of supporting 10,000 volts in the reverse blocking mode. The variation of the on-state voltage drop with high-level lifetime in the drift region for this case as predicted by the analytical model is provided in Fig. 7.14. It can be observed that the on-state voltage drop begins to increase when the lifetime is reduced below 10 microseconds. This is due to the relatively low diffusion length for holes in silicon carbide. The three components of the on-state voltage drop are also shown in the figure. The voltage drop at the Schottky contact is independent of the lifetime. Due to the large Schottky barrier height of 2.95 eV used in the analytical model, this contribution to the on-state voltage drop is much larger than for the silicon device. The voltage drop at the $N/N^+$ interface increases slightly when the lifetime becomes larger than 1 microsecond but makes only a small contribution to the total on-state voltage drop. The most significant increase in the voltage drop occurs for the middle region when the lifetime is reduced below 10 microseconds. It is therefore necessary to achieve lifetime values approaching 10 microseconds to obtain a low on-state voltage drop in the silicon carbide MPS rectifier. The measured lifetimes in the drift region of silicon carbide power devices have improved from less than 100 nanoseconds to 3 microseconds as the quality of the material has improved[5]. The on-state voltage drop of the 10 kV silicon carbide MPS rectifier is found to be just under 3 volts for a drift region lifetime of 10 microseconds.

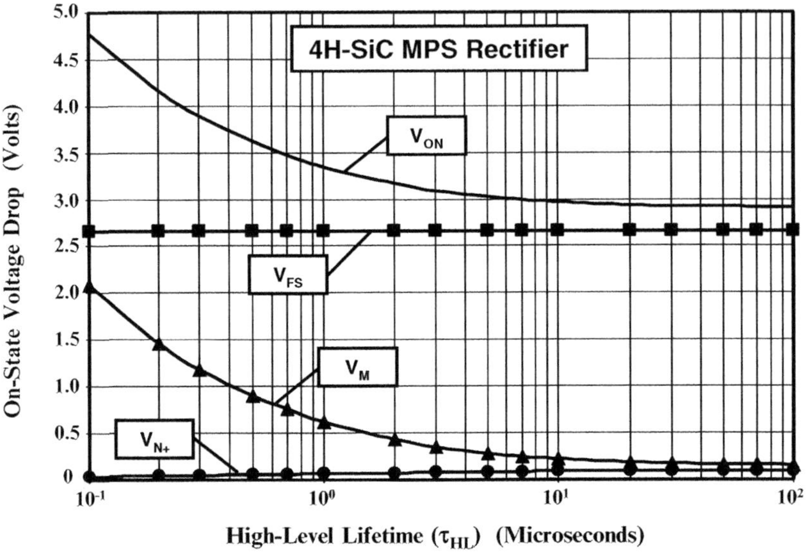


Fig. 7.14 Voltage Drops in the 10 kV Silicon Carbide MPS Rectifier.

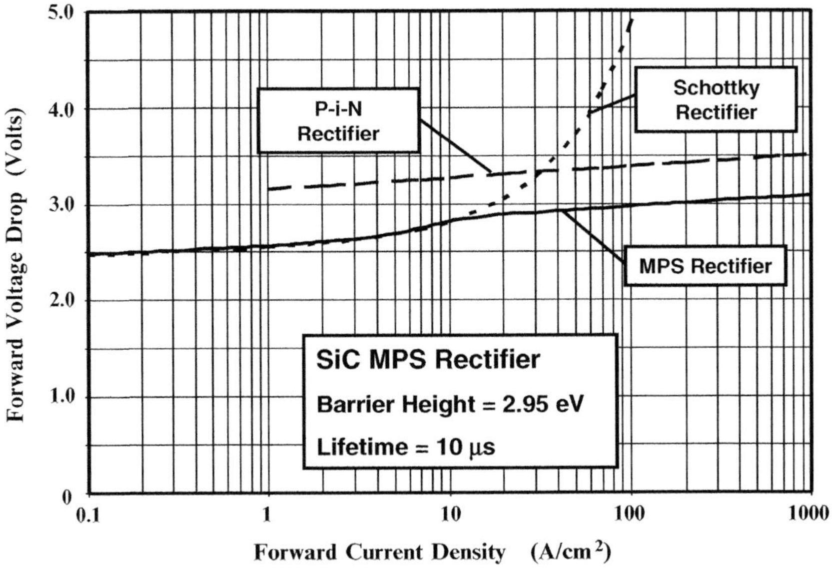


Fig. 7.15 On-state Characteristics for the 10 kV SiC MPS Rectifier with Lifetime of 10 microseconds in the Drift Region.

At lower current densities, the on-state voltage drop for the silicon carbide MPS rectifier resembles that for the Schottky rectifier as demonstrated in Fig. 7.15 for current density up to 10 A/cm$^2$. At larger forward current densities, the injected carrier density in the drift region exceeds the background doping concentration leading to high level injection conditions. In this mode of operation, the injected carrier concentration in the drift region increases in proportion to the current density resulting in a constant voltage drop across the drift region for the analytical model without recombination in the N$^+$ end region. From Eq. [7.71], an expression for the on-state current density can be derived:

$$J_{FC} = \sqrt{\frac{2qAT^2D_nN_D}{p}\frac{(d/L_a)}{\tanh(2d/L_a)}}e^{-\frac{q(\phi_{BN}+V_M)}{2kT}}e^{\frac{qV_{ON}}{2kT}} \quad [7.72]$$

The current flow is observed to become proportional to ($qV_{ON}/2kT$), similar to that observed for the P-i-N rectifier under high-level injection conditions, as observed in Fig. 7.15. The on-state voltage drop for the silicon carbide MPS rectifier is found to be lower than that for the silicon carbide P-i-N rectifier for the drift region lifetime of 10 microseconds.

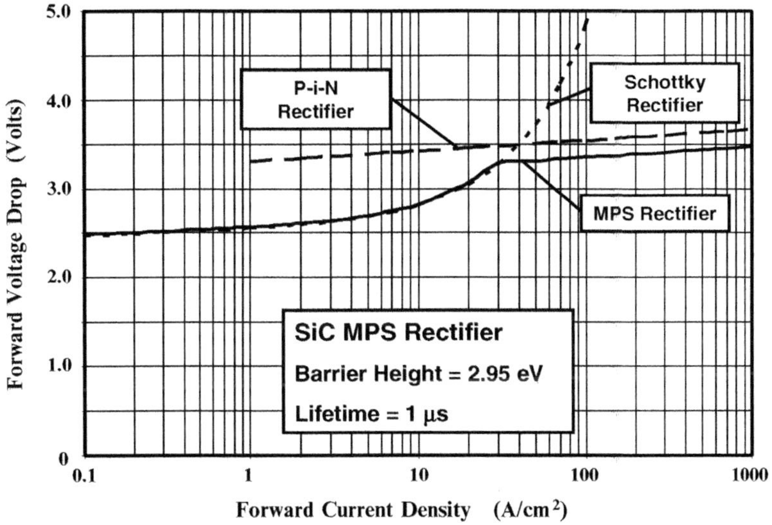


Fig. 7.16 On-state Characteristics for the 10 kV SiC MPS Rectifier with Lifetime of 1 microsecond in the Drift Region.

The impact of reducing the lifetime to 1 microsecond in the drift region on the on-state characteristics for the 10 kV silicon carbide MPS rectifier is shown in Fig. 7.16. The on-state voltage drop for the silicon carbide MPS rectifier increases by more than that for the P-i-N rectifier making its on-state voltage drop close that for the P-i-N rectifier. However, the stored charge in the silicon carbide MPS rectifier is smaller making the switching performance superior to that for the P-i-N rectifier.

The on-state voltage drop for the silicon carbide MPS rectifier is also determined by the barrier height of the Schottky metal. According to the analytical model, a reduction in the on-state voltage drop of the silicon carbide MPS rectifier can be achieved by reducing the barrier height as illustrated in Fig. 7.17 where the characteristics of devices with two barrier heights can be compared. The analytical model predicts a decrease in on-state voltage drop that is equal to the reduction of the barrier height. This behavior is similar to that observed in the Schottky rectifiers whose characteristics are shown in the figure by the dashed lines.

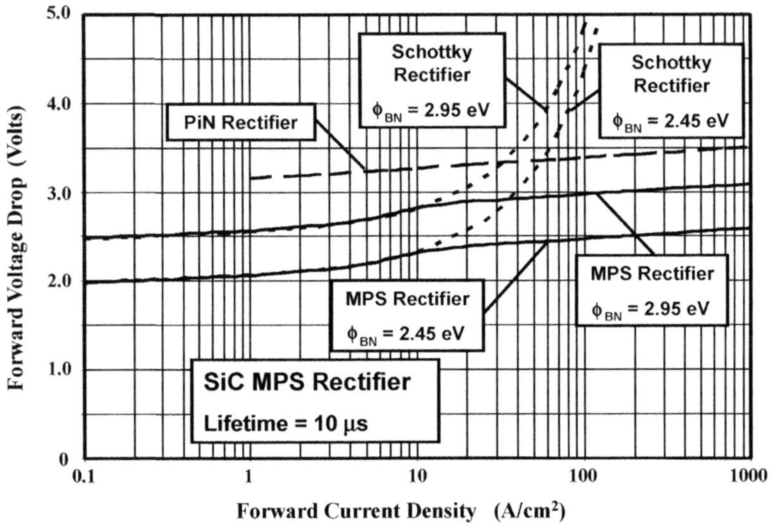


Fig. 7.17 On-state Characteristics for 10 kV SiC MPS Rectifiers with Different Schottky Barrier Heights.

In the presence of recombination in the end region, the on-state voltage drop for the silicon carbide MPS rectifier can be computed by utilizing the same equation [7.64] previously derived for the silicon MPS rectifier with the parameters appropriate to silicon carbide:

$$V_{ON} = \phi_{BN} + \frac{kT}{q}\ln\left(\frac{J_{FC}p}{AT^2d}\right) + \left\{\frac{2d}{(\mu_n + \mu_p)\sqrt{\dfrac{D_{pN+}J_{FC}}{qL_{pN+}N_{DN+}\tanh(W_{N+}/L_{pN+})}}} - \frac{kT}{2q}\right\}\ln\left(\frac{2d}{x_J}\right) + \frac{kT}{2q}\ln\left[\frac{J_{FC}L_{pN+}\tanh(W_{N+}/L_{pN+})}{qN_{DN+}D_{pN+}}\right] \quad [7.73]$$

Since the middle region voltage drop increases with increasing current density, the on-state voltage drop increases at a faster rate than for the analytical model without the end region recombination.

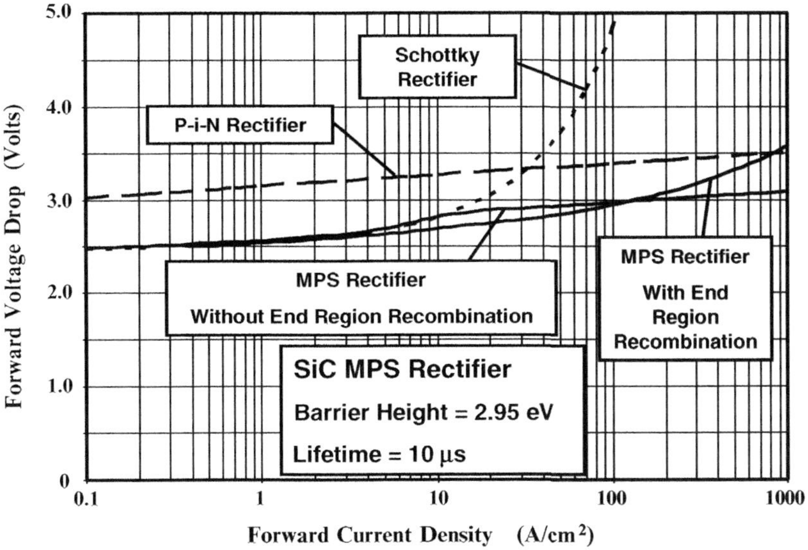


Fig. 7.18 On-State Characteristic for the 10 kV SiC MPS Rectifier after accounting for End Region Recombination.

The on-state characteristic for the silicon carbide MPS rectifier obtained by using the above analytical model with end region recombination is shown in Fig. 7.18 for the case of a lifetime of 10 microseconds in the drift region. For comparison, the on-state characteristic for a silicon carbide P-i-N rectifier with a lifetime of 10 microseconds in the drift region is also shown in the figure by the dashed line. In addition, the on-state characteristic for the Schottky rectifier with no modulation of the drift region is included in the figure for discussion. In order to compare the models with and without recombination in the end region, the characteristics obtained without end region recombination is also displayed in this figure. It can be observed that the slope of the on-state characteristic is not as steep with end region recombination. However, the on-state voltage drop at a current density of 100 A/cm$^2$ is the same with both analytical models. It is worth pointing out that, when end region recombination is accounted for in the P-i-N rectifier, its on-state voltage drop also increases more rapidly than shown in this figure[1].

## Simulation Example

In order to gain further insight into the physics of operation for the silicon carbide MPS rectifier, the results of two-dimensional numerical simulations are provided in this section for a structure designed for supporting 10,000 volts. For this case, a drift region with doping concentration of 2 x 10$^{15}$ cm$^{-3}$ was used with a thickness of

80 microns. The $P^+$ and $N^+$ end-regions were uniformly doped with a dopant concentration of 1 x $10^{19}$ $cm^{-3}$ and a depth of about 1 micron. In all cases, it was assumed that $\tau_{p0} = \tau_{n0}$. The influence of band-gap narrowing, auger recombination, and carrier-carrier scattering was included during the numerical simulations. The on-state characteristics were obtained for various values for the lifetime ($\tau_{p0}$ and $\tau_{n0}$). The impact of changes in the Schottky barrier height on the on-state characteristics was also analyzed to check the validity of the analytical models.

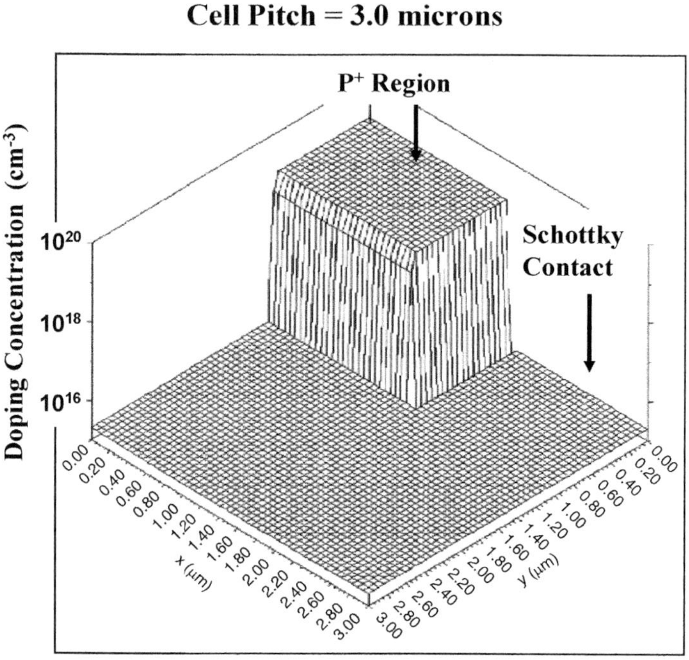


Fig. 7.16E Doping distribution in the Baseline 10 kV SiC MPS Rectifier Structure.

A three-dimensional view of the doping concentration in the upper portion of the baseline 10 kV silicon carbide MPS device structure with cell pitch of 3 microns is provided in Fig. 7.16E. The $P^+$ region is located on the upper left-hand-side. It is formed by using a 1 micron deep ion-implanted region from a 1.5 micron wide window in the cell. No diffusion of the dopant is assumed to occur during the ion-implant annealing process due to the small diffusion coefficients for dopants in silicon carbide. The Schottky contact has a width of 1.5 microns because there is no lateral diffusion of the $P^+$ region. Consequently, the $P^+$ region and the Schottky contact have an equal area in the baseline silicon carbide MPS rectifier structure used for the simulations.

The on-state characteristic for the baseline silicon carbide MPS rectifier structure obtained from the numerical simulations is shown in Fig. 7.17E for the case of lifetime ($\tau_{p0}$ and $\tau_{n0}$) of 10 microseconds in the drift region and a work function of 6.7 for the Schottky contact (corresponding to a Schottky barrier height of 3 eV). Several distinct regimes of operation are apparent in the shape of the characteristics. In order to relate these to the behavior of the inherent Schottky and P-i-N rectifiers within the MPS rectifier structure, the current flow through the contact to the $P^+$ region and the Schottky contact are also shown in the figure. It can be observed that current flow through the Schottky contact is dominant throughout the characteristics justifying the development of the analytical model at the Schottky contact. The on-state voltage drop at a current density of 100 A/cm$^2$ obtained using the numerical simulation is 3.02 volts. The value of 3 volts predicted by the analytical models with and without end region recombination is in excellent agreement providing validation of the analytical models.

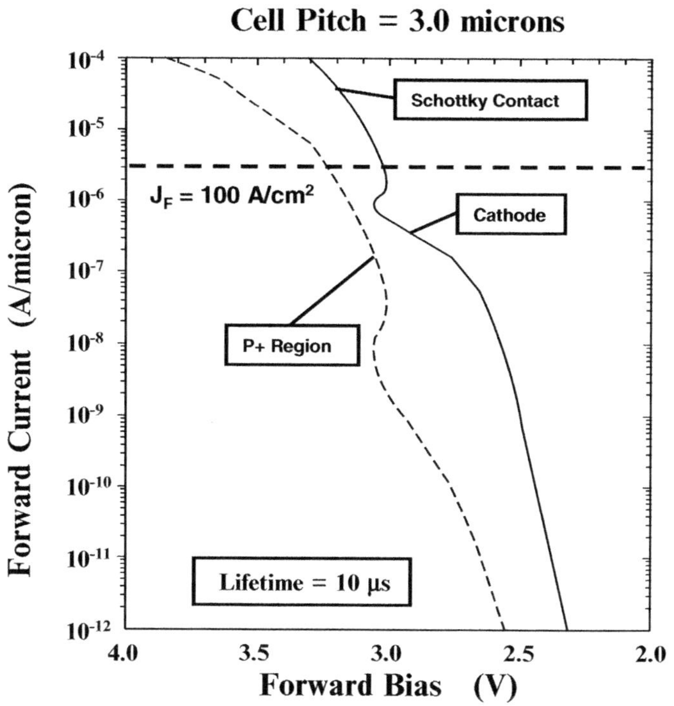


Fig. 7.17E On-State Characteristic of the 10 kV SiC Baseline MPS Rectifier.

The current distribution within the silicon carbide MPS rectifier can also be confirmed by examination of the current flow lines in the on-state. This is shown in Fig. 7.18E at an on-state current density of 100 A/cm$^2$. It can be observed that most of the current flow occurs into the Schottky contact. This validates the basic premise of the MPS rectifier concept which is based upon a current flow via the Schottky contact with sufficient injection from the P-N junction to provide conductivity modulation of the drift region. In the above figure, the depletion region at the P-N junction has also been delineated. It can be observed that the depletion width at the P-N junction is negligible in width due to the forward bias across the junction. This justifies the assumptions used in determination of the area of the Schottky contact for current flow in the analytical model.

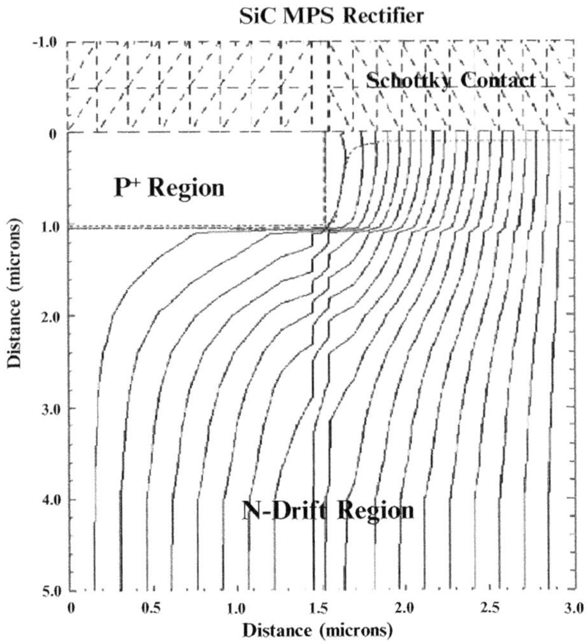


Fig. 7.18E Current Distribution within the 10 kV SiC Baseline MPS Rectifier.

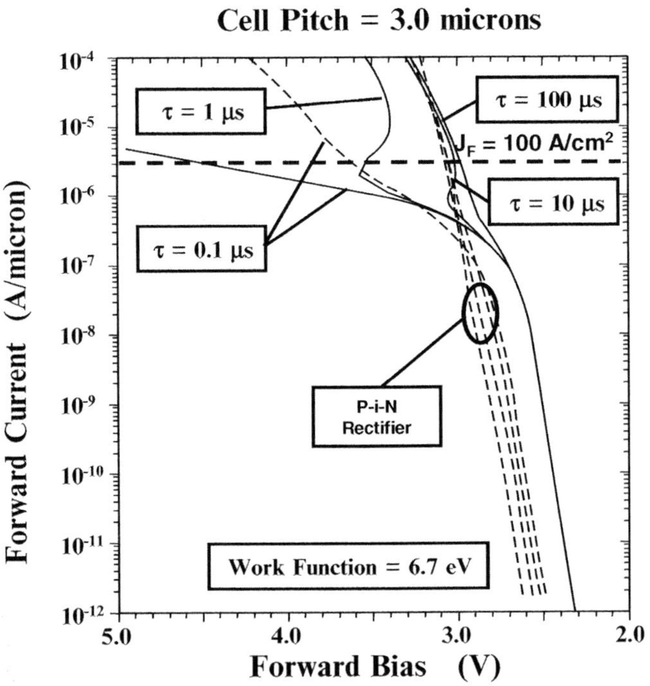


Fig. 7.19E On-state Characteristic of 10 kV SiC MPS and P-i-N Rectifiers.

The characteristics of the silicon carbide MPS rectifiers (solid lines) with various lifetime values in the drift region are compared in Fig. 7.19E with those of P-i-N rectifiers (dashed lines) made using the same drift region parameters. The P-i-N rectifier characteristics in the figure were obtained by simulations performed with a $P^+$ region across the entire cell width of 3 microns using the same surface concentration and junction depth as used in the MPS rectifier. It can be observed that there is a snap-back in the characteristics for the silicon carbide MPS rectifier when the lifetime is below 100 microseconds. However, for the case of lifetimes of 10 and 100 microseconds, the silicon carbide MPS rectifier exhibits a lower on-state voltage drop than the P-i-N rectifier at an on-state current density of 100 A/cm$^2$. This demonstrates the premise that the silicon carbide MPS rectifier can exhibit superior on-state characteristics to those of the silicon carbide P-i-N rectifier. When the lifetime in the drift region for the silicon carbide MPS rectifier is reduced to 1 microsecond, the on-state voltage drop exceeds that for the silicon carbide P-i-N rectifier with this lifetime. For the case of a lifetime of 0.1 microseconds, the bipolar mode is no longer observed in the silicon carbide MPS rectifier and its on-state voltage drop becomes large. For this low lifetime, the on-state voltage drop for the silicon carbide P-i-N rectifier also increases to 3.63 volts.

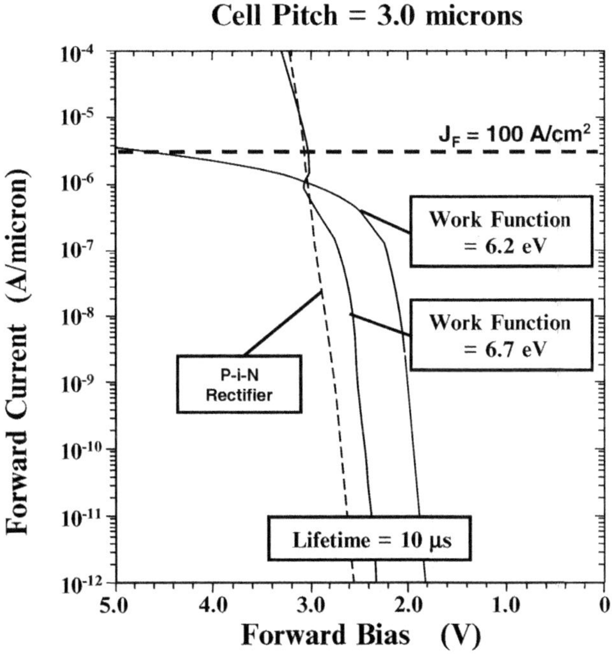


Fig. 7.20E On-state Characteristic of 10 kV SiC MPS Rectifiers.

The above on-state characteristics for the silicon carbide MPS rectifier were obtained using a rather large work function of 6.7 eV corresponding to a barrier height of 3 eV. When the work function of the Schottky contact is reduced to 6.2 eV, injection from the P-N junction gets suppressed and the device structure exhibits unipolar conduction up to an on-state voltage drop of 5 volts as shown in Fig. 7.20E. The bipolar mode of operation in the silicon carbide MPS rectifier can be produced in this case by reducing the width of the Schottky contact as demonstrated later in this section.

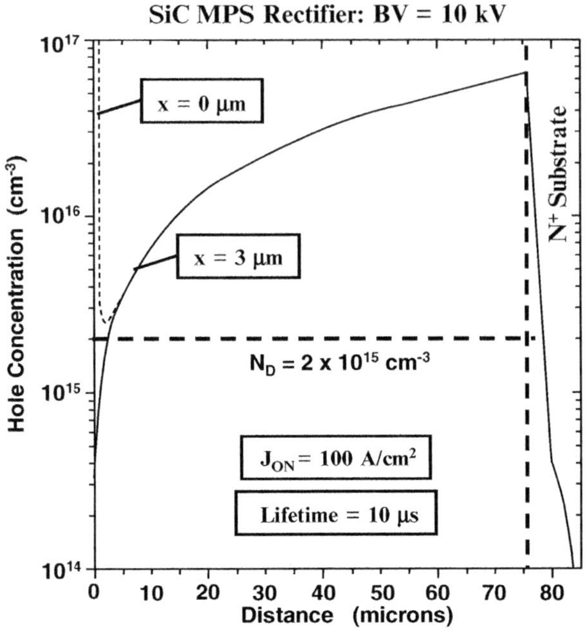


Fig. 7.21E Carrier Distribution within the 10 kV SiC MPS Rectifier.

The carrier distribution within the baseline silicon carbide MPS rectifier is shown in Fig. 7.21E for the case of a lifetime ($\tau_{p0}$ and $\tau_{n0}$) of 10 microsecond in the drift region at an on-state current density of 100 A/cm$^2$. Here, the hole concentration is shown with a solid line at the Schottky contact (x = 3 microns) and by the dotted line at the P-N junction (x = 0 microns). It can be seen that high-level injection conditions prevail in the drift region because the injected carrier concentration is much greater than the background doping concentration except in close proximity of the Schottky contact. The hole concentration profile obtained from the simulations is similar to that illustrated in Fig. 7.3 at the Schottky contact. In spite of injection of holes from the P-N junction, it can be observed from the figure that the hole distribution below the junction is very similar to that at the Schottky contact. This justifies utilization of a one-dimensional model for deriving the carrier distribution in the silicon carbide MPS rectifier structure. Due to the high-level injection conditions, the hole and electron concentrations are equal in magnitude throughout the drift region. From this figure, significant injection of holes can also be observed in the $N^+$ substrate region. It is therefore appropriate to include end region recombination when analyzing the current flow in the silicon carbide MPS rectifier.

The injection carrier concentration in the drift region reduces when the lifetime is reduced in the silicon carbide MPS rectifier structure. This is illustrated in Fig. 7.22E where the hole concentration profile obtained by using numerical simulations is shown for three lifetime values for the 10 kV baseline silicon carbide

MPS rectifier. The maximum hole concentration located at the interface between the drift region and the $N^+$ substrate reduces as the lifetime is reduced resulting in less modulation of the drift region. This behavior is modeled quite well by the analytical model with recombination in the drift region (see Fig. 7.13).

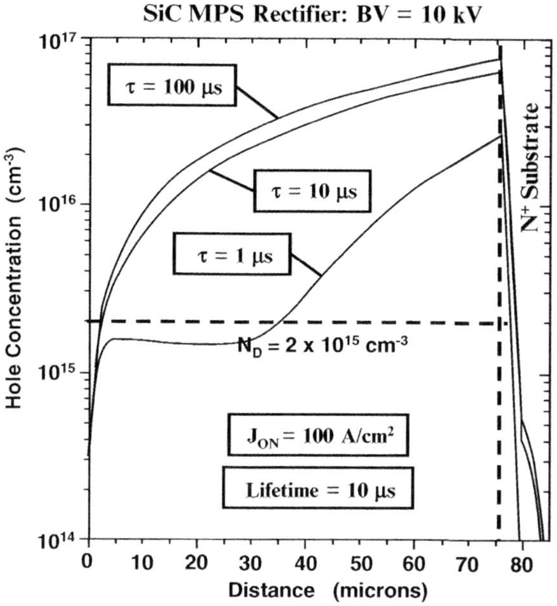


Fig. 7.22E Carrier Distribution within 10 kV SiC MPS Rectifiers.

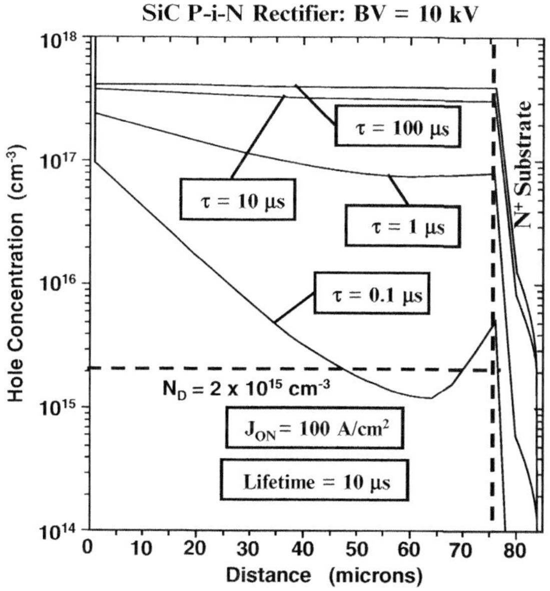


Fig. 7.23E Carrier Distribution within the 10 kV SiC P-i-N Rectifier.

For comparison, it is instructive to examine the hole carrier concentration profile for the silicon carbide P-i-N rectifier with the same range of lifetime values in the drift region. The hole concentration profiles obtained using the numerical simulations are shown in Fig. 7.23E. It can be observed that for lifetime values of 1, 10 and 100 microseconds, there is strong conductivity modulation of the entire drift region with a large amount of stored charge. This degrades the turn-off switching losses. When the lifetime is reduced to 0.1 microseconds, the stored charge is reduced and conductivity modulation does not extend through-out the drift region. This results in the increase in on-state voltage drop for the silicon carbide P-i-N rectifier as previously shown in Fig. 7.19E.

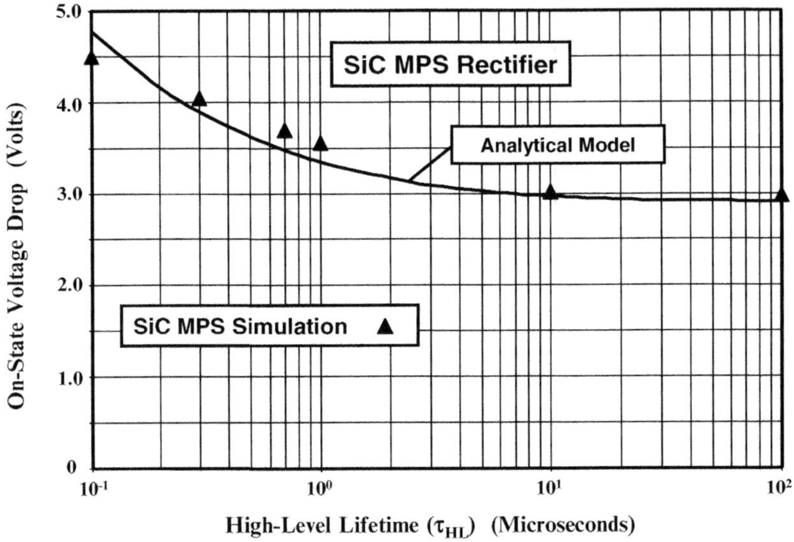


Fig. 7.24E On-State Voltage Drop for 10 kV SiC MPS Rectifiers.

The increase in the on-state voltage drop in the silicon carbide MPS rectifier observed with the numerical simulations when the lifetime is reduced is quite well predicted by the analytical model that takes into account recombination in the drift region. This is demonstrated in Fig. 7.24E where the on-state voltage drop predicted by the analytical model with recombination in the drift region is compared with the values obtained from the numerical solutions. These results provide validity for the analytical model developed in this chapter for the silicon carbide MPS rectifier structure.

The analytical model developed in this section for the MPS rectifier indicates that the carrier distribution profile in the drift region is linear in shape (see Eq. [7.30]) if recombination in the drift region can be neglected. It is worth verifying that this conclusion is also applicable for the silicon carbide structure by using the results of numerical simulations of the baseline silicon carbide MPS rectifier structure. The hole concentration profiles for this structure are shown in Fig. 7.25E for the case of three values for the lifetime using a linear scale for the vertical axis. It can be observed that the profile is close to linear in shape when the lifetime is large (10 and 100 microseconds). Although the hole concentration reduces at the interface between the drift region and the $N^+$ substrate when the lifetime is reduced, the profile can still be approximated by a straight line. A linear carrier profile can therefore be used as a good approximation when computing the stored charge in the drift region, voltage drop across the drift region, and for development of turn-off switching model for the silicon carbide MPS rectifier structure. When the lifetime is reduced to 1 microsecond, the hole concentration profile is no longer linear because high-level injection conditions do not prevail across the entire drift region. As discussed subsequently, such low lifetime values are not necessary for the silicon carbide MPS rectifier because of the good reverse recovery characteristics even when the lifetime is 10 microseconds.

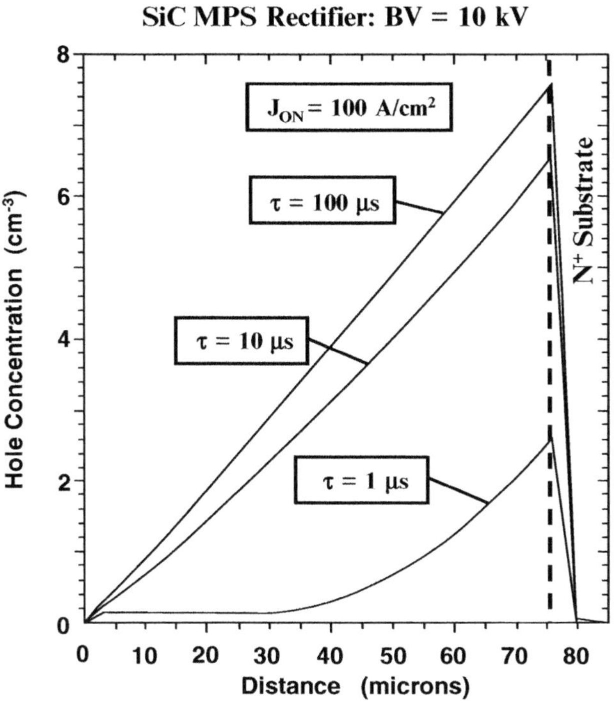


Fig. 7.25E Carrier Distribution within the 10 kV SiC MPS Rectifier.

For the silicon carbide baseline MPS rectifier structure discussed above with an ion-implant window (s) of 1.5 microns, it was found that a large Schottky barrier height is required to ensure bipolar mode of operation in the on-state. This problem can be overcome by reducing the size of the Schottky contact within the 3 micron cell. This can be illustrated by enlarging the ion implant window (s) to 2.5 microns while keeping the same cell pitch (p) of 3 microns. The on-state characteristics for this structure obtained using numerical simulations are shown in Fig. 7.26E for the case of barrier heights ranging from 1.0 to 3.0 eV. Once the device enters the bipolar mode of operation, the on-state voltage drop becomes essentially independent of the barrier height. At an on-state current density of 200 A/cm$^2$, the on-state voltage drop for this structure is 3.05 volts which is less than that for the P-i-N rectifier with the same lifetime in the drift region.

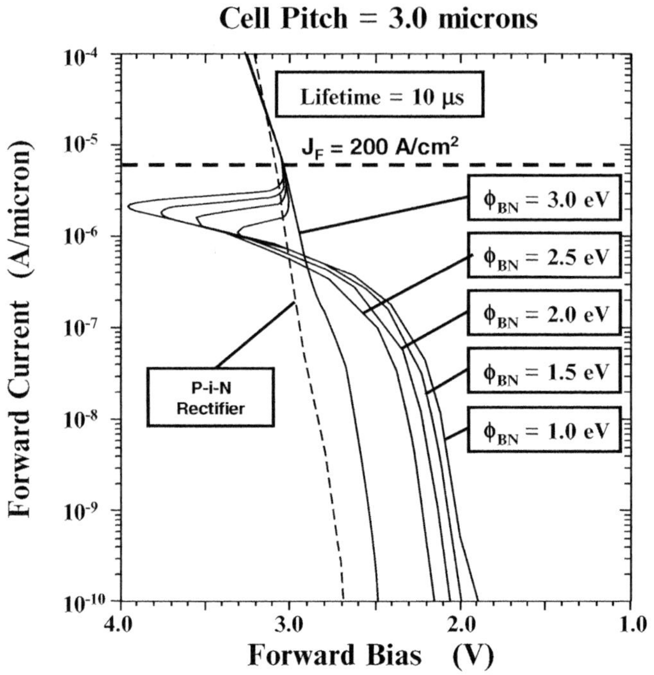


Fig. 7.26E On-state Characteristic of 10 kV SiC MPS Rectifiers.

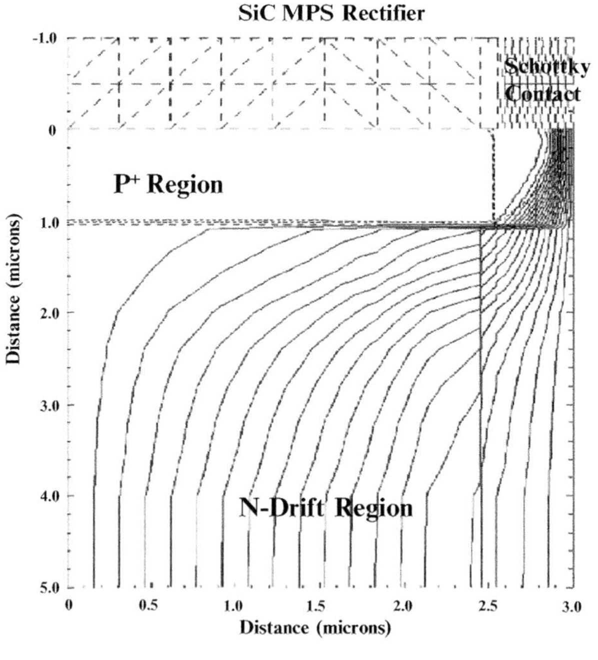


Fig. 7.27E Current Distribution within the 10 kV SiC MPS Rectifier.

The current distribution within the silicon carbide MPS rectifier with cell pitch (p) of 3 microns and ion implant window (s) of 2.5 microns is provided in Fig. 7.27E at an on-state current density of 200 A/cm$^2$. It can be observed that all the current is flowing via the Schottky contact despite its small area within the cell. This confirms that this cell structure is still operating as an MPS structure. A further confirmation of the operating physics can be obtained by examination of the hole concentration profile within the structure. The hole concentration profiles at the Schottky contact (x = 3 microns) and the middle of the P-N junction (x = 0 microns) are shown in Fig. 7.28E at an on-state current density of 200 A/cm$^2$. These profiles resemble those for the previous baseline silicon carbide MPS structure (see Fig. 7.21E) and those predicted by the analytical model for the MPS rectifier (see Fig. 7.13). Note that the hole carrier concentration in Fig. 7.13 was obtained at a current density of 100 A/cm$^2$ while that shown in Fig. 7.28E is obtained at a current density of 200 A/cm$^2$ leading to a larger concentration at the interface between the drift region and the N$^+$ substrate in this case. These results demonstrate that it is feasible to achieve the bipolar mode of operation in the silicon carbide high voltage MPS rectifier structure with Schottky barrier heights that are typical for metals such as nickel and gold.

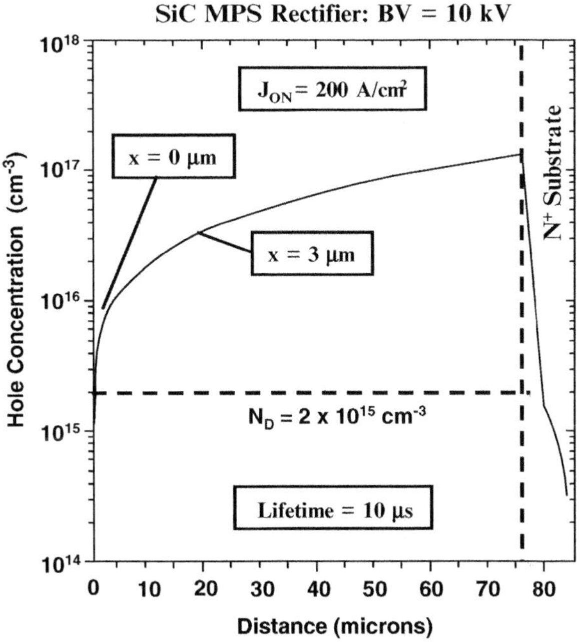


Fig. 7.28E Carrier Distribution within the 10 kV SiC MPS Rectifier.

## 7.3 Reverse Blocking

The reverse leakage current in the MPS rectifier is determined by current transport at the Schottky contact. In the case of the silicon MPS rectifier, the reduced electric field at the Schottky contact due to shielding by the P-N junction suppresses the barrier lowering effect. In the case of the silicon carbide MPS rectifier, the reduced electric field not only decreases the barrier lowering but also mitigates the influence of thermionic field emission. In both cases, the relatively large barrier height favored in the MPS rectifier structure (to obtain good on-state characteristics) results in a low leakage current. The impact of the reduction of the electric field at the Schottky contact in the MPS rectifier structure on the leakage current is similar to that previously discussed for the JBS rectifier. The presence of the P-N junction also diverts the avalanche multiplication current from the Schottky contact reducing the pre-breakdown multiplication effect.

### 7.3.1 Silicon MPS Rectifier: Reverse Leakage Model

In the case of the silicon MPS rectifier, the leakage current model must take into account the smaller Schottky contact area within the cell and the influence of the smaller electric field generated at the Schottky contact due to the shielding from the P-N junction. The leakage current for the silicon MPS rectifier is therefore given by:

$$J_L = \left(\frac{p - s - x_J}{p}\right) AT^2 \exp\left(-\frac{q\phi_b}{kT}\right).\exp\left(\frac{q\beta\Delta\phi_{bMPS}}{kT}\right) \qquad [7.74]$$

where $\beta$ is a constant to account for the smaller barrier lowering closer to the P-N junction as discussed previously for the JBS rectifier. In contrast with the Schottky rectifier, the barrier lowering for the silicon MPS rectifier is determined by the reduced electric field $E_{MPS}$ at the contact:

$$\Delta\phi_{bMPS} = \sqrt{\frac{qE_{MPS}}{4\pi\varepsilon_S}} \qquad [7.75]$$

The electric field at the Schottky contact varies with distance away from the P-N junction. The highest electric field is observed at the middle of the Schottky contact with a progressively smaller value closer to the P-N junction. In an analytical model with a worst case scenario, it is prudent to use the electric field at the middle of the contact to compute the leakage current. Until the depletion regions from the adjacent P-N junctions produce a potential barrier under the Schottky contact, the electric field at the metal-semiconductor interface in the middle of the contact increases with the applied reverse bias voltage as in the case of the Schottky rectifier. A potential barrier is established by the P-N junctions after depletion of the drift region below the Schottky contact. The voltage at which the depletion regions from the adjacent junctions intersect under the Schottky contact is referred to as the *pinch-off voltage*. The pinch-off voltage ($V_P$) can be obtained from the device cell parameters:

$$V_P = \frac{qN_D}{2\varepsilon_S}(p - s - x_J)^2 - V_{bi} \quad [7.76]$$

Although a potential barrier begins to form after the reverse bias exceeds the pinch-off voltage, the electric field continues to rise at the Schottky contact due to encroachment of the potential to the Schottky contact. This problem is more acute for the silicon MPS rectifier than for the silicon carbide structure because of the open shape of the planar junction. In order to analyze the impact of this on the reverse leakage current, the electric field $E_{MPS}$ can be related to the reverse bias voltage by:

$$E_{MPS} = \sqrt{\frac{2qN_D}{\varepsilon_S}(\alpha V_R + V_{bi})} \quad [7.77]$$

where $\alpha$ is a coefficient used to account for the build up in the electric field after pinch-off.

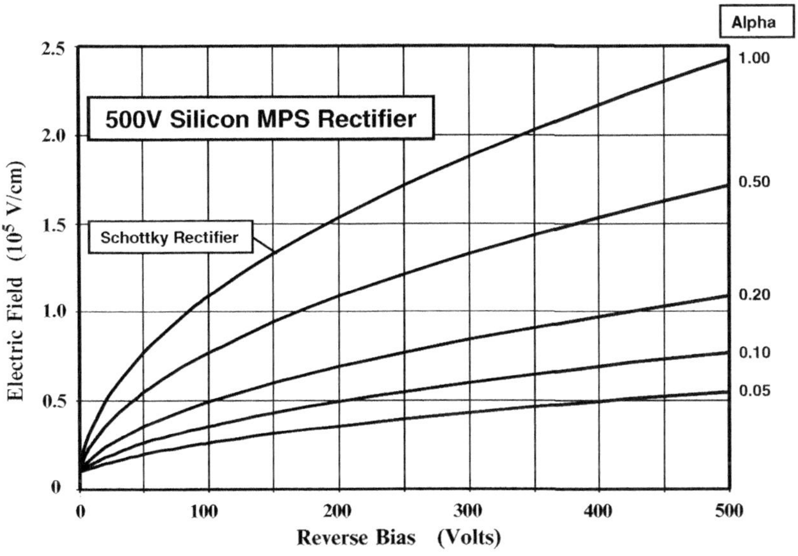


Fig. 7.19 Electric Field at the Schottky Contact for 500 V Silicon MPS Rectifiers.

As an example, consider the case of the 500 volt silicon MPS rectifier discussed earlier in the chapter with a cell pitch (p) of 3.0 microns, a $P^+$ region with dimension 's' of 0.5 microns, and a junction depth of 1 micron. The doping concentration in the drift region for this structure is 3.8 x $10^{14}$ $cm^{-3}$ to obtain a breakdown voltage of 500 volts. Due to the two-dimensional nature of the planar P-N junction in the silicon MPS rectifier structure, it is difficult to derive an analytical expression for alpha. However, the reduction of the electric field at the Schottky contact can be predicted by assuming various values for alpha in Eq. [7.77]. The results are shown in Fig. 7.19 for alpha values ranging between 0.05

and 1.00. An alpha of unity corresponds to the Schottky rectifier structure with no shielding. It can be observed that substantial reduction of the electric field at the Schottky contact is obtained as alpha is reduced. The electric field values for the silicon MPS rectifiers are smaller than those for the silicon JBS rectifier due to the lower doping concentration in the drift region for the high voltage structures.

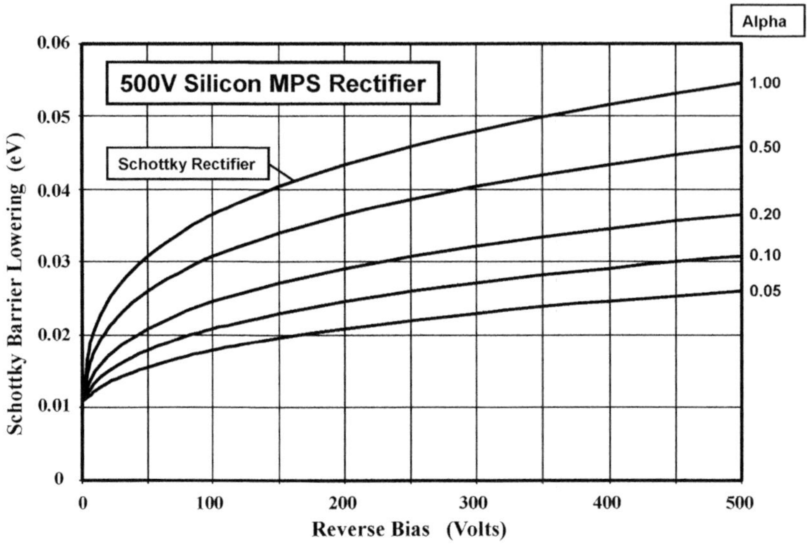


Fig. 7.20 Schottky Barrier Lowering in the 500 V Silicon MPS Rectifiers.

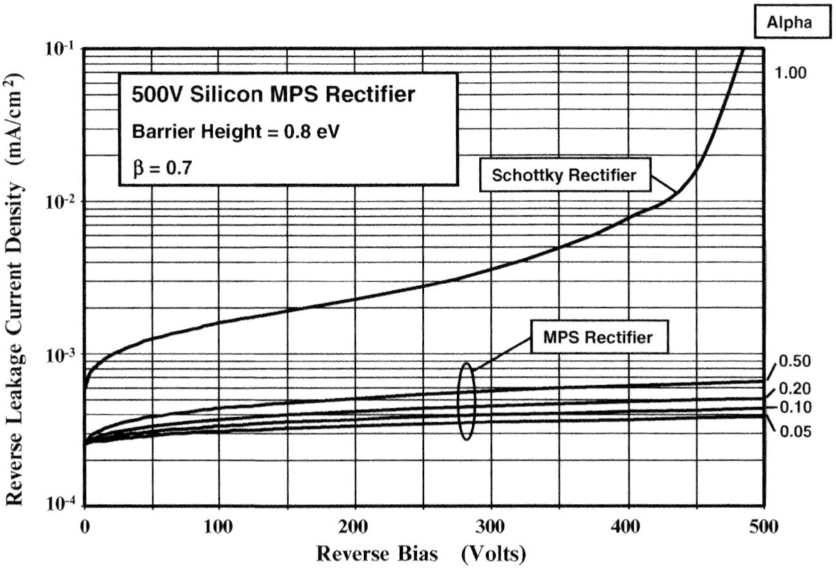


Fig. 7.21 Reverse Leakage Current for 500 V Silicon MPS Rectifiers with various Alpha coefficients.

The impact of the reduction of the electric field at the Schottky contact on the Schottky barrier lowering is shown in Fig. 7.20. Without the shielding by the P-N junction, a barrier lowering of 0.055 eV occurs in the Schottky rectifier. The barrier lowering is reduced to 0.037 eV with an alpha of 0.2 in the MPS rectifier structure. Although this may appear to be a small change, it has a large impact on the reverse leakage current. It is worth pointing out that the barrier lowering in the MPS rectifier is smaller than that observed in the JBS rectifier due to the smaller electric field values at the Schottky contact in the MPS rectifier.

The leakage current density for the 500 V silicon MPS rectifier structure computed by using the above analytical model is shown in Fig. 7.21. For these plots, a value of 0.7 was assumed for the constant $\beta$ based upon the results of the numerical simulations discussed below. For the MPS structure with pitch of 3.0 microns, an implant window (2s) of 1.0 microns, and junction depth of 1.0 micron, the Schottky contact area is reduced to 50 percent of the cell area. This results in a proportionate reduction of leakage current at low reverse bias voltages. The suppression of the Schottky barrier lowering and pre-breakdown multiplication, by the presence of the P-N junction, reduces the rate of increase in leakage current with increasing reverse bias. The net effect is a reduction in leakage current density by a factor of 360x when the reverse bias reaches 500 volts for the case of an alpha of 0.5. This demonstrates that a low reverse leakage current can be achieved with the MPS rectifier structure.

**Simulation Example**

In order to validate the above model for the reverse characteristics of the silicon MPS rectifier, the results of two-dimensional numerical simulations on a 500 V structure are described here. The structure had a drift region with a doping concentration of $3.8 \times 10^{14}$ cm$^{-3}$ and a thickness of 65 microns. The $P^+$ region had a depth of 1 micron with an ion-implant window (dimension s in Fig. 3.5) of 0.5 microns. The work function of the Schottky metal was chosen to obtain a barrier height of 0.9 eV.

A three dimensional view of the electric field distribution in the silicon MPS rectifier structure is shown in Fig. 7.29E. The Schottky contact is located on the lower right-hand-side in the figure with the $P^+$ region located at the top of the figure. A high electric field ($3 \times 10^5$ V/cm) is observed at the P-N junction. However, the electric field at the middle of the Schottky contact is reduced to $2 \times 10^5$ V/cm by the shielding of the P-N junction.

The increase in the electric field at the middle of the Schottky contact in the silicon MPS rectifier structure with cell pitch of 3.0 microns and $P^+$ diffusion window (s) of 0.5 microns is shown in Fig. 7.30E. For comparison, the growth of the electric field at the contact for the 500V Silicon Schottky rectifier is shown in Fig. 7.31E. From these figures, it is apparent that the electric field at the Schottky contact is suppressed in the MPS rectifier due to the incorporation of the P-N junction. An even greater suppression of the electric field at the Schottky contact can be obtained by reducing the width of the Schottky contact as shown in Fig. 7.32E for the MPS rectifier structure with the same cell pitch of 3 microns and a

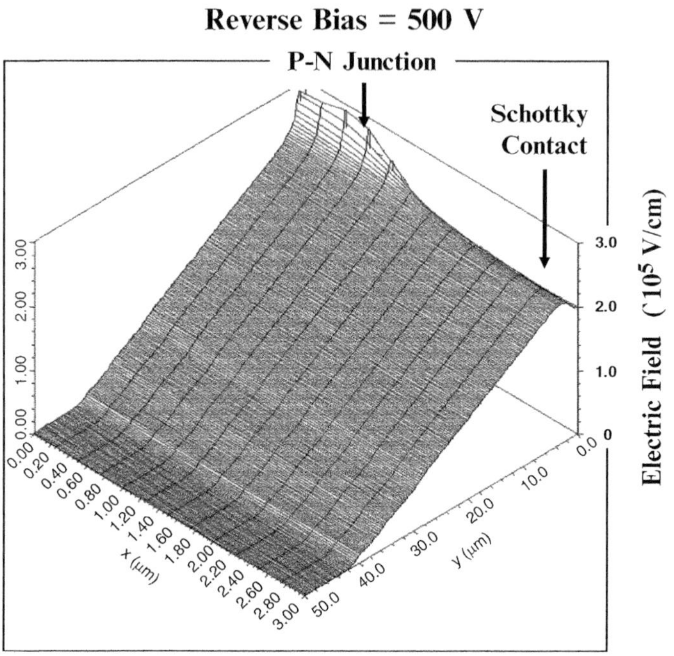


Fig. 7.29E Electric Field Distribution in a 500 V Silicon MPS Rectifier.

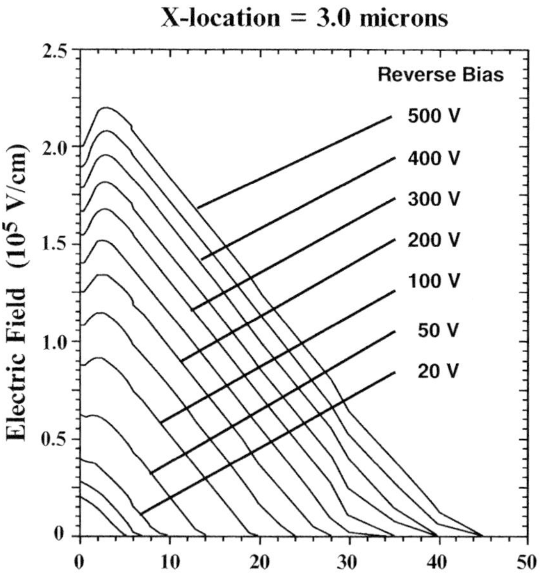


Fig. 7.30E Growth of the Electric Field at the Middle of the Schottky Contact in a 500 V Silicon MPS Rectifier.

wider $P^+$ region with diffusion window "s" of 1.5 microns. The electric field at the middle of the Schottky contact for this case is reduced to 1.4 x $10^5$ V/cm at a reverse bias of 500 volts when compared with 2.0 x $10^5$ V/cm for the MPS structure with diffusion window "s" of 0.5 microns.

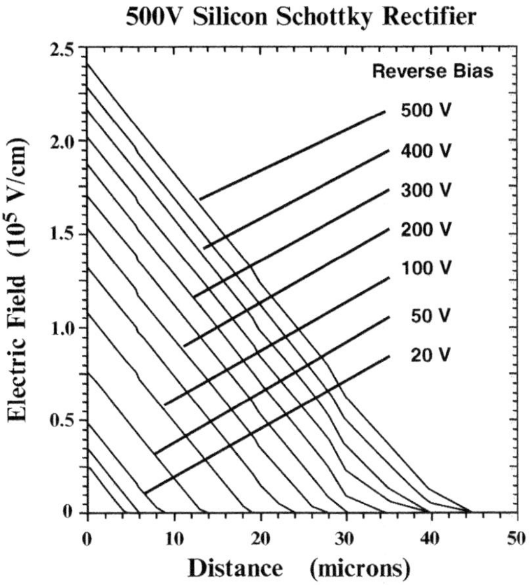


Fig. 7.31E Growth of the Electric Field at the Middle of the Contact in a 500 V Silicon Schottky Rectifier.

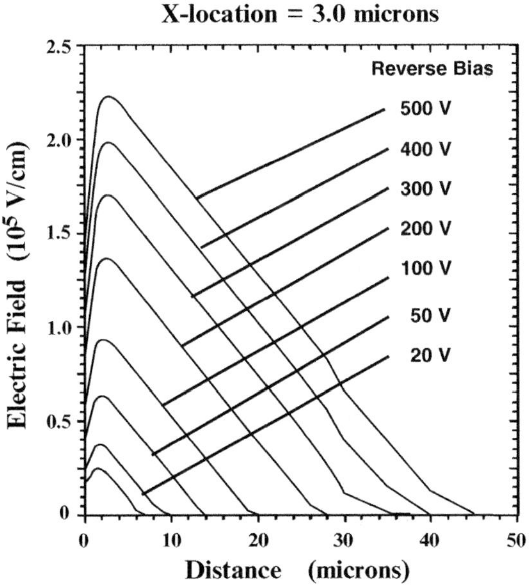


Fig. 7.32E Growth of the Electric Field at the Middle of the Schottky Contact in a 500 V Silicon MPS Rectifier.

The coefficient alpha, that governs the rate at which the electric field increases at the middle of the Schottky contact in the analytical model for the silicon MPS rectifier structure, can be extracted from the results of the two-dimensional numerical simulations. The increase in the electric field at the middle of the Schottky contact, obtained from the numerical simulations, is shown in Fig. 7.33E for the silicon MPS rectifiers with cell pitch (p) of 3.0 microns and diffusion windows (s) of 0.5 and 1.5 microns by the symbols. The results of calculations based upon using the analytical Eq. [7.77] are shown by the solid lines with the values for alpha adjusted to fit the results of the numerical simulations. The case with alpha of unity fits the Schottky rectifier quite well as expected. The value for alpha for the silicon MPS rectifier with diffusion window (s) of 0.5 microns is found to be 0.680 while that for a diffusion window (s) of 1.5 micron is 0.328. With these values of alpha, the analytical model accurately predicts the behavior of the electric field at the middle of the Schottky contact. It can therefore be used to compute the Schottky barrier lowering and leakage current in silicon MPS rectifiers with planar diffused junctions.

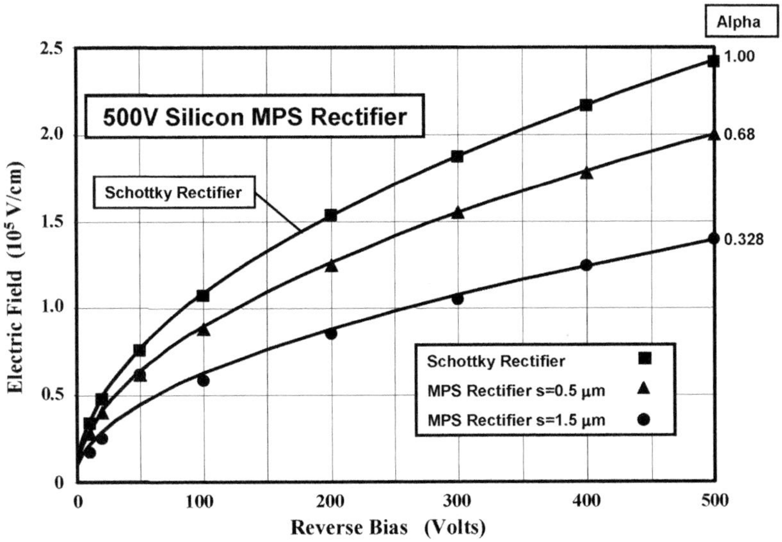


Fig. 7.33E Growth of the Electric Field at the Middle of the Schottky Contact for Silicon 500 V MPS Rectifiers.

### 7.3.2 Silicon Carbide MPS Rectifier: Reverse Leakage Model

The leakage current in the silicon carbide MPS rectifier can be calculated using the same approach as for the silicon MPS rectifier structure. Firstly, it is important to account for the smaller Schottky contact area in the MPS rectifier cell. Secondly, it is necessary to include Schottky barrier lowering while accounting for the smaller electric field at the Schottky contact due to shielding by the P-N junction. Third, the thermionic field emission current must be included while accounting for the smaller electric field at the Schottky contact due to shielding by the P-N junction. After making these adjustments, the leakage current for the silicon carbide MPS rectifier can be calculated by using:

$$J_L = \left(\frac{p-s}{p}\right) AT^2 \exp\left(-\frac{q\phi_b}{kT}\right).\exp\left(\frac{q\Delta\phi_{bMPS}}{kT}\right).\exp\left(C_T E_{MPS}^2\right) \quad [7.78]$$

where $C_T$ is a tunneling coefficient ($8 \times 10^{-13}$ $cm^2/V^2$ for 4H-SiC). In contrast to the Schottky rectifier, the barrier lowering for the MPS rectifier is determined by the reduced electric field $E_{MPS}$ at the contact:

$$\Delta\phi_{bMPS} = \sqrt{\frac{qE_{MPS}}{4\pi\varepsilon_S}} \quad [7.79]$$

As in the case of the silicon MPS structure, the electric field at the Schottky contact varies with distance away from the P-N junction. The highest electric field is observed at the middle of the Schottky contact with a progressively smaller value closer to the P-N junction. When developing an analytical model with a worst case scenario, it is prudent to use the electric field at the middle of the contact to compute the leakage current.

Until the depletion regions from the adjacent P-N junctions produce a potential barrier under the Schottky contact, the electric field at the metal-semiconductor interface in the middle of the contact increases with the applied reverse bias voltage as in the case of the Schottky rectifier. A potential barrier is established by the P-N junctions after depletion of the drift region below the Schottky contact. As in the case of the silicon MPS rectifier structure, the pinch-off voltage ($V_P$) can be obtained from the device cell parameters:

$$V_P = \frac{qN_D}{2\varepsilon_S}(p-s)^2 - V_{bi} \quad [7.80]$$

It is worth pointing out that the built-in potential for 4H-SiC is much larger than for silicon. Although the potential barrier begins to form after the reverse bias exceeds the pinch-off voltage, the electric field continues to rise at the Schottky contact due to encroachment of the potential to the Schottky contact. This problem is less acute for the silicon carbide structure than in the silicon MPS rectifier because of the rectangular shape of the P-N junction resulting from the very low diffusion coefficients for dopants in 4H-SiC. In order to analyze the impact of this on the reverse leakage current, the electric field $E_{MPS}$ can be related to the reverse bias voltage by:

$$E_{MPS} = \sqrt{\frac{2qN_D}{\varepsilon_S}(\alpha V_R + V_{bi})} \quad [7.81]$$

where $\alpha$ is a coefficient used to account for the build up in the electric field after pinch-off.

As an example, consider the case of the 10kV silicon carbide MPS rectifier discussed earlier in the chapter with a cell pitch (p) of 3.0 microns and a $P^+$ region with dimension 's' of 1.5 microns. The pinch-off voltage for this structure is only 1 volt for a drift region with doping concentration of $2 \times 10^{15}$ $cm^{-3}$. Due to the two-dimensional nature of the P-N junction in the MPS rectifier structure, it is difficult to derive an analytical expression for alpha. However, the reduction of the electric field at the Schottky contact can be predicted by assuming various values for alpha in Eq. [7.81]. The results are shown in Fig. 7.22 for alpha values ranging between 0.05 and 1.00. An alpha of unity corresponds to the Schottky rectifier structure with no shielding. It can be observed that substantial reduction of the electric field at the Schottky contact is obtained as alpha is reduced. It is worth pointing out that the electric field at the Schottky contact for the 10 kV silicon carbide MPS rectifier is smaller than for the 3 kV silicon carbide JBS rectifier due to the smaller doping concentration in the drift layer for the higher voltage structure.

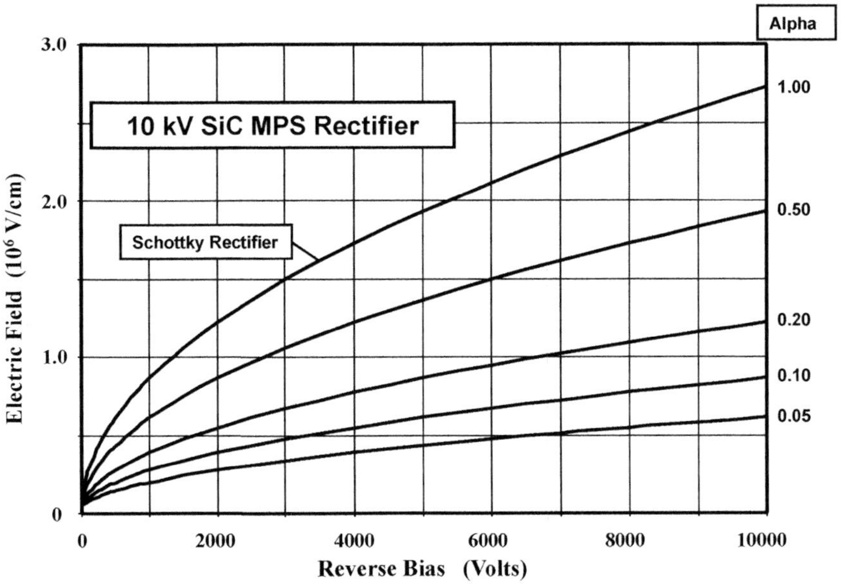


Fig. 7.22 Electric Field at the Schottky Contact for 10 kV Silicon Carbide MPS Rectifiers.

The impact of the reduction of the electric field at the Schottky contact, due to the shielding by the P-N junction in the silicon carbide MPS structure, on the Schottky barrier lowering is shown in Fig. 7.23. Without the shielding by the P-N junction, a barrier lowering of 0.20 eV occurs in the Schottky rectifier. This is much greater than for silicon devices due to the larger electric field at the contact. The barrier lowering is reduced to 0.13 eV with an alpha of 0.2 in the 4H-SiC MPS rectifier structure. These smaller values for alpha are appropriate for the silicon carbide structure because the rectangular shape of the P-N junction favors a stronger shielding of the Schottky contact.

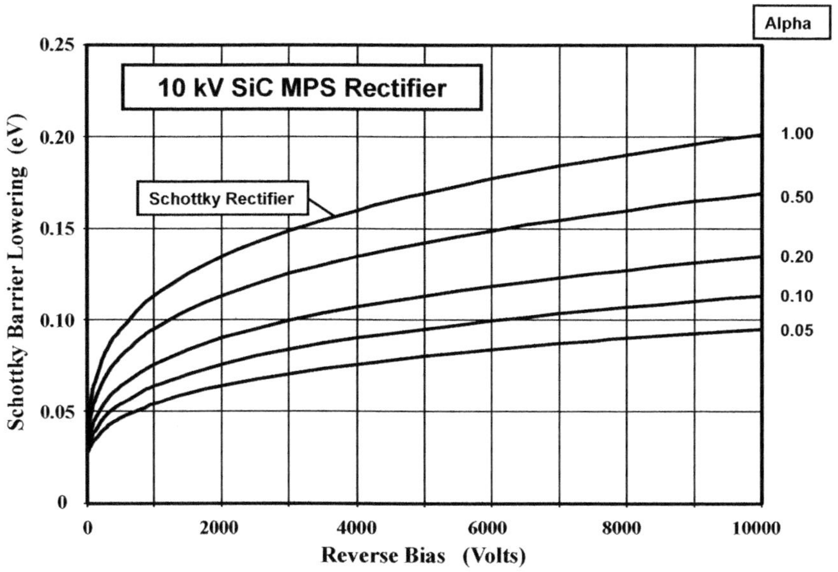


Fig. 7.23 Schottky Barrier Lowering in 10 kV Silicon Carbide MPS Rectifiers with various Alpha Coefficients.

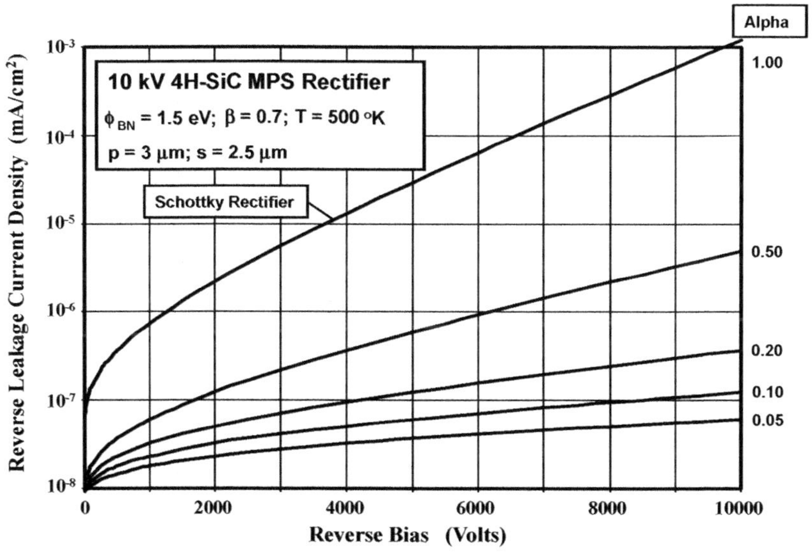


Fig. 7.24 Reverse Leakage Current for 10 kV Silicon Carbide MPS Rectifiers with various Alpha coefficients.

The larger barrier lowering for silicon carbide, in conjunction with the thermionic field emission current, results in an increase in leakage current by four-orders of magnitude when the voltage increases to 10000 volts for the 10 kV Schottky rectifier. This is illustrated in Fig. 7.24, as the plot with alpha of unity by using a barrier height of 1.5 eV at an ambient temperature of 500 °K. The reverse leakage current in the MPS rectifier is greatly reduced by the shielding provided by the P-N junction. This can be observed in Fig. 7.24 for the plots with various values for alpha for the 4H-SiC MPS rectifier structure with pitch of 3.0 microns, an implant window (s) of 2.5 microns, and junction depth of 1 micron. In this case, the Schottky contact area is reduced to only 16.7 percent of the cell area. This results in a proportionate reduction of leakage current at low reverse bias voltages. More importantly, the suppression of the electric field at the Schottky contact, by the presence of the P-N junction, greatly reduces the rate of increase in leakage current with increasing reverse bias. The net effect is a reduction in leakage current density by a factor of 250x when the reverse bias reaches 10 kV for the case of an alpha of 0.5 and a factor of 3,300x for an alpha of 0.2.

### Simulation Example

In order to validate the model for the reverse blocking characteristics for the silicon carbide MPS rectifier structure, a structure with breakdown voltage of 10,000 volts, using a drift region with doping concentration of 2 x $10^{15}$ $cm^{-3}$ and thickness of 80 microns, will be considered here. The two-dimensional numerical simulations were performed with several spacing between the $P^+$ regions by varying the implant window (2s) with a fixed pitch (p) of 3 microns and a depth of 1 micron for the $P^+$ region.

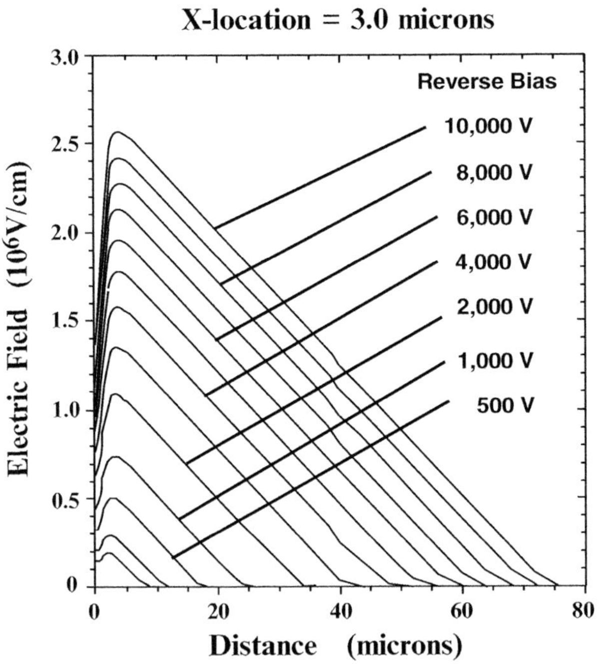


Fig. 7.34E Electric Field distribution in a 10 kV 4H-SiC MPS Rectifier.

The electric field profile at the center of the Schottky contact is shown in Fig. 7.34E for the case of a cell pitch (p) of 3 microns and a $P^+$ ion implant window (s) of 1.5 microns. It can be observed that the electric field at the surface under the contact is significantly reduced when compared the peak electric field in the bulk. At a reverse bias of 10,000 volts, the electric field at the Schottky contact is only $1.4 \times 10^6$ V/cm compared with $2.6 \times 10^6$ V/cm at the electric field maxima which occurs in the bulk at a depth of 2 microns.

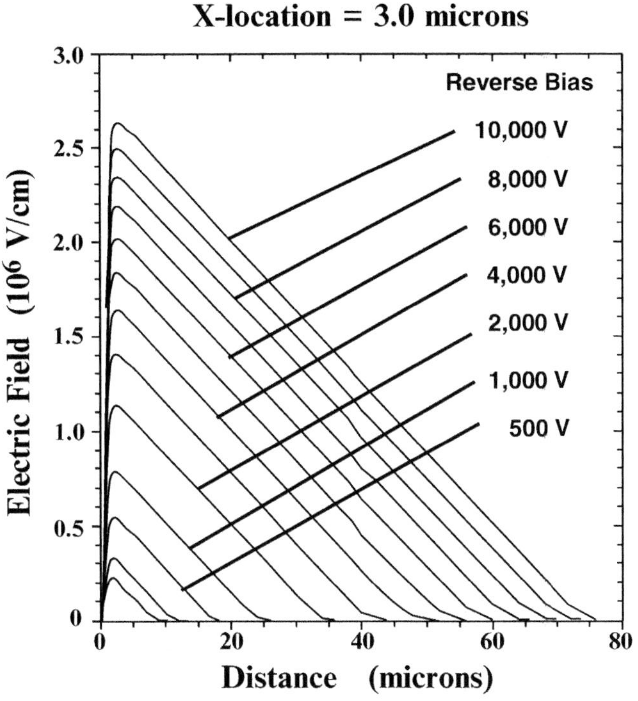


Fig. 7.35E Electric Field variation with Reverse Voltage in a 10 kV 4H-SiC MPS Rectifier.

It was previously demonstrated that the bipolar mode of operation in the silicon carbide MPS rectifier structure can be enhanced by using a small width for the Schottky contact. It was found that this design results in a low on-state voltage drop with reduced snap-back in the on-state characteristics. Fortunately, this design also produces a decrease in the electric field at the Schottky contact which is favorable for reducing the reverse leakage current. In order to illustrate this, the electric field profile at the center of the Schottky contact is shown in Fig. 7.35E for the case of a cell pitch (p) of 3 microns and an ion implant window (s) of 2.5 microns. It can be observed that the electric field at the surface under the contact is drastically reduced when compared the peak electric field in the bulk. At a reverse bias of 10,000 volts, the electric field at the Schottky contact is very small ($1.5 \times 10^5$ V/cm) compared with $2.6 \times 10^6$ V/cm at the electric field maxima which occurs in the bulk at a depth of 2 microns.

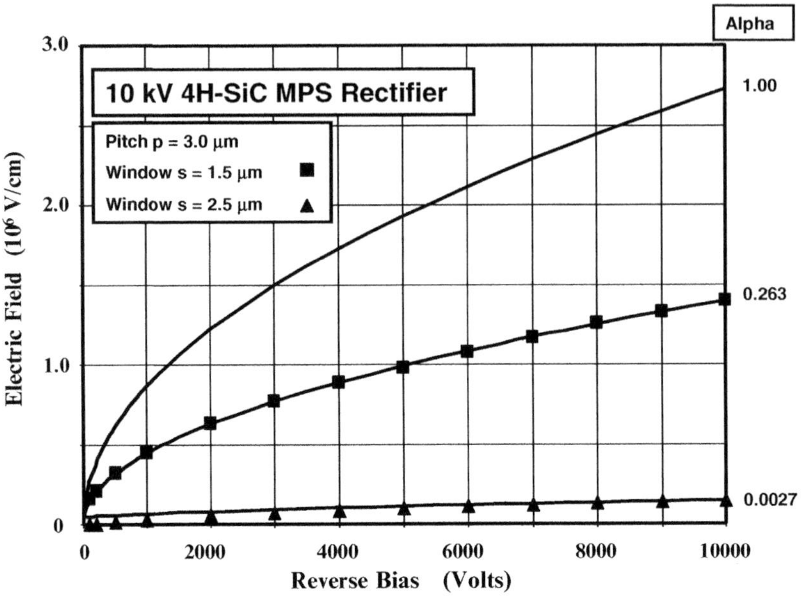


Fig. 7.36E Electric Field variation with Reverse Voltage in 10 kV 4H-SiC MPS Rectifiers.

The increase in the electric field at the Schottky contact with increasing reverse bias voltage, obtained from the numerical simulations, is plotted in Fig. 7.36E for two cases of the ion implant window. These data points are compared with the calculated values obtained using the analytical model as shown by the solid lines. Here, the value for alpha was adjusted to obtain a good match to the simulation data for each cell pitch. The value for alpha is smaller for the silicon carbide MPS rectifier structure when compared with the silicon MPS rectifier structure with the same junction depth and spacing 'd'. This is due to the rectangular shape for the junction in the silicon carbide structure in comparison with a planar cylindrical shape for the junction in the silicon structure. The rectangular shape for the junction produces a stronger potential barrier under the Schottky contact which suppresses the electric field at the contact to a greater degree. This is accounted for in the analytical model by a smaller value for the alpha coefficient in Eq. [7.81].

The smaller electric fields at the contact in the 10 kV 4H-SiC MPS rectifier structures greatly reduce the leakage current because of not only suppressing the Schottky barrier lowering effect but also suppressing the tunneling current. This reduction of the leakage current is shown in Fig. 7.37E for the two cell structures discussed above. With a ion implant window (s) of 1.5 microns, the leakage current is reduced by a factor of 600 at reverse blocking voltages near the breakdown voltage. With a ion implant window (s) of 2.5 microns, the leakage current is reduced by a factor of 80,000 at reverse blocking voltages near the breakdown voltage. Since the cell structure with the larger ion implant window was previously shown to produce excellent on-state characteristics, this value is preferable for the 4H-SiC MPS rectifier with a blocking voltage of 10,000 volts.

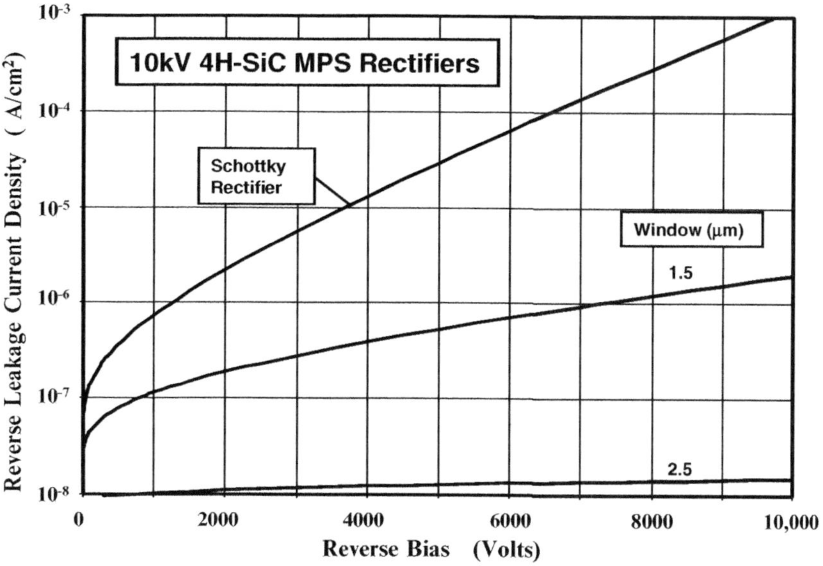


Fig. 7.37E Leakage Current Suppression in the 10 kV 4H-SiC MPS Rectifiers.

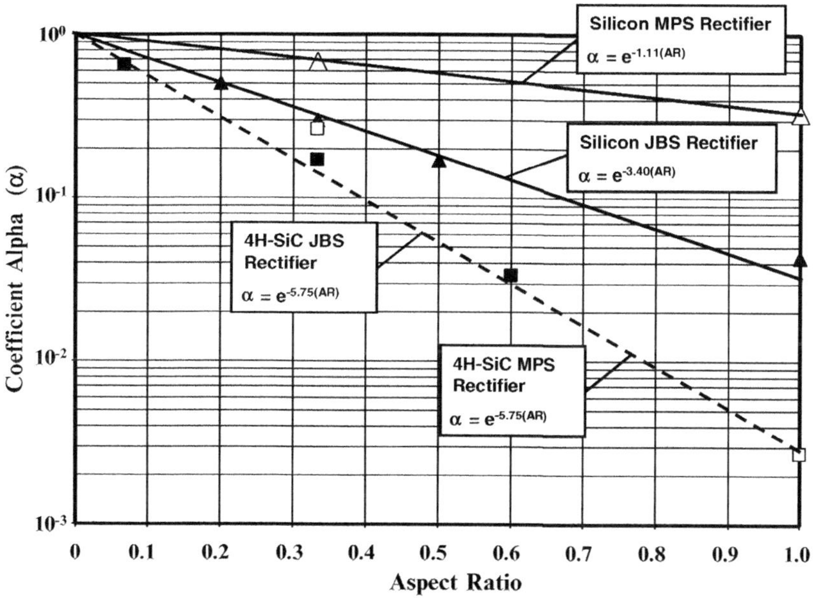


Fig. 7.38E Impact of Aspect Ratio on Alpha for MPS Rectifiers.

For the JBS rectifier, it was demonstrated in chapter 3 that the aspect ratio of the current conduction region located below the Schottky contact has a strong influence on the suppression of the electric field at the contact. This phenomenon is quantified by the coefficient alpha ($\alpha$) in Eq. (3.37) and Eq. (3.41) with the aspect ratio defined by Eq. [3.42]. The variation of the coefficient alpha ($\alpha$) with aspect ratio, obtained from the numerical simulations, is shown in Fig. 7.38E for silicon and silicon carbide MPS rectifiers by the open data points. The values for the silicon and silicon carbide JBS rectifiers (filled data points) obtained in chapter 3 are included in this figure for comparison. It can be observed that the coefficient alpha ($\alpha$) varies exponentially with the aspect ratio. It is smaller for the case of silicon carbide devices when compared with silicon devices with the same aspect ratio because the electric field is not suppressed to the same degree in silicon devices due to the cylindrical shape of the $P^+$ region when compared with the rectangular shape for silicon carbide devices. The variation of the alpha coefficient with aspect ratio for silicon MPS rectifiers is found to be different from that for the silicon JBS rectifiers. However, the variation of the alpha coefficient with aspect ratio for silicon carbide MPS rectifiers is found to be the same as that for the silicon carbide JBS rectifiers.

## 7.4 Switching Performance

Power rectifiers operate for part of the time in the on-state when the bias applied to the anode is positive and for the rest of the time in the blocking state when the bias applied to the anode is negative. During each operating cycle, the diode must be rapidly switched between these states to minimize power losses. Much greater power losses are incurred when the diode switches from the on-state to the reverse blocking state than when it is turned on. The presence of a large concentration of free carriers in the drift region during on-state current flow is responsible for the low on-state voltage drop of high voltage P-i-N and MPS rectifiers. The stored charge within the drift region of the power rectifier produced by the on-state current flow must be removed before it is able to support high voltages. This produces a large reverse current for a short time duration. This phenomenon is referred to as *reverse recovery*.

### 7.4.1 Stored Charge

It is instructive to compare the stored charge in the MPS rectifier structure with that in a P-i-N rectifier because this provides a relative measure of the energy loss that will occur during the turn-off transient. In the case of the P-i-N rectifier, the drift region has almost a uniform (average) carrier concentration given by Eq. [6.23]. Using this equation for the case of the 500 V silicon P-i-N rectifier example in the previous sections, the average carrier concentration is found to be $9 \times 10^{17}$ cm$^{-3}$ at an on-state current density of 100 A/cm$^2$ if the lifetime in the drift region is 10 microseconds. The total stored charge in the drift region based upon a uniform concentration of carriers is then found to be 1.01 mC/cm$^2$ by multiplying the carrier concentration and the thickness of the drift region. However, with a high lifetime in the drift region, the end-region recombination becomes dominant and the carrier concentration in the drift region is greatly reduced. From simulations, it is found that the average carrier concentration in the drift region is about $7 \times 10^{16}$ cm$^{-3}$ leading to a total stored charge of 0.078 mC/cm$^2$.

In contrast, the carrier distribution in the MPS rectifier has a triangular shape with a maximum value at the interface between the drift region and $N^+$ substrate. The maximum carrier concentration in the 500V silicon MPS can be obtained by using either Eq. [7.21] or Eq. [7.31]. For a lifetime of 10 microseconds in the drift region, the maximum carrier concentration is found to be 6 x $10^{16}$ $cm^{-3}$ at an on-state current density of 100 A/$cm^2$. The total stored charge in the drift region of the silicon MPS rectifier is then found to be 0.034 mC/$cm^2$. Consequently, the stored charge in the silicon MPS rectifier is about 2-times smaller than for the P-i-N rectifier in spite of a lower on-state voltage drop. Moreover, the carrier concentration for the MPS rectifier is zero at the upper junction during on-state operation. This allows the device to begin supporting a reverse voltage much faster than in the case of the P-i-N rectifier which shortens the reverse recovery process making the peak reverse recovery current of the MPS rectifier much smaller than that of the P-i-N rectifier. A similar reduction of the stored charge holds true for the silicon carbide MPS rectifier in comparison with the silicon carbide P-i-N rectifiers.

### 7.4.2 Reverse Recovery

As discussed previously in chapter 6, it is common-place to use power rectifiers with an *inductive load* with the current reducing at a constant ramp rate ('a'). A large *peak reverse recovery current* ($J_{PR}$) occurs due to the stored charge followed by the reduction of the current to zero. In the case of the P-i-N rectifier, the device

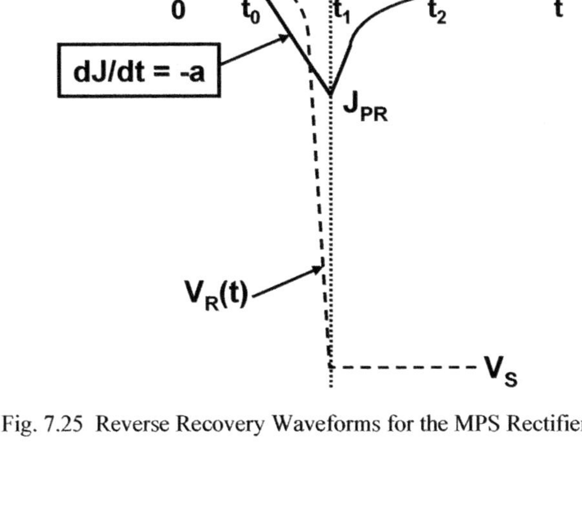


Fig. 7.25 Reverse Recovery Waveforms for the MPS Rectifier.

remains in its forward biased mode with a low on-state voltage drop even after the current reverses direction as illustrated in Fig. 6.8 due to the high minority carrier concentration at the junction in the initial on-state. Until the minority carrier concentration reduces to zero at time $t_1$ (in Fig. 6.8), the junction is unable to support a reverse blocking voltage. The voltage across the diode then rapidly increases to the supply voltage with the rectifier operating in its reverse bias mode. The current flowing through the rectifier in the reverse direction reaches a maximum value ($J_{PR}$) at time $t_2$ (in Fig. 6.8) when the reverse voltage becomes equal to the reverse bias supply voltage ($V_S$).

In the case of the MPS rectifier, the carrier concentration at the Schottky contact is zero in the on-state making the carrier concentration at the junction also close to zero. Consequently, this device is able to support a reverse blocking voltage immediately after the current reverses direction as illustrated in Fig. 7.25. The voltage across the diode then rapidly increases to the supply voltage with the rectifier operating in its reverse bias mode. The current flowing through the rectifier in the reverse direction reaches a maximum value ($J_{PR}$) at time $t_1$ (in Fig. 7.25) when the reverse voltage becomes equal to the reverse bias supply voltage ($V_S$). After this time, the remaining stored charge in the drift region is removed by the diffusion of carriers to the $N^+$ cathode and the space charge region.

An analytical model for the reverse recovery process for the turn-off of the MPS rectifier under a constant rate of change of the current (*current ramp-rate*) can be created by assuming that the initial concentration of the free carriers in the drift region can be linearized as illustrated in Fig. 7.26. As shown in the figure, the initial carrier distribution established by the on-state current flow is linear with the concentration increasing from zero at x = –d to the concentration of $p_M$ at x = +d:

$$p(x) = n(x) = \left(\frac{x+d}{2d}\right) p_M \qquad \textbf{[7.82]}$$

The hole and electron concentrations in the drift region can be assumed to be equal in the on-state and during the turn-off transient because of charge neutrality. In the case of finite lifetime in the drift region, the maximum carrier concentration $p_M$ is given by Eq. [7.21]:

$$p_M = n_M = \frac{J_{ON} L_a}{2qD_n} \tanh\left(\frac{2d}{L_a}\right) \qquad \textbf{[7.83]}$$

where $J_{ON}$ is the on-state current density.

During the first phase of the turn-off process, the current flow reduces from the on-state current density to zero. Since the current flow remains in the forward direction, the Schottky contact and the P-N junction remain in forward bias during this time interval. The hole and electron concentrations remain close to zero at the Schottky contact and the P-N junction to satisfy the boundary conditions at the Schottky contact (as in the case of the on-state analysis). However, electrons are extracted from the cathode side during this time interval. The current flowing at the cathode due to the diffusion of electrons is given by:

$$J = 2qD_n\left(\frac{dn}{dx}\right)_{x=+d} \qquad [7.84]$$

If the carrier distribution during the first phase is linearized as illustrated in Fig. 7.26, the slope of the carrier profile near the cathode is given by:

$$\left(\frac{dn}{dx}\right)_{x=+d} = \left(\frac{dp}{dx}\right)_{x=+d} = \frac{J(t)}{2qD_n} = \frac{J_{ON} - at}{2qD_n} \qquad [7.85]$$

where 'a' is the current ramp rate. At the end of the first phase (time $t_0$ in Fig. 7.25), the current becomes equal to zero leading to a zero slope for the carrier profile as illustrated in Fig. 7.26. As in the case of the turn-off analysis for the P-i-N rectifier, it will be assumed that the carrier profile pivots around a fixed point located at a distance 'b' from the interface between the drift region and the $N^+$ substrate.

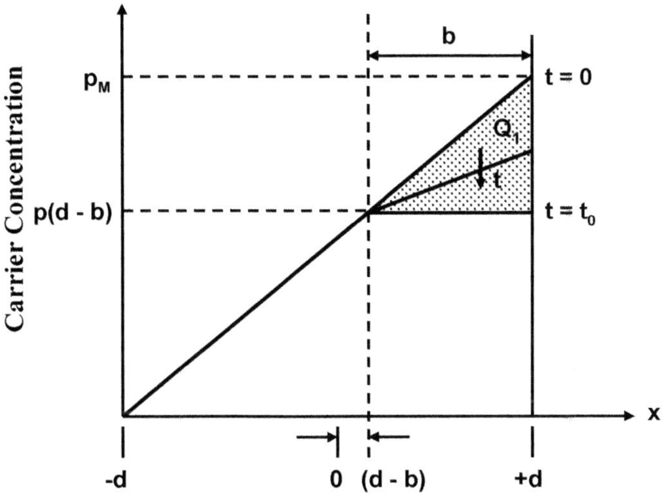


Fig. 7.26 Carrier Distribution Profiles in the MPS Rectifier during the First Phase of the Reverse Recovery Process.

The distance 'b' in Fig. 7.26 can be obtained by relating the charge $Q_1$ removed during the first phase to the current flow. Note that the x-values are defined from the center of the drift region as shown in Fig. 7.26. The hole concentration at x = (d - b) can be assumed to remain the same as during the on-state operation if recombination is neglected during the turn-off transient. This assumption is justified because the turn-off time is much shorter than the lifetime in the drift region. The change in the stored charge within the drift region during the first phase can be obtained from the shaded area, indicated by $Q_1$, in the figure:

$$Q_1 = \frac{qb}{2}\left[p_M - p(d-b)\right] \tag{7.86}$$

Using Eq. [7.82] for the initial carrier profile, the hole concentration at x = (d – b) is given by:

$$p(d-b) = \left(\frac{2d-b}{2d}\right)p_M \tag{7.87}$$

Substituting this into Eq. [7.86] yields:

$$Q_1 = \frac{qp_M}{4d}b^2 \tag{7.88}$$

According to the charge control principle, this charge can be related the current flow during the turn-off transient from t = 0 to t = $t_0$:

$$Q_1 = \int_0^{t_0} J(t)\,dt = \int_0^{t_0} (J_{ON} - at)\,dt = J_{ON}t_0 - \frac{at_0^2}{2} = \frac{J_{ON}^2}{2a} \tag{7.89}$$

because the time $t_0$ at which the current crosses zero is given by:

$$t_0 = \frac{J_{ON}}{a} \tag{7.90}$$

Combining the above relationships:

$$b = \sqrt{\frac{2d}{qap_M}}J_{ON} \tag{7.91}$$

The distance 'b' can therefore be calculated from the on-state current density and the ramp rate 'a'.

Once the current density becomes negative at the start of the second phase of the turn-off process, the MPS rectifier immediately begins to support a reverse voltage as illustrated in Fig. 7.25 with the formation of a space-charge region $W_{SC}(t)$ at the $P^+$/N junction that expands with time as illustrated in Fig. 7.27. The expansion of the space-charge region is achieved by extraction of the stored charge in the drift region in the vicinity of the junction resulting in the reverse current continuing to increase after time $t_0$. It can be assumed that the initial hole distribution in the vicinity of the junction does not change during the second phase of the turn-off process in the conductivity modulated portion of the N-base region because the lifetime in the drift region is much greater than the switching time interval. Consequently, the concentration of holes at the edge of the space-charge region ($p_e$) increases during the turn-off process as the space-charge width increases:

$$p_e(t) = p_M\left[\frac{W_{SC}(t)}{2d}\right] \tag{7.92}$$

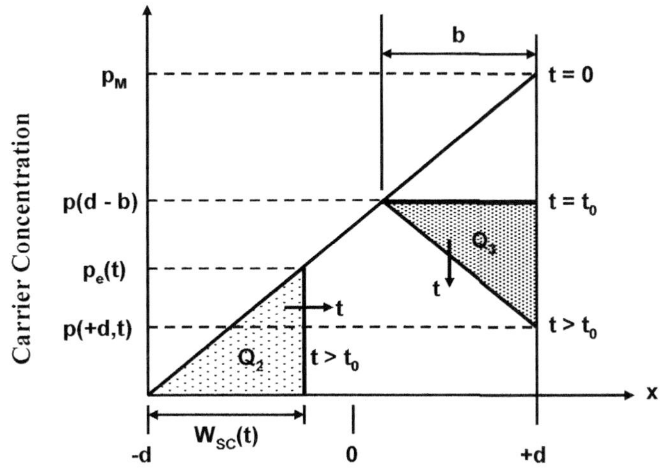


Fig. 7.27 Carrier Distribution Profiles in the MPS Rectifier during the Second

According to the charge-control principle, the charge removed by the expansion of the space-charge layer must equal the charge removed due to collector current flow:

$$J(t) = qp_e(t)\frac{dW_{SC}(t)}{dt} = qp_M\left[\frac{W_{SC}(t)}{2d}\right]\frac{dW_{SC}(t)}{dt} \quad [7.93]$$

by using Eq. [7.92]. The collector current density increases linearly at the ramp rate during the second phase of the turn-off process. Consequently:

$$qp_M\left[\frac{W_{SC}(t)}{2d}\right]\frac{dW_{SC}(t)}{dt} = at \quad [7.94]$$

where 'a' is the ramp rate. Integrating this equation and applying the boundary condition of zero width for the space-charge layer at time zero provides the solution for the evolution of the space-charge region width with time:

$$W_{SC}(t) = \sqrt{\frac{2da}{qp_M}}t \quad [7.95]$$

According to this analysis, the space-charge layer expands towards the right-hand-side at a constant rate as indicated by the horizontal time arrow in Fig. 7.27.

The collector voltage supported by the MPS rectifier structure is related to the space charge layer width by:

$$V_R(t) = \frac{q(N_D + p_{SC})W_{SC}^2(t)}{2\varepsilon_S} = \left(\frac{N_D + p_{SC}}{p_M}\right)\left(\frac{da}{\varepsilon_S}\right)t^2 \quad [7.96]$$

The hole concentration in the space-charge layer can be related to the collector current density under the assumption that the carriers are moving at the saturated drift velocity in the space-charge layer:

$$p_{SC}(t) = \frac{J_R(t)}{qv_{sat,p}} = \frac{at}{qv_{sat,p}} \tag{7.97}$$

In the MPS rectifier, the hole concentration in the space-charge increases during the voltage rise-time because the current density is increasing. The analytical model for turn-off of the MPS rectifier structure predicts an increase in the collector voltage approximately quadratically with time (see Eq. [7.96]).

The end of the second phase of the turn-off process occurs when the collector voltage reaches the reverse bias supply voltage ($V_S$). This time interval ($t_A$ in Fig. 7.25) can be obtained by making the reverse bias voltage equal to the supply voltage in Eq. [7.96]:

$$t_A = \sqrt{\frac{\varepsilon_S p_M V_S}{ad(N_D + p_{SC})}} \tag{7.98}$$

According to the analytical model, the voltage rise-time is proportional to the square root of the reverse bias supply voltage and inversely proportional to square root of the ramp time.

The width of the space-charge layer at the end of the voltage transient can be obtained by using the collector supply voltage:

$$W_{SC}(t_A) = \sqrt{\frac{2\varepsilon_S V_{CS}}{q(N_D + p_{SC})}} \tag{7.99}$$

The width of the space-charge layer at the end of the second phase depends upon the reverse bias supply voltage and the peak reverse recovery current (via $p_{SC}$).

Since the end of the second phase occurs when the reverse bias across the MPS rectifier becomes equal to the supply voltage ($V_S$):

$$t_1 = t_0 + t_A = t_0 + \sqrt{\frac{\varepsilon_S p_M V_S}{ad(N_D + p_{SC})}} \tag{7.100}$$

The hole concentration ($p_{SC}$) in the space charge layer is a function of time because of the increasing reverse current density. However, its magnitude is typically much smaller than the doping concentration ($N_D$) in the drift region. Consequently, the time ($t_1$) at which the peak reverse recovery current occurs can be computed using:

$$t_1 = \frac{J_{ON}}{a} + \sqrt{\frac{\varepsilon_S p_M V_S}{adN_D}} \tag{7.101}$$

This expression indicates that the time for the end of the second phase is reduced with increasing ramp rate and increased with increasing reverse bias supply voltage.

The time taken to reach the peak reverse recovery current after the current crosses zero is defined in Fig. 7.25 as $t_A$. Using this value of time, the peak reverse recovery current can be obtained:

$$J_{PR} = at_A = \sqrt{\frac{a\varepsilon_S p_M V_S}{d(N_D + p_{SC})}} \quad [7.102]$$

Based upon this expression, it can be concluded that the peak reverse recovery current will increase with increasing ramp rate and increasing reverse bias supply voltage.

It is worth pointing out that electrons are also removed from the vicinity of the interface between the drift region and the $N^+$ substrate during the second phase of the turn-off process. This is illustrated in Fig. 7.27 by the shaded area marked $Q_3$. The slope of the carrier profile in the neutral region on the right-hand-side (RHS) is also related to the reverse current density at any time instant because the current is sustained by the diffusion of electrons towards the $N^+$ substrate:

$$\left(\frac{dn}{dx}\right)_{RHS} = \left(\frac{dp}{dx}\right)_{RHS} = \frac{J_R(t)}{2qD_n} = -\frac{at}{2qD_n} \quad [7.103]$$

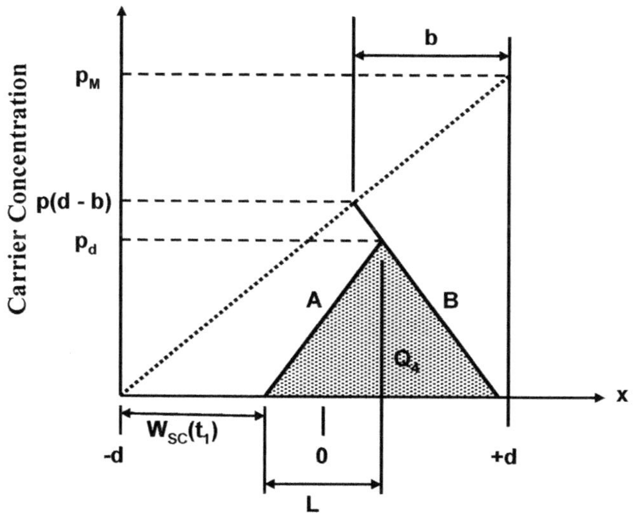


Fig. 7.28 Stored Charge within the MPS Rectifier at the end of the Second Phase.

As discussed above, carriers are extracted from the drift region during the second phase by an expanding space charge layer on the left hand side (LHS) and by diffusion of electrons on the right-hand-side. In addition, holes also diffuse towards the space charge layer from the neutral region. In order to account for the reduction in the carrier density due to this process, a linearized carrier profile for the left-hand-side, as shown in Fig. 7.28 by the line marked A, can be used. The positive slope of this line is related to the peak reverse recovery current flowing at the end of the second phase:

$$\left(\frac{dn}{dx}\right)_{LHS} = \left(\frac{dp}{dx}\right)_{LHS} = \frac{J_{PR}}{2qD_p} \qquad [7.104]$$

Using a carrier concentration of zero for this line at $x = [W_{SC}(t_1) - d]$ with the above positive slope yields an equation for line A:

$$p_{LHS}(x) = \frac{J_{PR}}{2qD_p}\left[x + d - W_{SC}(t_1)\right] \qquad [7.105]$$

Similarly, the carrier profile on the right-hand-side of the drift region at the end of the second phase can be derived by using the negative slope given by Eq. [7.103] at time $t_1$ and the carrier concentration $p(d - b)$:

$$p_{RHS}(x) = p(d-b) - \frac{J_{PR}}{2qD_n}\left[x - (d-b)\right] \qquad [7.106]$$

The intersection point for these lines provides the maximum carrier density in the drift region at the beginning of the third phase:

$$p_d = \frac{D_n p(d-b)}{D_n + D_p} + \frac{J_{PR}}{2q}\left[\left(\frac{d-b}{D_n + D_p}\right) - \frac{D_n\left[d - W_{SC}(t_1)\right]}{D_p(D_n + D_p)} + \frac{d - W_{SC}(t_1)}{D_p}\right] \qquad [7.107]$$

The second term in this expression is much smaller in magnitude than the first term.

During the third phase of the turn-off process, the reverse current reduces at an exponential rate as illustrated in Fig. 7.25 while the voltage supported by the MPS rectifier remains constant at the supply voltage. The stored charge within the drift region after the end of the second phase is illustrated in Fig. 7.28 by the shaded area marked $Q_4$. This charge must be removed during the third phase of the turn-off process by the diffusion of the free carriers towards the space charge boundary on the left-hand-side and towards the $N^+$ substrate on the right-hand-side. As in the case of the non-punch-through P-i-N rectifier structure[1], the current in the MPS rectifier reduces as governed by the diffusion of the excess holes remaining in the drift region towards the edges of the space-charge region in a time frame that is much shorter than the recombination lifetime. Since the voltage is supported across the space-charge region during this process, the electric field can be assumed to be small during the diffusion of the excess carriers in the un-depleted region. The drift component of the current in the continuity equation for the excess carriers can therefore be neglected. By using the same approach as used for the non-punch-through P-i-N rectifier, the reverse recovery current for the MPS rectifier during the third phase can be obtained:

$$J_R(t) = J_{PR}\left(\frac{t_1}{t}\right)^3 e^{\frac{L^2}{4D_p}\left(\frac{1}{t}-\frac{1}{t_1}\right)} \frac{\left[1+\cot\mathrm{anh}\left(L^2/4D_p t\right)\right]}{\left[1+\cot\mathrm{anh}\left(L^2/4D_p t_1\right)\right]} \qquad \textbf{[7.108]}$$

The utility of the analytical model can be illustrated by performing the analysis of the reverse recovery for a specific silicon MPS rectifier structure. Consider the case of a silicon MPS rectifier designed to support 500 volts with a drift region thickness of 70 microns and doping concentration of $3.8 \times 10^{14}$ cm$^{-3}$. The reverse recovery process in this structure is analyzed here using the analytical solutions for various ramp rates and lifetime values. In all cases, the reverse recovery is assumed to begin with on-state operation at a current density of 100 A/cm$^2$.

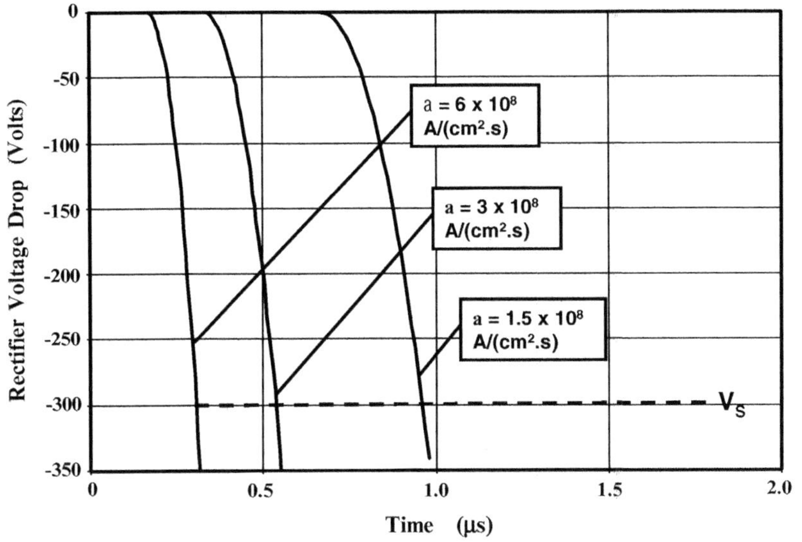


Fig. 7.29 Analytically Calculated Reverse Recovery Voltage Waveforms for the Silicon 500 V MPS Rectifier Structure.

The calculated reverse recovery current waveforms for various cases of the ramp rate are shown in Fig. 7.29. It can be seen that the start of the increase in the reverse voltage is delayed with reduced ramp rate because the current crosses zero at a later time. The time $t_0$ at which current crosses zero and the junction becomes reverse biased increases from 0.167 microseconds to 0.333 microseconds to 0.667 microseconds as the ramp rate decreases from $6 \times 10^8$ A/cm$^2$-s to $3 \times 10^8$ A/cm$^2$-s to $1.5 \times 10^8$ A/cm$^2$-s. The reverse voltage supported by the MPS rectifier then increases quadratically with time. The time at which the voltage reaches a reverse supply voltage of 300 volts is indicated in the figure by the dashed line. The time $t_1$ at which the reverse bias voltage reaches a supply voltage of 300 volts increases from 0.319 microseconds to 0.549 microseconds to 0.972 microseconds as the ramp rate decreases from 6 x $10^8$ A/cm$^2$-s to 3 x $10^8$ A/cm$^2$-s to 1.5 x $10^8$ A/cm$^2$-s. This point in the voltage waveforms defines the end of the second phase of the reverse recovery process.

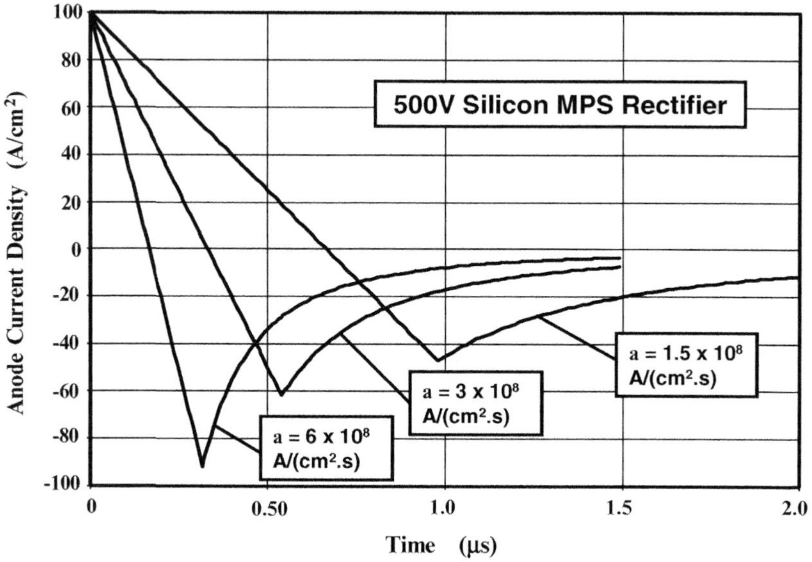


Fig. 7.30 Analytically Calculated Current Waveforms for the 500V Silicon MPS Rectifier during the Reverse Recovery Process using various Ramp Rates.

The current waveforms calculated for the 500 V silicon MPS rectifier using the analytical solutions are shown in Fig. 7.30 for the case of the three ramp rates. The peak reverse recovery current occurs at the end of the second phase. The peak reverse recovery current densities predicted by the analytical model are 92, 62, and 47 A/cm$^2$, respectively, for the three cases of the ramp rate. After the second phase, the reverse current decays to zero at an exponential rate.

The analytical model can also be utilized to examine the influence of the minority carrier lifetime on the reverse recovery process. Consider the case of the same 500-V MPS rectifier structure switched off from an on-state current density of 100 A/cm$^2$ at a ramp rate of 3 x $10^8$ A/cm$^2$s. The voltage waveforms predicted by the analytical model for the reverse recovery process with lifetime values of 2 and 20 microseconds in the drift region are shown in Fig. 7.31. The model predicts a slightly faster rate of increase in the anode voltage during the second phase when the lifetime is reduced. The current flow during the reverse recovery process is shown in Fig. 7.32 for the case of the two lifetime values. The peak reverse recovery current density predicted by the analytical model reduces from 62 to 56 A/cm$^2$ when the lifetime is reduced from 10 to 1 microseconds. A relatively small reduction of the peak reverse recovery current is predicted by model.

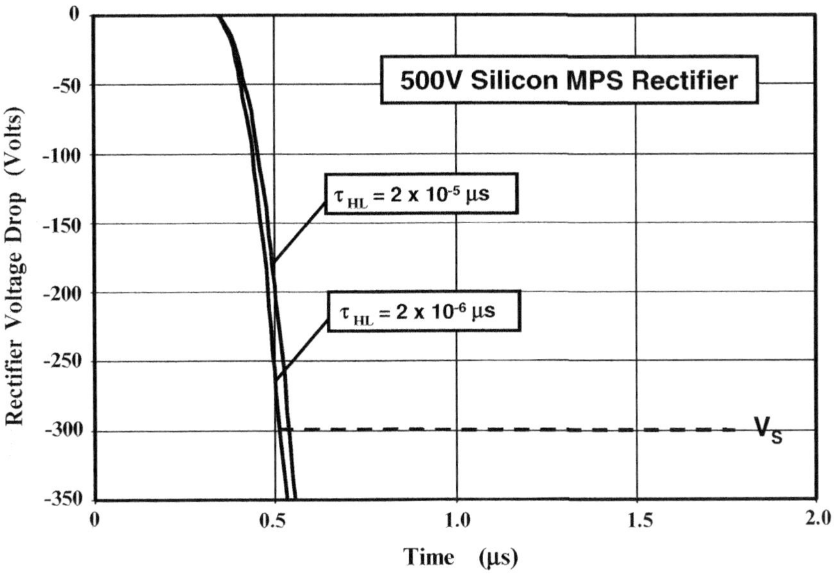


Fig. 7.31 Analytically Calculated Voltage Waveforms for the 500V Silicon MPS Rectifier during the Reverse Recovery Process for various Lifetime Values.

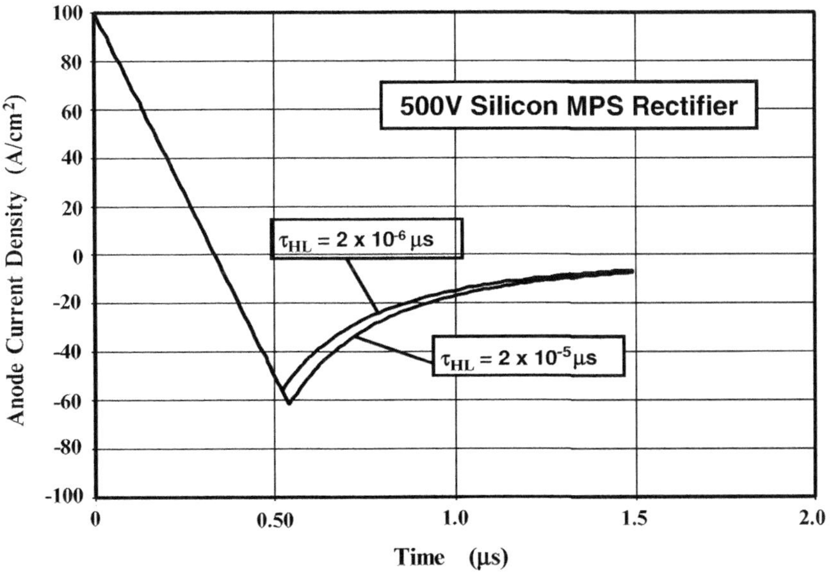


Fig. 7.32 Analytically Calculated Current Waveforms for the 500V Silicon MPS Rectifier during the Reverse Recovery Process for various Lifetime Values.

**Simulation Example**

In order to validate the above model for the reverse recovery transient in the MPS rectifier, the results of numerical simulations on a 500 V silicon MPS rectifier structure are described here. The structure had a drift region thickness of 65 microns with a doping concentration of 3.8 x $10^{14}$ $cm^{-3}$. The cathode current was ramped down from an on-state current density of 100 A/cm2 using various values of negative ramp rates. In addition, the impact of changing the lifetime was examined for comparison with the analytical model.

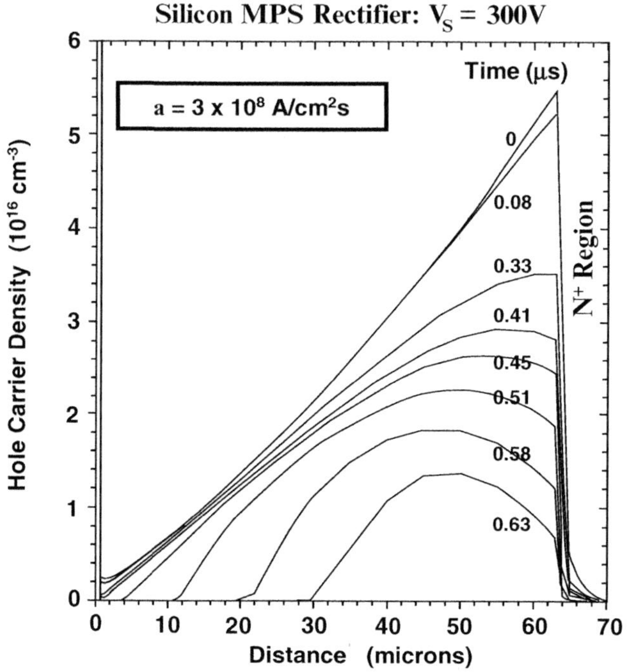


Fig. 7.39E Carrier Distribution in a 500V Silicon MPS Rectifier during Phase 1 and 2 of the Reverse Recovery Transient.

First consider the case of a negative ramp rate of 3 x $10^8$ A/cm$^2$s with a lifetime ($\tau_{p0}$ and $\tau_{n0}$) value of 10 microsecond in the drift region. The maximum carrier concentration in the drift region (at the interface between the drift region and the $N^+$ substrate) under steady state conditions with an on-state current density of 100 A/cm$^2$is found to be 5.5 x $10^{16}$ $cm^{-3}$ as shown in Fig. 7.39E. The carrier concentration profile exhibits a zero slope at time t = 0.33 μs, at the interface between the drift region and the $N^+$ substrate, corresponding to time t = $t_0$ in the analytical model (see Fig. 7.30). The carrier concentration at the junction remains close to zero throughout this time interval. After this time, a space charge layer begins to form at the junction and expands towards the right hand side. The silicon MPS structure was found to support a reverse bias of 300 volts when the time reached 0.63 μs. At this time, the space charge layer width is 30 microns. The value predicted by the analytical model (using Eq. [7.95]) is also 30 microns in excellent agreement with the simulations. It can also be observed from Fig. 7.39E that carriers are being extracted from the right-hand-side by diffusion during the voltage rise-time. This process reduces the peak concentration of the remaining holes in the drift region well below the steady-state hole concentration. This was accounted for when developing the analytical model (see Fig. 7.28).

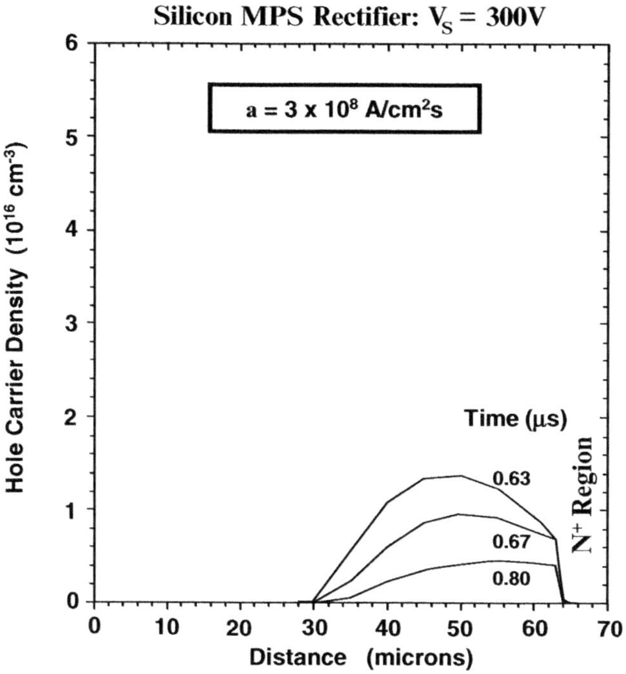


Fig. 7.40E Carrier Distribution in a 500V Silicon MPS Rectifier during Phase 3 of the Reverse Recovery Transient.

The carrier distribution during the third phase of the reverse recovery process obtained from the numerical simulations is shown in Fig. 7.40E. The carrier distribution has a Gaussian shape which decreases with time due to removal of the carriers by diffusion into the space charge region on the left-hand-side and the $N^+$ substrate on the right-hand-side. This validates the assumptions used when developing the analytical model for the decay of the reverse recovery current during the third phase.

The voltage and current waveforms obtained with the aid of the numerical simulations of the reverse recovery for the silicon MPS rectifier are shown in Fig. 7.41E and 7.42E, respectively. The waveforms for three cases of the ramp rate have the same features predicted by the analytical model (see Fig. 7.29 and Fig. 7.30). In the simulations, the voltage increases at a quadratic rate as expected from the analytical model and the peak reverse currents obtained using the analytical model are in good agreement with those observed in the simulations. In addition, the reverse recovery current decays exponentially after reaching the peak value as predicted by the analytical model based upon the diffusion of the carriers and not at the rate of recombination for a lifetime of 10 microseconds.

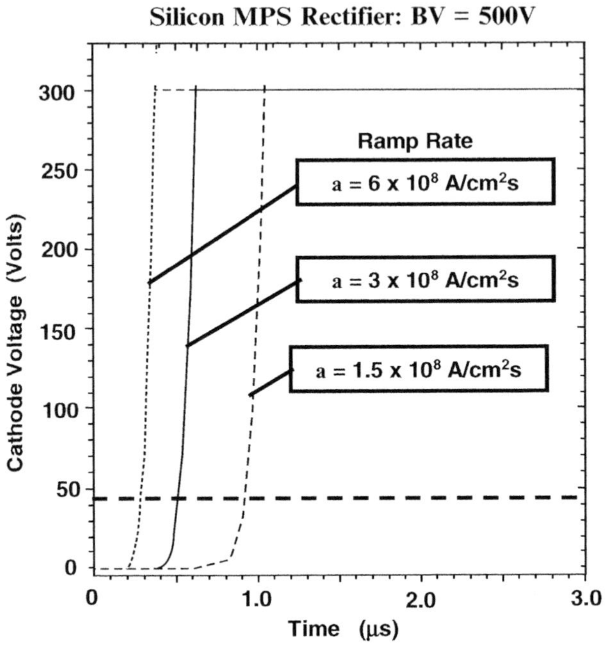


Fig. 7.41E Voltage Waveforms for a 500V Silicon MPS Rectifier during the Reverse Recovery Transient with various Ramp Rates.

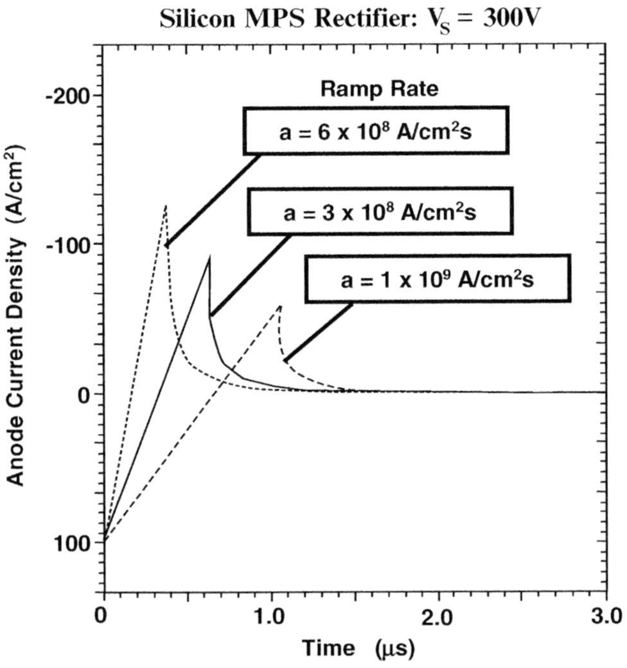


Fig. 7.42E Current Waveforms for a 500V Silicon MPS Rectifier during the Reverse Recovery Transient with various Ramp Rates.

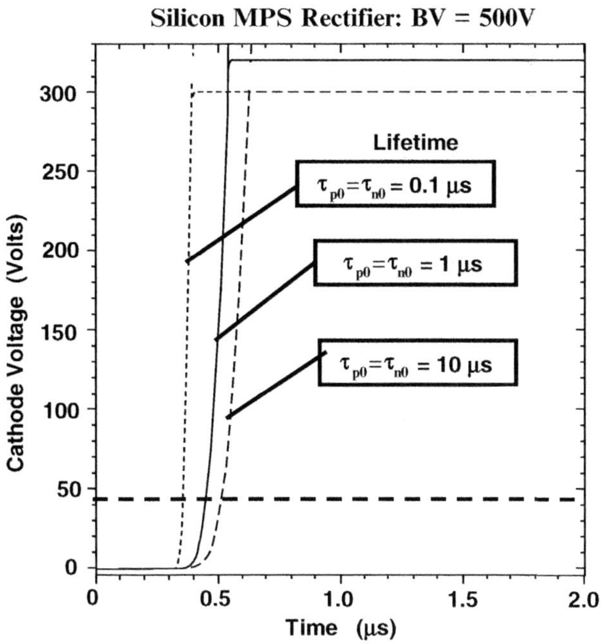


Fig. 7.43E Voltage Waveforms for a 500V Silicon MPS Rectifier during the Reverse Recovery Transient with various Lifetime Values.

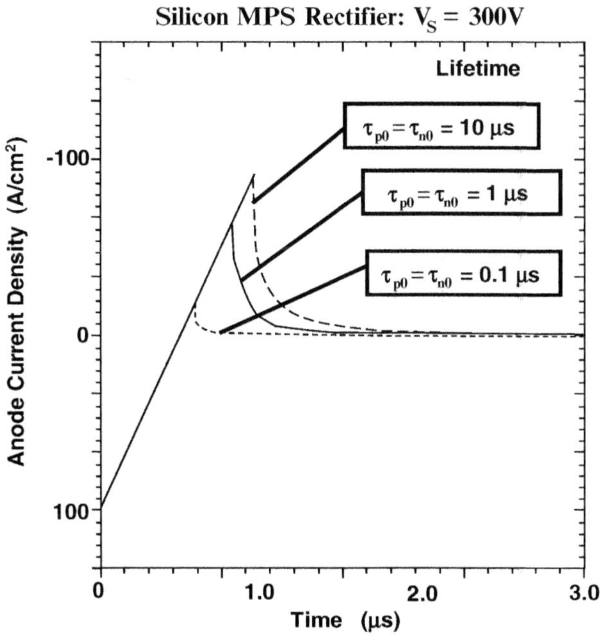


Fig. 7.44E Current Waveforms for a 500V Silicon MPS Rectifier during the Reverse Recovery Transient with various Lifetime Values.

The validity of the analytical model can also be examined by observation of the impact of changes in the lifetime on the reverse recovery process. In order to illustrate this, the lifetime was reduced from a baseline value of 10 microseconds while performing the reverse recovery at ramp rate of 3 x $10^8$ A/cm$^2$s. The diode voltage and current waveforms obtained using the numerical simulations are shown in Fig. 7.43E and 7.44E, respectively. The waveforms shown in the above figures have the same features as the waveforms obtained using the analytical model (see Fig. 7.31 and 7.32). The peak reverse recovery current decreases with a reduction of the lifetime as predicted by the model. However, the reduction in the peak reverse recovery current observed with the simulations is greater than calculated using the analytical model. This is due to neglecting the recombination during the reverse recovery in the analytical model. This assumption is adequate when the lifetime is 10 microseconds in the drift region. However, when the lifetime is reduced to 1 microsecond, it becomes comparable to the time duration of the transient. Furthermore, when the lifetime is reduced to 0.1 microseconds, the entire drift region does not operate under high level injection conditions making the analytical model inappropriate for this case.

### 7.4.3 Silicon Carbide MPS Rectifier: Reverse Recovery

The analytical model described in the previous section for the reverse recovery process in MPS rectifiers can also be applied to silicon carbide devices. Since silicon carbide JBS rectifiers offer excellent on-state characteristics with negligible stored charge, these structures are preferable for rectifiers with reverse blocking voltages of up to 5000 volts. It is appropriate to consider the bipolar mode of operation in the MPS rectifier only for silicon carbide devices with blocking voltages well above 5000 volts. However, as discussed in section 7.2, due to the large built-in potential for the P-N junction in silicon carbide when compared with silicon, it is necessary to use Schottky contacts with large barrier heights and sufficiently small area to ensure injection from the junction to provide the desired conductivity modulation of the drift region.

Consider the case of a silicon carbide MPS rectifier designed to support 10,000 volts with a drift region thickness of 80 microns and doping concentration of 2 x $10^{15}$ cm$^{-3}$ whose on-state characteristics were discussed in section 7.2. The reverse recovery process for this structure is analyzed here for various ramp rates and lifetime values in the drift region. It is first worth comparing the stored charge in this silicon carbide MPS rectifier with that in the P-i-N rectifier under the same on-state current density. The stored charge in the 10 kV silicon carbide MPS rectifier can be computed from the carrier profile shown in Fig. 7.13 at a current density of 100 A/cm$^2$. Although not obvious from the figure, the hole concentration is approximately linear in shape with a maximum concentration of 6.2 x $10^{16}$ cm$^{-3}$ at the interface between the drift region and the N$^+$ substrate. The stored charge computed using this triangular distribution is 39.7 microCoulombs per cm$^2$. In comparison, the average stored charge in the 10 kV silicon carbide P-i-N rectifier computed by using Eq. [6.23] is 7.8 x $10^{17}$ cm$^{-3}$. Numerical simulations of the 10 kV silicon carbide P-i-N rectifier indicate that the average hole concentration in the drift region is reduced to 3.5 x $10^{17}$ $cm^{-3}$ due to the presence of end region recombination (see Fig. 7.23E). Using this value for the average carrier concentration, the stored charge in the 10kV silicon carbide P-i-N rectifier is found to be 448 microCoulombs per $cm^2$. The stored charge in the 10kV silicon carbide P-i-N rectifier is therefore more than 10-times larger than in the silicon carbide MPS rectifier even though their on-state voltage drops are approximately equal.

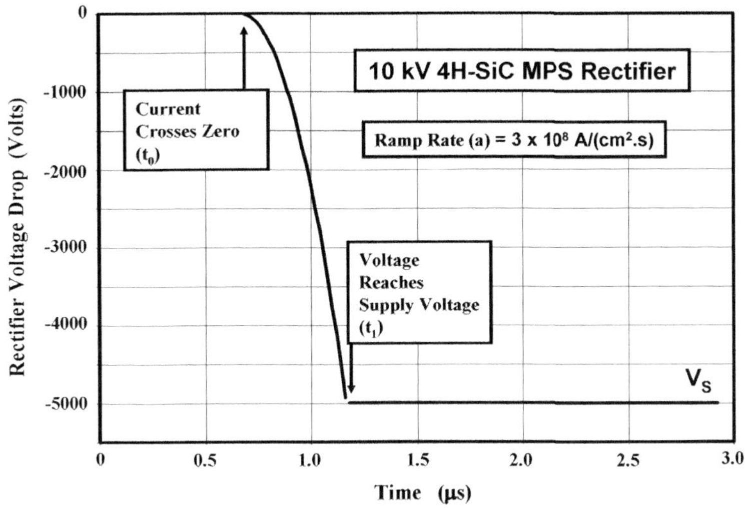


Fig. 7.33 Analytically Calculated Reverse Recovery Voltage Waveform for the Silicon Carbide 10 kV MPS Rectifier Structure.

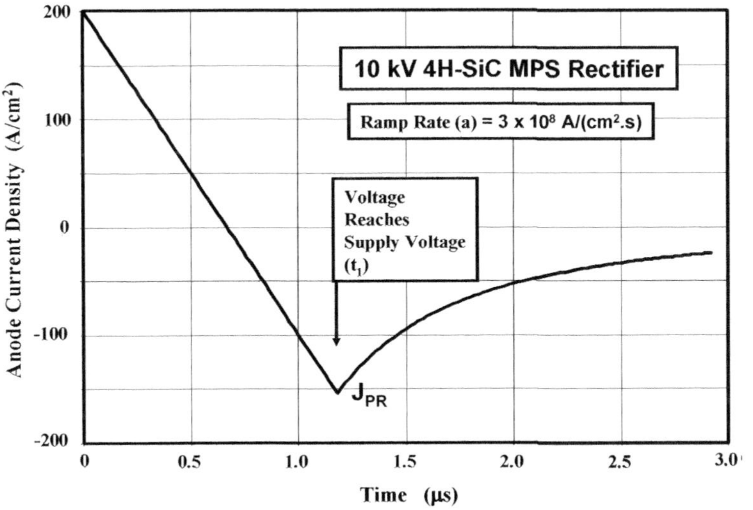


Fig. 7.34 Analytically Calculated Reverse Recovery Current Waveform for the Silicon Carbide 10 kV MPS Rectifier Structure.

The calculated reverse recovery voltage and current waveforms for the silicon carbide 10 kV MPS rectifier structure are shown in Fig. 7.33 and Fig. 7.34 for the case of a ramp rate of 3 x $10^8$ A/cm$^2$-s. An on-state current density of 200 A/cm$^2$ was used in this example. The time $t_0$ at which current crosses zero is 0.667 microseconds. The reverse voltage supported by the silicon carbide MPS rectifier then increases quadratically with time. The time at which the voltage reaches a reverse supply voltage of 5000 volts is indicated in the figure. The time $t_1$ at which the reverse bias voltage reaches a supply voltage of 5000 volts is 1.18 microseconds for the ramp rate of 3 x $10^8$ A/cm$^2$-s. This point in the voltage waveforms defines the end of the second phase of the reverse recovery process.

The peak reverse recovery current occurs at the end of the second phase. The peak reverse recovery current density predicted by the analytical model is 154 A/cm$^2$ for ramp rate of 3 x $10^8$ A/cm$^2$-s. After the second phase, the reverse current decays to zero at an exponential rate that is much faster than the recombination rate dictated by a lifetime of 10 microseconds in the drift region because the model is based upon extraction of the stored charge by diffusion.

## Simulation Example

In order to validate the above model for the reverse recovery transient in the MPS rectifier, the results of numerical simulations on a 10 kV 4H-SiC MPS rectifier structure are described here. The structure had a drift region thickness of 80 microns with a doping concentration of 2.0 x $10^{15}$ cm$^{-3}$. The cathode current was ramped down from an on-state current density of 200 A/cm$^2$ using a ramp rate of 3 x $10^8$ A/cm$^2$-s with a lifetime ($\tau_{p0}$ and $\tau_{n0}$) value of 10 microsecond in the drift region. The maximum carrier concentration in the drift region (at the interface between the drift region and the N$^+$ substrate) under steady state conditions with an on-state current density of 200 A/cm$^2$is found to be 1.33 x $10^{17}$ cm$^{-3}$ as shown in Fig. 7.45E. This is in excellent agreement with the value computed using Eq. [7.21] providing validation for the on-state model. The carrier concentration profile exhibits a zero slope at time t = 0.74 $\mu$s, at the interface between the drift region and the N$^+$ substrate, approximately corresponding to time t = $t_0$ in the analytical model (see Fig.7.34). The carrier concentration at the junction remains close to zero throughout this time interval. After this time, a space charge layer begins to form at the junction and expands towards the right hand side. The 10kV 4H-SiC MPS rectifier structure was found to support a reverse bias of 5000 volts when the time reached 1.26 $\mu$s. At this time, the space charge layer width is 52 microns. The value predicted by the analytical model (using Eq. [7.95]) is 52.4 microns in excellent agreement with the simulations. It can also be observed from Fig. 7.45E that carriers are being extracted from the right-hand-side by diffusion during the voltage rise-time. This process reduces the peak concentration of the remaining holes in the drift region well below the steady-state hole concentration. This was accounted for when developing the analytical model (see Fig. 7.28).

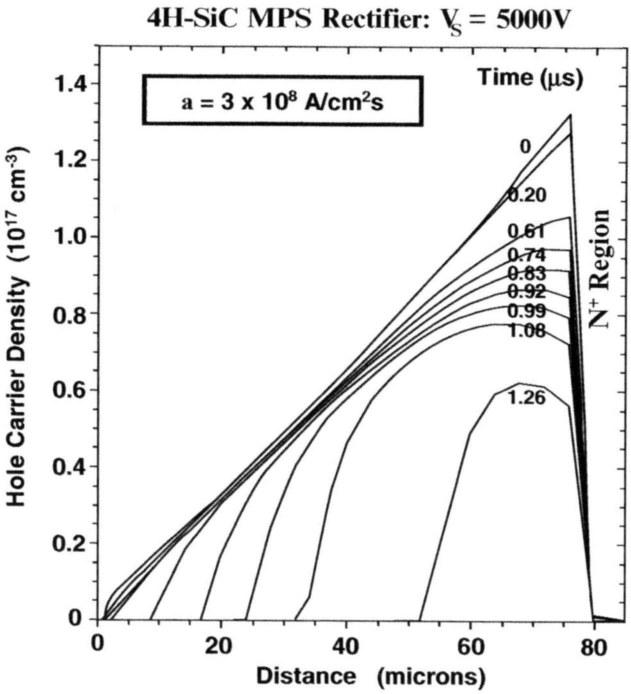


Fig. 7.45E Carrier Distribution in a 10 kV 4H-SiC MPS Rectifier during Phase 1 and 2 of the Reverse Recovery Transient.

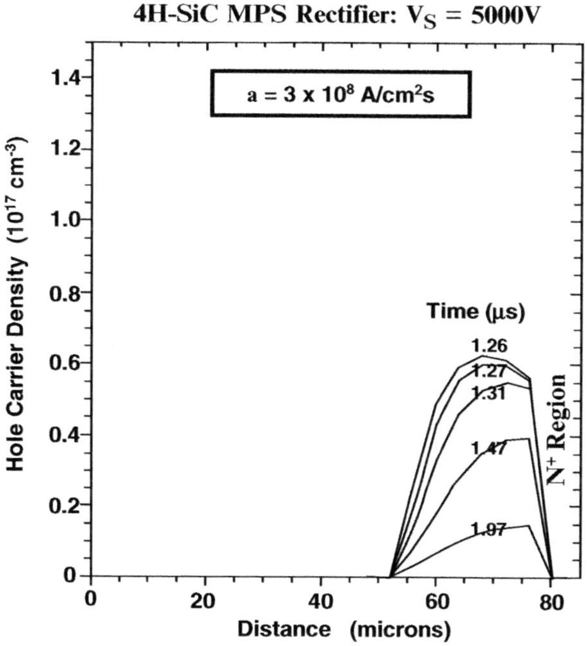


Fig. 7.46E Carrier Distribution in a 10 kV 4H-SiC MPS Rectifier during Phase 3 of the Reverse Recovery Transient.

The carrier distribution during the third phase of the reverse recovery process for the 10 kV 4H-SiC MPS rectifier structure obtained from the numerical simulations is shown in Fig. 7.46E. The carrier distribution has a Gaussian shape which decreases with time due to removal of the carriers by diffusion into the space charge region on the left-hand-side and the $N^+$ substrate on the right-hand-side. This validates the assumptions used when developing the analytical model for the decay of the reverse recovery current during the third phase.

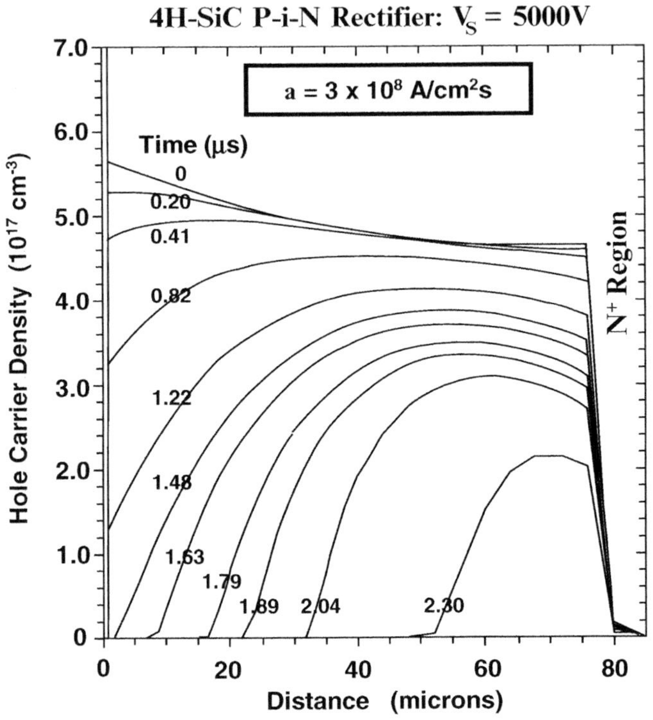


Fig. 7.47E Carrier Distribution in a 10 kV 4H-SiC P-i-N Rectifier during Phase 1, 2 and 3 of the Reverse Recovery Transient.

In order to quantify the improvement in the reverse recovery behavior derived by using the silicon carbide MPS rectifier structure, the results of numerical simulations performed on a 10 kV silicon carbide P-i-N rectifier are included in this section. This P-i-N rectifier structure has the same drift region parameters with the $P^+$ anode region extending across the entire width of the 3 micron wide cell structure. The hole concentration profiles within the 4H-SiC P-i-N rectifier structure are shown in Fig. 7.47E at various time intervals during the turn-off transient. The initial carrier profile at an on-state current density of 200 A/cm$^2$ has an average value of about 5 x $10^{17}$ cm$^{-3}$. During the first phase of the turn-off process (see Fig. 6.8 for the phases for the P-i-N rectifier), the slope of the carrier profile on the left-hand-side reduces due to the decreasing current density until it becomes zero at 0.667 microseconds. Once the current becomes negative during the second phase, the carrier concentration on the left-hand-side reduces until the junction becomes reverse biased at a time of 1.48 microseconds. This long delay before the P-i-N rectifier can support a reverse bias voltage is responsible for the large peak reverse recovery current. A space charge layer forms after this time during the third phase and takes a width of about 52 microns when the voltage reaches the supply voltage of 5000 volts at time ($t_2$) of 2.3 microseconds. As in the case of the MPS rectifier some charge removal also occurs from the right-hand-side of the structure due to the diffusion of electrons into the $N^+$ cathode region.

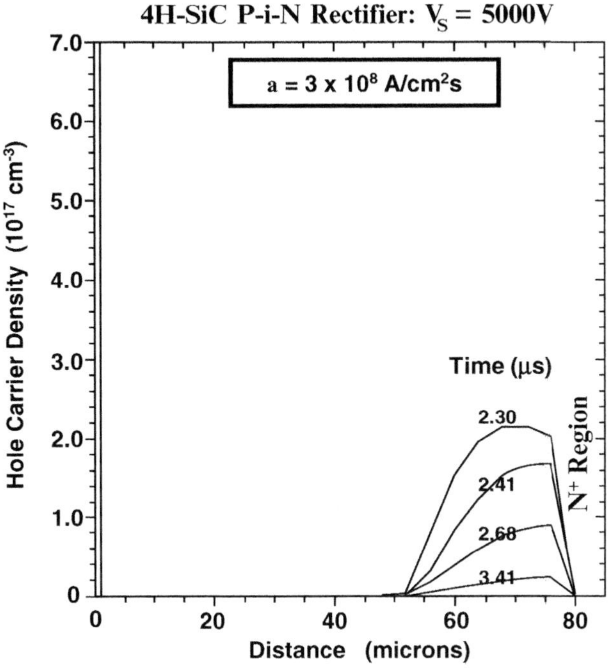


Fig. 7.48E Carrier Distribution in a 10 kV 4H-SiC P-i-N Rectifier during Phase 4 of the Reverse Recovery Transient.

The carrier distribution during the fourth phase (as defined in Fig. 6.8) of the reverse recovery process for the 10 kV 4H-SiC P-i-N rectifier structure obtained from the numerical simulations is shown in Fig. 7.48E. The carrier distribution has a Gaussian shape which decreases with time due to removal of the carriers by diffusion into the space charge region on the left-hand-side and the $N^+$ substrate on the right-hand-side. This validates the assumptions used when developing the analytical model for the decay of the reverse recovery current in the P-i-N rectifier during the fourth phase.

The voltage and current waveforms obtained with the aid of the numerical simulations of the reverse recovery for the 10 kV silicon carbide MPS and P-i-N rectifiers are compared in Fig. 7.49E and Fig. 7.50E, respectively. In both cases, a ramp rate of 3 x $10^8$ A/cm$^2$-s was used for the reverse recovery and a reverse bias supply of 5000 volts was assumed.

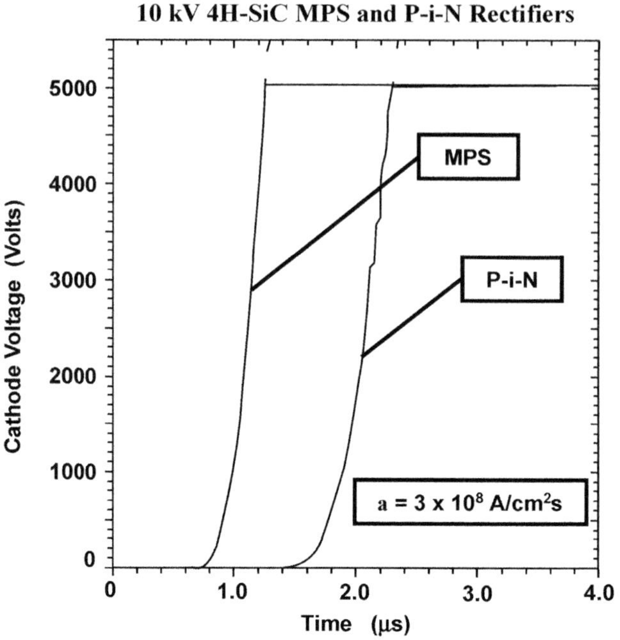


Fig. 7.49E Voltage Waveforms for the 10 kV 4H-SiC MPS and P-i-N Rectifiers during the Reverse Recovery Transient.

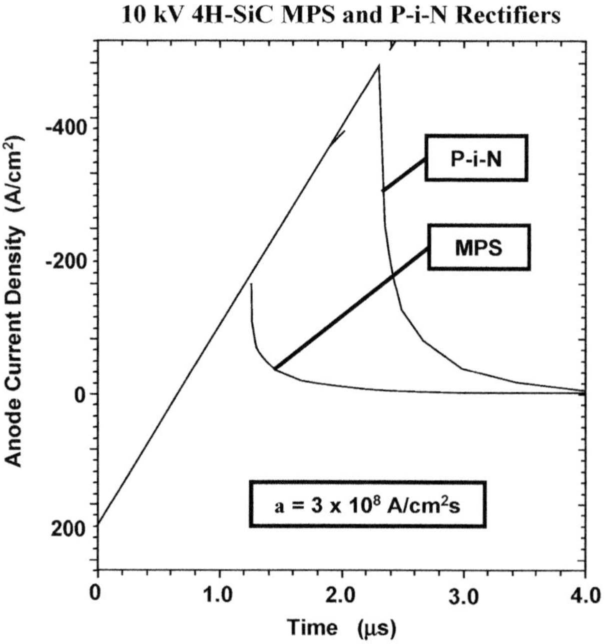


Fig. 7.50E Current Waveforms for the 10 kV 4H-SiC MPS and P-i-N Rectifiers during the Reverse Recovery Transient.

It can be observed in Fig. 7.49E that the numerical simulations of the silicon carbide MPS rectifier show the voltage increasing at a quadratic rate immediately after the current crosses zero as expected from the analytical model. In contrast, the voltage for the silicon carbide P-i-N rectifier begins to rise at a much later time due to the large stored charge in the vicinity of the P-N junction. It can be observed from Fig. 7.50E that the peak reverse recovery current for the MPS rectifier is 178 $A/cm^2$ which is in good agreement with the predictions of the analytical model. In contrast, the peak reverse recovery current for the P-i-N rectifier is 493 $A/cm^2$ which is nearly 3-times larger than for the MPS rectifier. From the current waveforms for the two devices in Fig. 7.50E, the reverse recovery time for the MPS rectifier is found to be 1.3 microseconds in comparison with 2.3 microseconds for the P-i-N rectifier. These results indicate that a substantial improvement in the reverse recovery power loss can be achieved by utilizing the MPS rectifier structure for silicon carbide high voltage rectifiers while simultaneously obtaining a low on-state voltage drop. Since the lifetime control processes for silicon carbide devices have not yet been developed, the MPS structure provides an elegant alternative to creating high performance, high voltage rectifiers.

## 7.5 MPS Rectifier Trade-Off Curves

In the previous section, it was demonstrated that the peak reverse recovery current and the turn-off time can be reduced by reducing the minority carrier lifetime in the drift region of the silicon MPS rectifier structure. This enables reduction of the power losses during the switching transient. However, the on-state voltage drop of the silicon MPS rectifier increases when the minority carrier lifetime is reduced, which produces an increase in the power dissipation during on-state current flow. For power system applications, it is desirable to reduce the total power dissipation produced in the rectifiers to maximize the power conversion efficiency. This also reduces the heat generated within the power devices maintaining a lower junction temperature which is desirable to prevent thermal runaway and reliability problems. To minimize the power dissipation, it is common-place to perform a trade-off between on-state and switching power losses for power rectifiers by developing trade-off curves.

| Lifetime ($\tau_{P0}$ & $\tau_{N0}$) | $V_{ON}$ (100 $A/cm^2$) | $t_{rr}$ ($\mu s$) | $J_{PR}$ ($A/cm^2$) | $Q_{RR}$ ($\mu C/cm^2$) |
|---|---|---|---|---|
| 10 | 0.870 | 0.50 | 86.7 | 21.7 |
| 1.0 | 0.958 | 0.392 | 60 | 11.8 |
| 0.1 | 1.645 | 0.137 | 13.3 | 0.91 |

Fig. 7.35 On-state Voltage Drop and Reverse Recovery Parameters obtained by Numerical Simulations of the 500V Silicon MPS Rectifiers.

In order to compare the performance of the silicon MPS rectifier with that of the silicon P-i-N rectifier, the on-state voltage drop and reverse recovery characteristics were obtained for the 500 volt structures using numerical simulations. The data acquired from the simulations are tabulated in Fig. 7.35 and Fig. 7.36 for the silicon MPS and P-i-N rectifier, respectively, by varying the lifetime in the drift region. For the same lifetime in the drift region, the peak reverse recovery current, the reverse recovery time, and the reverse recovery charge are substantially smaller for the silicon MPS rectifier when compared with the silicon P-i-N rectifier. This is consistent with the stored charge in the silicon MPS rectifier being about half that in the P-i-N rectifier during on-state operation. These results obtained from the numerical simulations are consistent with the experimental results reported in the literature for 1000 volt silicon rectifiers[6].

| **Lifetime ($\tau_{P0}$ & $\tau_{N0}$)** | **$V_{ON}$ (100 A/cm$^2$)** | **$t_{rr}$ (μs)** | **$J_{PR}$ (A/cm$^2$)** | **$Q_{RR}$ (μC/cm$^2$)** |
|---|---|---|---|---|
| **10** | **0.847** | **0.697** | **140** | **48.8** |
| **1.0** | **0.857** | **0.512** | **105** | **27.9** |
| **0.1** | **0.996** | **0.167** | **37** | **3.09** |

Fig. 7.36 On-state Voltage Drop and Reverse Recovery Parameters obtained by Numerical Simulations of the 500V Silicon P-i-N Rectifiers.

| **Device** | **$V_{ON}$ (200 A/cm$^2$)** | **$t_{rr}$ (μs)** | **$J_{PR}$ (A/cm$^2$)** | **$Q_{RR}$ (μC/cm$^2$)** |
|---|---|---|---|---|
| **MPS** | **3.05** | **1.3** | **178** | **116** |
| **P-i-N** | **3.10** | **2.3** | **493** | **566** |

Fig. 7.37 On-state Voltage Drop and Reverse Recovery Parameters obtained by Numerical Simulations of the 10 kV 4H-SiC MPS and P-i-N Rectifiers.

In order to compare the performance of the silicon carbide MPS rectifier with that of the P-i-N rectifier, the on-state voltage drop and reverse recovery characteristics obtained for the 10 kV structures using numerical simulations are provided in Fig. 7.37. For the same lifetime in the drift region, the peak reverse recovery current, the reverse recovery time, and the reverse recovery charge are all substantially smaller for the 4H-SiC MPS rectifier when compared with the P-i-N rectifier. This is consistent with the stored charge in the P-i-N rectifier being about ten-times larger than in the MPS rectifier during on-state operation. At the same time, the on-state voltage drop for the MPS rectifier is slightly smaller than that for the P-i-N rectifier. These improved characteristics can be achieved with a Schottky contact with a practical barrier height as long as the area for the Schottky contact is made sufficiently small to allow the P-N junction to inject carriers in the on-state.

## 7.6 Summary

The physics of operation of the MPS rectifier has been analyzed in this chapter. Analytical expressions have been derived for the on-state and blocking state, as well as the reverse recovery transients. At on-state current levels, the injected minority carrier density in the drift region exceeds the relatively low doping concentration required to achieve high breakdown voltages. This high level injection in the drift region modulates its conductivity producing a reduction in the on-state voltage drop. Unlike the P-i-N rectifier, the carrier concentration at the junction is close to zero with a maximum value at the interface between the drift region and the $N^+$ cathode region. Due to the presence of the Schottky contact with a lower barrier for current flow when compared with the P-N junction, the on-state voltage drop for the MPS rectifier can be smaller than that for the P-i-N rectifier.

The MPS rectifier can support a large voltage in the reverse blocking mode by appropriate choice of the doping concentration and thickness of the drift region. The leakage current in the reverse direction is larger than that for the P-i-N rectifier due to the thermionic emission current across the Schottky contact. However, this current can be made small by using a large Schottky barrier height and a small area for the Schottky contact.

As in the case of the P-i-N rectifier, the switching of the MPS rectifier from the on-state to the reverse blocking state is accompanied by a significant current flow in the reverse direction. However, the peak reverse recovery current and reverse recovery time are smaller than those observed in the P-i-N rectifier. This is especially true for the silicon carbide devices. Since a lifetime control process has not yet been developed for silicon carbide bipolar devices, the MPS structure offers an elegant approach to producing high voltage power rectifiers with low on-state voltage as well as much smaller reverse recovery power losses. In addition, the current density flowing through the P-N junction in the silicon carbide MPS rectifier is low because it is a small fraction of on-state current density. This reduced current density at the P-N junction in the silicon carbide MPS rectifier structure suppresses the highly undesirable increase in on-state voltage drop observed during prolonged on-state operation at high current density of silicon carbide P-i-N rectifiers[7].

## References

[1] B.J. Baliga, "Fundamentals of Power Semiconductor Devices", Springer-Scientific, New York, 2008.

[2] B.J. Baliga and E. Sun, "Comparison of Gold, Platinum, and Electron Irradiation for Controlling Lifetime in Power Rectifiers", IEEE Transactions on Electron Devices, Vol. ED-24, pp. 685-688, 1977.

---

[3] B.J. Baliga and S. Krishna, "Optimization of Recombination Levels and their Capture Cross-sections in Power Rectifiers and Thyristors", Solid State Electronics, Vol. 20, pp. 225-232, 1977.

[4] B.J. Baliga, "Analysis of the High-Voltage Merged P-i-N/Schottky (MPS) Rectifier", IEEE Electron Device Letters, Vol. EDL-8, pp. 407-409, 1987.

[5] M.E. Levinshtein, T.T. Mnatsakanov, P.A. Ivanov, R. Singh, K.G. Irvine, J. Palmour, "Carrier lifetime measurements in 10 kV 4H-SiC diodes", Electronics Letters, Volume 39, Issue 8, Page(s):689 – 691, 17 April 2003.

[6] S.L. Tu and B.J. Baliga, "Controlling the Characteristics of the MPS Rectifier by variation of the Area of the Schottky Region", IEEE Transactions on Electron Device, Vol. 40, pp. 1307-1315, 1993.

[7] A. Hefner, et al, "Recent Advances in High-Voltage, High-Frequency, Silicon-Carbide Power Devices", IEEE 41$^{st}$ Industrial Application Society Conference, Vol. 1, pp. 330-337, 2006.

# Chapter 8

# SSD Rectifiers

In the previous chapter, it was demonstrated that the trade-off between the on-state and reverse recovery power loss can be greatly improved by merging the physics of the P-i-N rectifier and the Schottky rectifier[1] to create the MPS rectifier structure illustrated in Fig. 7.1. In this structure, the drift region is designed using the same criteria as used for P-i-N rectifiers in order to support the desired reverse blocking voltage. The device structure contains a P-N junction over a portion under the metal contact and a Schottky contact for the remaining portion. During on-state current flow in the MPS rectifier, most of the current flows via Schottky contact with some modulation of the drift region by injection of holes into the drift region. The on-state voltage drop can be less than that of the P-i-N rectifier under these conditions. At the same time, the carrier distribution in the MPS rectifier has a low value at the anode side making the reverse recovery process much superior to that for the P-i-N rectifier. One drawback of the MPS rectifier structure is the larger leakage current when compared with the P-i-N rectifier due to thermionic emission current flow at the Schottky contact. The leakage current can be reduced to acceptable levels by using a relatively large barrier height and the shielding provided by the P-N junction.

An alternative to the MPS rectifier structure that produces a similar injected carrier distribution in the on-state is the Static-Shielding-Diode (SSD) structure[2] under appropriate design conditions. The SSD rectifier structure is illustrated in Fig. 8.1. It contains a deep $P^+$ region with a high doping concentration as in the case of the MPS rectifier over a portion of the anode region. The remainder of the cell contains a lightly-doped, shallower P-type region. The electrical performance of the SSC rectifier structure is strongly dependent on the doping profile for the lightly-doped P-type region. On the one extreme, if the doping concentration of the P-type region is made very large as in the case of the deeper $P^+$ region, the SSD

B.J. Baliga, *Advanced Power Rectifier Concepts*, DOI: 10.1007/978-0-387-75589-2_8, 

rectifier characteristics approach those exhibited by the P-i-N rectifier. On the other extreme, if the doping concentration of the P-type region is made very small, the SSD rectifier characteristics approach those exhibited by the MPS rectifier due to the poor injection efficiency at the junction between the P-type region and the drift region. The electrical characteristics of the SSD rectifier can be tailored by adjusting the doping concentration and thickness of the P-type region as discussed in this chapter. While choosing the doping concentration in the P-type region to optimize the trade-off between the on-state voltage drop and the reverse recovery characteristics, it is also necessary to ensure that the depletion region does not punch through the lightly-doped P-type region to the anode contact. If the depletion region in the P-type region extends to the anode contact, a large leakage current occurs in the SSD rectifier structure because of the high surface recombination velocity at the ohmic contact.

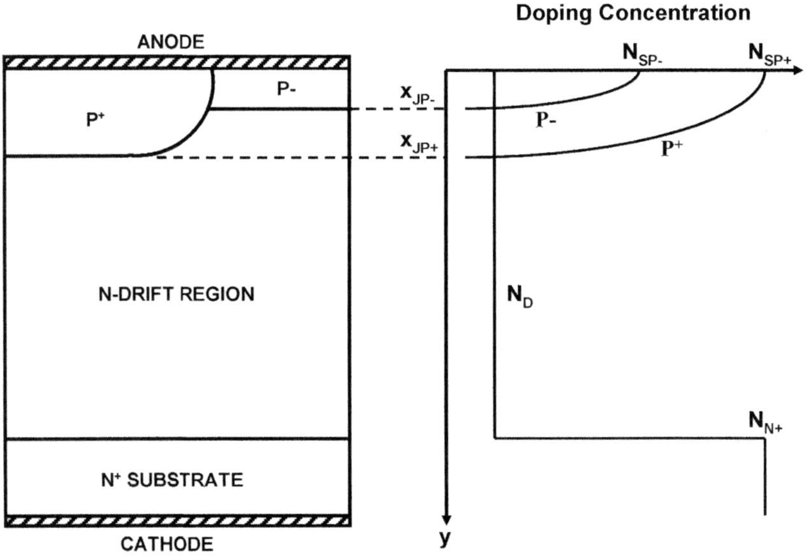


Fig. 8.1 The SSD Rectifier Structure.

## 8.1 Device Physics

The basic structure for the SSD rectifier is shown in Fig. 8.1. The anode region consists of two P-type regions – one with a high doping concentration and the second with a low doping concentration. The on-state and reverse recovery characteristics for the SSD rectifier are controlled by adjusting the doping concentration at the surface of the lightly-doped P- region ($N_{SP-}$) and its junction depth ($x_{JP-}$). By utilizing a low doping concentration in the P- region, the injection efficiency at the junction between the P- region and the drift region can be reduced. With a sufficiently small injection efficiency, the carrier concentration at this junction becomes close to zero. The injected carrier profile in the drift region then becomes similar to that observed in the MPS rectifier under high-level injection conditions over most of the drift region.

The on-state current flow in the SSD rectifier can be expected to occur via the $P^+$/N junction and the P-/N junction. As the forward bias voltage is increased, the P-N junctions begin to inject a significant concentration of holes into the drift region. The drift region then operates under high-level injection conditions as in the case of the P-i-N rectifier. Since the resistance of the drift region is reduced by conductivity modulation, a large current flow can occur with an on-state voltage drop that is similar to that for a P-i-N rectifier.

The carrier concentration at the P-/N junction remains low in the SSD rectifier because of the low injection efficiency at this junction. The resulting carrier profile is superior to that observed in the P-i-N rectifier in terms of charge removal during the turn-off transient. The SSD rectifier exhibits a smaller peak reverse recovery current and charge leading to reduced switching power loss. Moreover, the trade-off between the on-state voltage drop and reverse recovery power loss can be generated by changing the doping concentration in the P- region. A further refinement of the trade-off curve can be achieved by using the lifetime control techniques commonly used for the P-i-N rectifier.

### 8.1.1 High Level Injection Conditions

As the forward bias applied to the SSD rectifier is increased, the injected minority carrier concentration from the P-N junction also increases in the drift region until it ultimately exceeds the background doping concentration ($N_D$) in the drift region resulting in *high level injection.* When the injected hole concentration in the drift region becomes much greater than the background doping concentration, charge neutrality requires that the concentrations for electrons and holes become equal:

$$n(x) = p(x) \quad \textbf{[8.1]}$$

The large concentration of free carriers reduces the resistance of the drift region resulting in *conductivity modulation* of the drift region. As in the case of the P-i-N and MPS rectifiers, conductivity modulation of the drift region is beneficial for allowing the transport of a high current density through lightly doped drift regions with a low on-state voltage drop.

The carrier distribution within the drift region for the SSD rectifier is different from that observed for the P-i-N and MPS rectifiers because of the presence of the lightly-doped P-region. When the lifetime in the drift region is large, it is possible to derive the carrier distribution along the line A-A in Fig. 8.2 while neglecting recombination in the drift region. This is equivalent to the case of a very large value for the ambipolar diffusion length ($L_a$) in Eq. [7.11] leading to the expression:

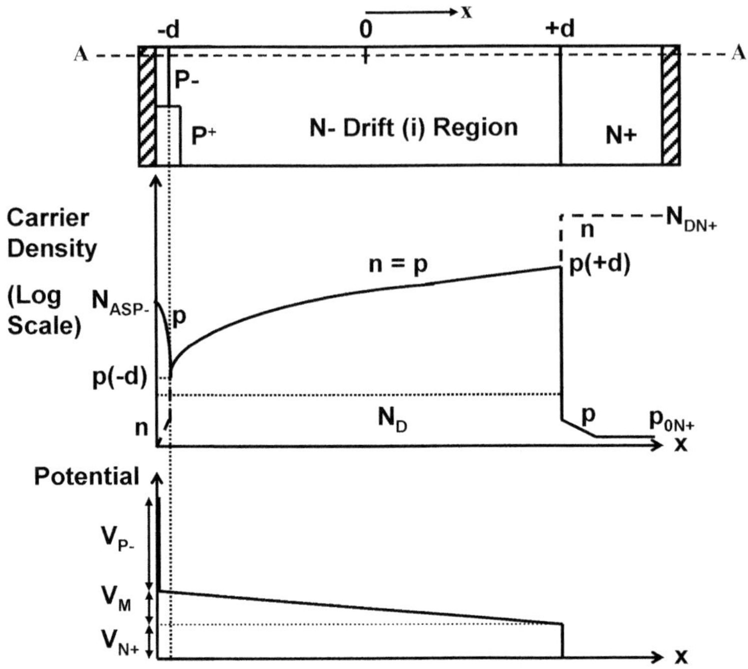


Fig. 8.2 Carrier and Potential Distribution under High-Level Injection Conditions for the SSD Rectifier.

$$\frac{d^2p}{dx^2} = 0 \qquad \textbf{[8.2]}$$

The solution for the carrier concentration governed by Eq. [8.2] is a linear carrier distribution given by:

$$p(x) = Ax + B \qquad \textbf{[8.3]}$$

with the constants A and B determined by the boundary conditions for the N-drift region.

At the junction between the N-drift region and the $N^+$ cathode region (located at x = +d in Fig. 8.2), the total current flow occurs exclusively by electron transport:

$$J_T = J_n(+d) \qquad \textbf{[8.4]}$$

and

$$J_p(+d) = 0 \qquad \textbf{[8.5]}$$

Using these equations:

$$J_T = 2qD_n\left(\frac{dn}{dx}\right)_{x=+d} = 2qD_n\left(\frac{dp}{dx}\right)_{x=+d} \qquad [8.6]$$

The above boundary condition can be used to obtain the constant A in Eq. [8.3]:

$$A = \frac{J_T}{2qD_n} \qquad [8.7]$$

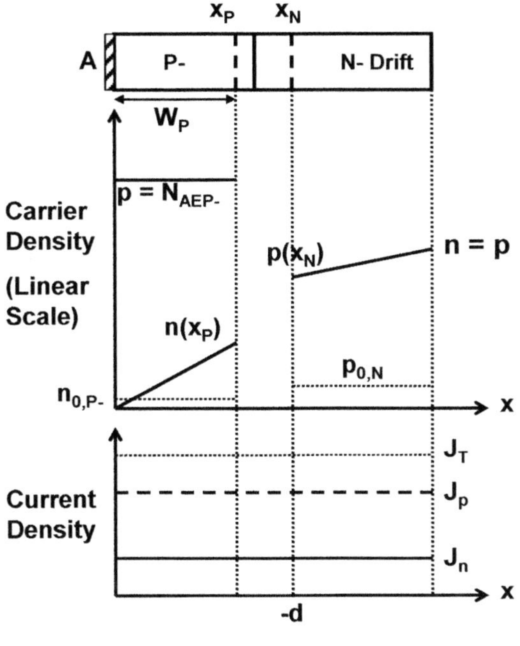


Fig. 8.3 Carrier Distribution at the Junction between the Lightly-Doped P- Region and the N- Drift Region for the SSD Rectifier.

The second boundary condition for the SSD rectifier occurs at the junction between the N-drift region and the lightly-doped P- region (located at $x = -d$ in Fig. 8.2). Here, the hole concentration becomes determined by the finite injection efficiency of the junction. The carrier distribution at this junction is illustrated in Fig. 8.3. The boundaries at the junction are labeled $x_P$ and $x_N$. Note that the boundary $x_N$ corresponds to $x = -d$ in Fig. 8.2. The figure also shows the current density at the junction indicating that a portion of the total current is due to the injection of electrons into the lightly-doped P- region. The electron current contribution can be derived from the injected carrier profile in the P- region. Since the width ($W_P$) of this region is small and its doping concentration is moderate, the diffusion length for electrons in the P- region can be assumed to be much greater than its width. The electron concentration then varies linearly with distance as illustrated in Fig. 8.3 from $n(x_P)$ at the junction to zero at the ohmic contact.

Using the Boltzmann quasi-equilibrium condition[3]:

$$\frac{p(x_N)}{p(x_P)} = \frac{n(x_P)}{n(x_N)} = e^{-(q\Delta\psi/kT)} \quad \textbf{[8.8]}$$

where $\Delta\psi$ is the potential barrier across the junction under forward bias operation. Assuming low-level injection conditions in the P- region, the hole concentration is equal to the effective acceptor doping concentration ($N_{AEP-}$):

$$p(x_P) = N_{AEP-} \quad \textbf{[8.9]}$$

The electron and hole concentrations at the junction in the N- drift region are equal due to high-level injection conditions:

$$n(x_N) = p(x_N) \quad \textbf{[8.10]}$$

Using these relationships in Eq. [8.8] yields:

$$n(x_P) = \frac{p^2(x_N)}{N_{AEP-}} \quad \textbf{[8.11]}$$

The diffusion current due to electrons in the P- region under low-level injection conditions is then given by:

$$J_n(x_P) = \frac{qD_n n(x_P)}{W_P} = \frac{qD_n p^2(x_N)}{W_P N_{AEP-}} \quad \textbf{[8.12]}$$

Since the electron current is continuous through the depletion region:

$$J_n(x_N) = J_n(x_P) = \frac{qD_n p^2(x_N)}{W_P N_{AEP-}} \quad \textbf{[8.13]}$$

Due to high-level injection conditions in the N- drift region, the hole current density is given by:

$$J_p(x_N) = 2qD_p\left(\frac{dp}{dx}\right)_{x=-d} = 2qD_p A = \frac{D_p}{D_n} J_T \quad \textbf{[8.14]}$$

by using Eq. [8.7]. Since the sum of the electron and hole current densities must equal the total on-state current density:

$$\frac{D_p}{D_n} J_T + \frac{qD_n p^2(x_N)}{W_P N_{AEP-}} = J_T \quad \textbf{[8.15]}$$

Solving for the hole concentration:

$$p(x_N) = p(-d) = \sqrt{\frac{W_B N_{AEP-} J_T}{qD_n}\left(1 - \frac{D_p}{D_n}\right)} \quad \textbf{[8.16]}$$

This expression indicates that the hole concentration at the junction between the lightly-doped P- region and the drift region increases as the square root of the current density. It also indicates that the carrier concentration at the junction between the lightly-doped P- region and the drift region will increase with increasing net charge (product of doping and width) in the lightly-doped P- region.

Using the above hole concentration at x = –d in Eq. [8.3] provides a solution for the constant B:

$$B = \frac{J_T d}{2qD_n} + \sqrt{\frac{W_B N_{AEP-} J_T}{qD_n}\left(1 - \frac{D_p}{D_n}\right)} \quad \textbf{[8.17]}$$

The hole (and electron) carrier concentration in the N- drift region is then given by substitution of the constants A and B into Eq. [8.3]:

$$\mathrm{p(x) = n(x)} = \frac{\mathrm{J_T}}{\mathrm{2qD_n}}(\mathrm{x + d}) + \mathrm{p(x_N)} \quad \textbf{[8.18]}$$

where x extends from –d to +d across the drift region. The carrier distribution described by this equation has a maximum value at the interface between the drift region and the $N^+$ substrate with a magnitude of:

$$\mathrm{p(+d) = n(+d)} = \frac{\mathrm{J_T d}}{\mathrm{qD_n}} + \mathrm{p(x_N)} \quad \textbf{[8.19]}$$

and reduces monotonically when proceeding towards the lightly-doped P- region in the negative x direction. The concentration becomes equal to $p(x_N)$ at the P- region as required to satisfy the boundary condition used to derive the expression.

As a particular example, the carrier distributions calculated by using Eq. [8.19] are shown in Fig. 8.4 for a silicon SSD rectifier with drift region thickness of 70 microns corresponding to a breakdown voltage of 500 volts using various values for the surface concentration of the lightly-doped P- region. In performing these computations, it was assumed that the effective doping concentration ($N_{AEP-}$) of the P- region is one-half of the surface doping concentration ($N_{ASP-}$) of the P- region. It can be observed that the carrier concentration at the junction between the lightly-doped P- region and the N- drift region increases significantly with increasing surface concentration of the P- region. For the lowest surface concentration of the P- region, the carrier profile resembles that for the MPS rectifier structure. For the largest surface concentration for the P- region, the carrier profile begins to resemble that for the P-i-N rectifier. Note that for this analytical model, the carrier concentration also increases at the interface between the N- drift region and the $N^+$ substrate with increasing surface concentration of the P- region.

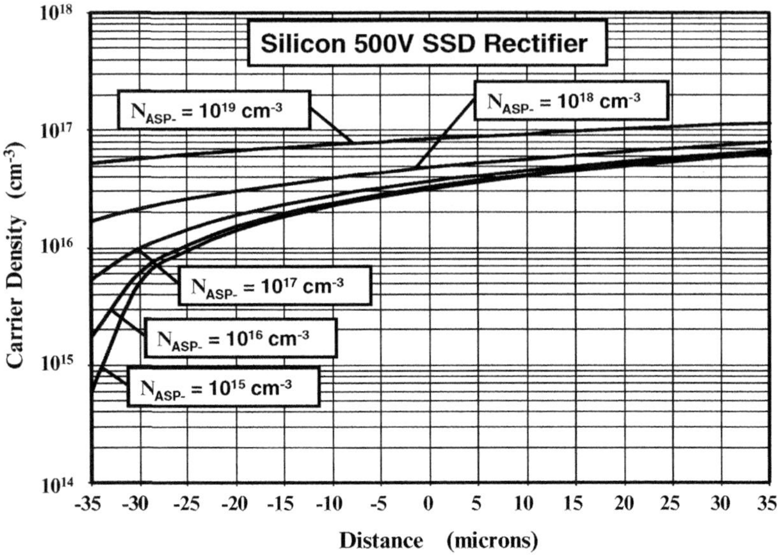


Fig. 8.4 Carrier Distribution for the Silicon SSD Rectifier with various Surface Doping Concentrations for the P- Region.

### 8.1.2 On-State Voltage Drop

The on-state voltage drop for the SSD rectifier can be obtained by summing the voltage drops along the path marked 'A-A' in Fig. 8.2 through the lightly-doped region. The total voltage drop along this path consists of the voltage drop across the junction between the P- region and the drift region, the voltage drop across the drift region (middle region voltage $V_M$), and the voltage drop at the interface with the $N^+$ substrate ($V_{N+}$):

$$V_{ON} = V_{P-} + V_M + V_{N+} \qquad \textbf{[8.20]}$$

The voltage drop at the junction between the P- region and the drift region is given by[3]:

$$V_{P-} = \frac{kT}{q}\ln\left[\frac{p(-d)}{p_{0N}}\right] = \frac{kT}{q}\ln\left[\frac{p(-d)N_D}{n_i^2}\right] \qquad \textbf{[8.21]}$$

where $p_{0N}$ is the minority carrier concentration in the drift region in equilibrium, which has been related to the doping concentration ($N_D$). This voltage drop is smaller in the SSD rectifier than for the P-i-N rectifier because the injected carrier concentration p(-d) is much smaller at the junction.

Analysis of the voltage drop across the drift (middle) region can be performed by integration of the electric field. The electric field in the drift region can be obtained from the carrier distribution. The hole and electron currents flowing in the drift region are given by:

$$J_p = q\mu_p\left(pE - \frac{kT}{q}\frac{dp}{dx}\right) \quad [8.22]$$

and

$$J_n = q\mu_n\left(nE + \frac{kT}{q}\frac{dn}{dx}\right) \quad [8.23]$$

The total current at any location in the drift region is constant and given by:

$$J_T = J_p + J_n \quad [8.24]$$

Combining these relationships:

$$E(x) = \frac{J_T}{q(\mu_n + \mu_p)n} - \frac{kT}{2qn}\frac{dn}{dx} \quad [8.25]$$

The charge neutrality condition n(x) = p(x) was utilized when deriving this expression.

In the case of the SSD rectifier, it was found in the previous section that the carrier distribution has a linear shape when recombination in the drift region is neglected. The case with no recombination in the drift region is treated here with the carrier profile given by Eq. [8.18]. Substituting this profile in Eq. [8.25] yields:

$$E(x) = \left[\frac{2D_n}{(\mu_n + \mu_p)} - \frac{kT}{2q}\right]\frac{J_T}{\left[(x+d)J_T + 2qD_n p(x_N)\right]} \quad [8.26]$$

The voltage drop in the drift (middle) region can be obtained by integration of the electric field across the drift region:

$$V_M = \left[\frac{2D_n}{(\mu_n + \mu_p)} - \frac{kT}{2q}\right]\ln\left(\frac{2J_T d + 2qD_n p(x_N)}{2qD_n p(x_N)}\right) \quad [8.27]$$

For the case of a SSD rectifier with a drift region thickness (2d) of 70 microns and surface concentration for the P- region of 1 x $10^{16}$ $cm^{-3}$, the voltage drop across the middle region at room temperature is found to be 0.09 volts at an on-state current density of 100 $A/cm^2$. This voltage drop is much smaller than the voltage drop of 8.47 volts for an un-modulated drift region with doping concentration of 3.8 x $10^{14}$ $cm^{-3}$ and thickness of 70 microns.

The voltage drop across the interface between the drift region and the $N^+$ substrate can be determined from the carrier density at x = +d:

$$V_{N+} = \frac{kT}{q} \ln\left[\frac{n(+d)}{N_D}\right] \quad [8.28]$$

The carrier concentration in the drift region at this interface is given by Eq. [8.19] without recombination in the drift region. The on-state voltage drop for the 500 V silicon SSD rectifier with a drift region width of 70 microns at an on-state current density of 100 A/cm$^2$ is then found to be 0.79 volts by summing the voltage drops across the P-/N junction, the middle region, and the N/$N^+$ interface. This is less than that for the P-i-N rectifier.

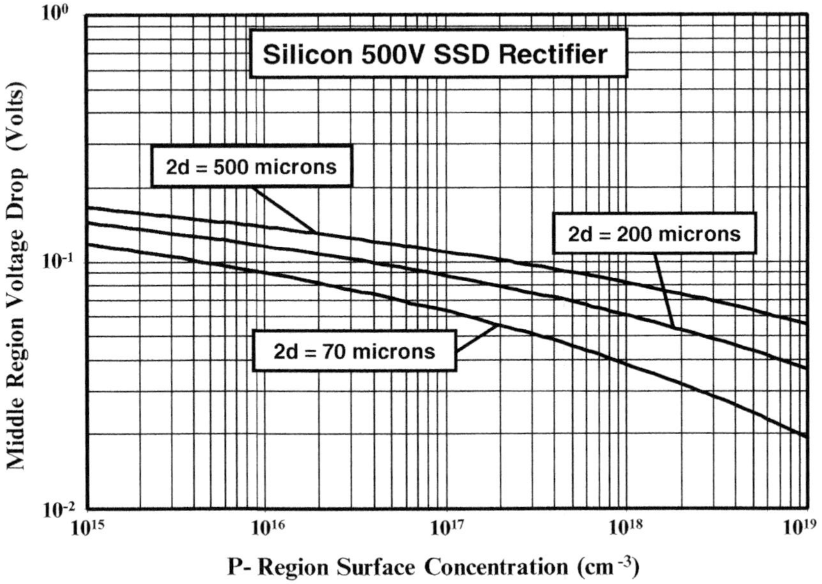


Fig. 8.5 Middle Region Voltage Drop for 500V Silicon SSD Rectifiers.

The voltage drop in the drift (middle) region can be obtained by using Eq. [8.27]. According to this expression, the voltage drop across the drift (middle) region is dependent on the on-state current density and the thickness of the drift region. The voltage drop is also determined by the doping concentration of the lightly-doped P- region via the term $p(x_N)$. The values for the middle region voltage drop for silicon SSD rectifiers with various drift region thicknesses at an on-state current density of 100 A/cm$^2$ are shown in Fig. 8.5 as a function of the surface concentration of the P- region. In making this graph, the effective doping concentration of the P- region was assumed to be half the surface concentration. The middle region voltage drop at an on-state current density of 100 A/cm$^2$ is found to be about 0.1 volts.

The on-state voltage drop for the SSD rectifier can be computed by utilizing the three components discussed above. Combining the voltage drop given by Eq. [8.21] across the P-/N junction and Eq. [8.28] across the N/N$^+$ interface:

$$V_{P-} + V_{N+} = \frac{kT}{q} \ln\left[\frac{p(x_N)n(+d)}{n_i^2}\right] \quad [8.29]$$

Using this in Eq. [8.20] with Eq. [8.27] yields:

$$V_{ON} = \frac{kT}{q} \ln\left(\frac{p(x_N)n(+d)}{n_i^2}\right) + \left\{\frac{2D_n}{(\mu_n + \mu_p)} - \frac{kT}{2q}\right\} \ln\left[\frac{2J_T d + 2qD_n p(x_N)}{2qD_n p(x_N)}\right] \quad [8.30]$$

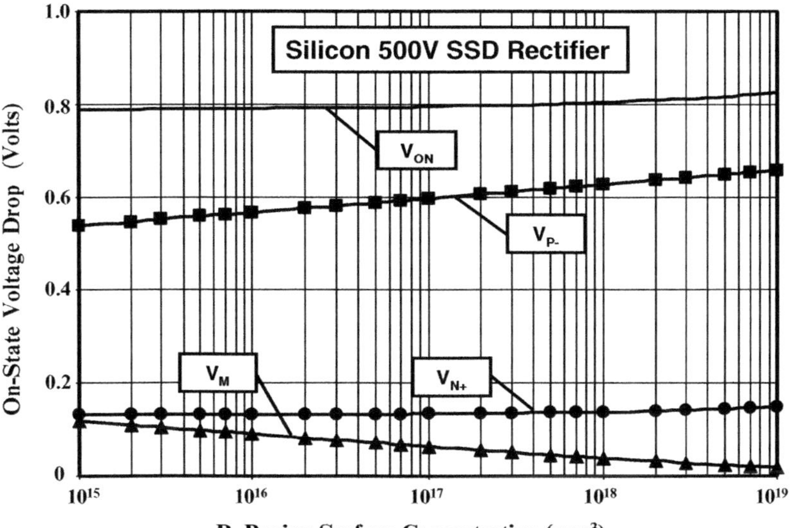


Fig. 8.6 Voltage Drops in the 500V Silicon SSD Rectifier.

The on-state voltage drop for the SSD rectifier is a function of the doping concentration of the P- region via the hole concentration $p(x_N)$ at the P-/N junction. As an example, consider the case of a silicon SSD rectifier with a drift region doping concentration of $3.8 \times 10^{14}$ cm$^{-3}$ and thickness of 70 microns capable of supporting 500 volts in the reverse blocking mode. The variation of the on-state voltage drop with surface concentration of the P- region for this case as predicted by the analytical model is provided in Fig. 8.6. It can be observed that the on-state voltage drop increases slightly as the doping concentration of the P- region is increased. In order to get a better understanding of this behavior the three components of the on-state voltage drop are also shown in the figure. The voltage drop at the $N/N^+$ interface increases only slightly with increasing doping concentration of the P- region. However, the voltage drop at the P-/N junction increases significantly with increasing doping concentration of the P- region due to the larger hole concentration in the drift region at the junction. Concomitantly, the voltage drop in the drift (middle) region goes down with increasing doping concentration of the P- region because of the larger hole concentration within the drift region. The change in these two voltage drops is nearly equal resulting in the total on-state voltage drop increasing only slightly with increasing doping concentration of the P- region.

### 8.1.3 Forward Conduction Characteristics

The on-state i-v characteristics for the SSD rectifier can be predicted by using Eq. [8.30]. The on-state characteristics for the 500V silicon SSD rectifier are shown in Fig. 8.7 for the case of various surface concentration values for the P- region. At low on-state current densities, the on-state voltage drop is significantly smaller for the SSD rectifier structure with the smaller surface concentration for the P- region. At higher on-state current densities, the on-state voltage drops for all the cases become nearly equal. At an on-state current density of 100 A/cm$^2$, the on-state voltage drop of the SSD rectifier is found to be about 0.8 volts which is smaller than that for the P-i-N rectifier. This demonstrates that it is possible to obtain a smaller on-state voltage drop in the SSD rectifier while also obtaining a smaller stored charge in the drift region.

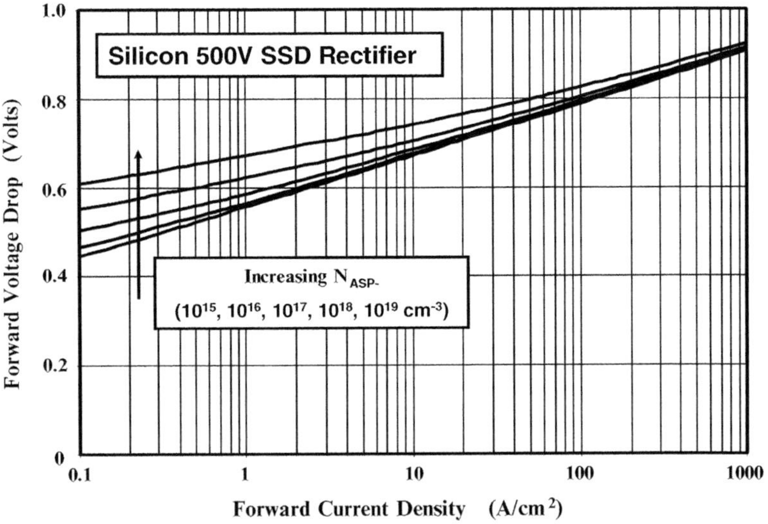


Fig. 8.7 On-State Characteristics for the 500V Silicon SSD Rectifier with Surface Doping Concentration of P- Region as a Parameter.

### 8.1.4 Injection into the N$^+$ End-Region

In the case of P-i-N rectifiers, recombination in the end regions (P$^+$ and N$^+$ regions) has been shown to have a strong impact on the on-state characteristics especially when the lifetime in the drift region is large. In the presence of recombination in the end regions, the carrier concentration in the drift region no longer increases in proportion to the on-state current density. The voltage drop in the middle region is then no longer constant but increases as the current density is increased. This produces a significant increase in the on-state voltage drop. A similar effect can be expected to occur in the SSD rectifier when the lifetime in the drift region is large. An analytical model that takes the recombination in the end region (N$^+$ region for the SSD rectifier) into account is derived in this section.

With the presence of recombination in the end-region for the SSD rectifier, the total current flow must accommodate for the recombination of carriers in the N$^+$ end-region. In the analysis presented below, it will be assumed that recombination in the N$^+$ end region is dominant. Due to the high doping concentration in the N$^+$ end-region, the injected minority carrier density is well below the majority carrier density even during operation at very high on-state current densities. The current corresponding to the N$^+$ end-region can therefore be analyzed using low-level injection theory under the assumption of a uniform doping concentration in the region. Under low-level injection conditions:

$$J_{N+} = J_{SN+} e^{\frac{qV_{N+}}{kT}} \qquad \textbf{[8.31]}$$

where $J_{SN+}$ are referred to as the *saturation current density* for the heavily doped N$^+$ end-region.

The injected carrier concentrations on the two sides of the N$^+$/N junction are related under quasi-equilibrium conditions by:

$$p_{N+}(+d).n_{N+}(+d) = p(+d).n(+d) \qquad \textbf{[8.32]}$$

Under low-level injection conditions within the N$^+$ end-region:

$$n_{N+}(+d) = n_{0N+} \qquad \textbf{[8.33]}$$

and

$$p_{N+}(+d) = p_{0N+} e^{\frac{qV_{N+}}{kT}} \qquad \textbf{[8.34]}$$

Using these relationships in Eq. [8.32]:

$$p(+d).n(+d) = p_{0N+}.n_{0N+} e^{\frac{qV_{N+}}{kT}} = n_{ieN+}^2 e^{\frac{qV_{N+}}{kT}} \qquad \textbf{[8.35]}$$

where $n_{ieN+}$ is the effective intrinsic carrier concentration in the N$^+$ cathode region including the influence of band-gap narrowing. Due to high-level injection conditions in the drift region p(+d) = n(+d), leading to:

$$e^{\frac{qV_{N+}}{kT}} = \left[\frac{n(+d)}{n_{ieN+}}\right]^2 \quad [8.36]$$

Using this expression in Eq. [8.31]:

$$J_{N+} = J_{SN+}\left[\frac{n(+d)}{n_{ieN+}}\right]^2 \quad [8.37]$$

Recombination in the drift region can be neglected when the lifetime in the drift region is large, allowing equating the total current density to the recombination current given by Eq. [8.37] for the $N^+$ end-region. The carrier concentration at the interface between the drift region and the $N^+$ region is then given by:

$$n(+d) = n_{ieN+}\sqrt{\frac{J_T}{J_{SN+}}} \quad [8.38]$$

From this equation, it can be concluded that the carrier concentration in the drift region will increase as the square root of the total current density if the end region recombination becomes dominant. Under these circumstances, the middle region voltage drop increases more rapidly with increasing current density resulting in an increase in the total on-state voltage drop. It is worth pointing out that the concentration at the interface between the drift region and the $N^+$ substrate is independent of the doping concentration of the P- region in this model.

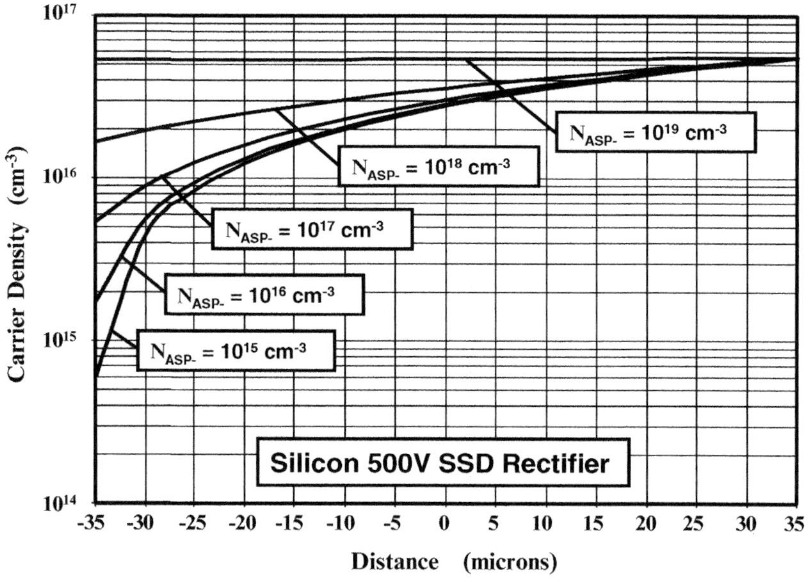


Fig. 8.8 Carrier Distribution for the Silicon SSD Rectifier with various Surface Doping Concentrations for the P- Region when End-region Recombination is Dominant.

If recombination in the drift region is neglected, the carrier concentration profile determined by solving Eq. [8.2] is given by Eq. [8.3] with the concentration at x = –d given by Eq. [8.16] for $p(x_N)$ and the concentration at x = +d given by Eq. [8.38]. Using these boundary conditions:

$$p(x) = n(x) = \frac{n_{ieN+}}{2}\sqrt{\frac{J_T}{J_{SN+}}}\left(\frac{x}{d}+1\right)+\frac{p(x_N)}{2}\left(1-\frac{x}{d}\right) \qquad \textbf{[8.39]}$$

where x extends from –d to +d across the drift region.

As an example, the carrier distributions calculated by using Eq. [8.39] are shown in Fig. 8.8 for a silicon SSD rectifier with drift region thickness of 70 microns (corresponding to a breakdown voltage of 500 volts) using various values for the surface concentration of the lightly-doped P- region. In performing these computations, it was assumed that the effective doping concentration ($N_{AEP-}$) of the P- region is one-half of the surface doping concentration ($N_{ASP-}$) of the P- region. It can be observed that the carrier concentration at the junction between the lightly-doped P- region and the N- drift region increases significantly with increasing surface concentration of the P- region. For the lowest surface concentration of the P- region, the carrier profile resembles that for the MPS rectifier structure. For the largest surface concentration for the P- region, the carrier profile begins to resemble that for the P-i-N rectifier. Note that for this analytical model, the carrier concentration at the interface between the N- drift region and the $N^+$ substrate does not change with increasing surface concentration of the P- region. Its value is consequently smaller than observed with the analytical model without end-region recombination (see Fig. 8.4).

The on-state voltage drop for the SSD rectifier structure can be analyzed in the presence of $N^+$ end-region recombination by summing the voltage at the P-/N junction, the middle region, and the N/$N^+$ interface. The voltage drop at the P-/N junction can be computed by using Eq. [8.21]. The voltage drop across the middle region can be determined by obtaining the electric field. Substituting Eq. [8.39] into Eq. [8.25] for the electric field yields:

$$E(x) = \left\{\frac{J_T}{q(\mu_n+\mu_p)} - \frac{kT\left[n(+d)-p(x_N)\right]}{4qd}\right\} \left\{\frac{2d}{\left[n(+d)-p(x_N)\right]x+\left[n(+d)+p(x_N)\right]d}\right\} \qquad \textbf{[8.40]}$$

The voltage drop in the drift (middle) region can be obtained by integration of the electric field across the drift region:

$$V_M = \left[\frac{2J_T d}{q(\mu_n+\mu_p)\left[n(+d)-p(x_N)\right]} - \frac{kT}{2q}\right]\ln\left[\frac{n(+d)}{p(x_N)}\right] \qquad \textbf{[8.41]}$$

The voltage drop across the interface between the drift region and the $N^+$ substrate also gets altered when recombination in the end region becomes dominant. This voltage drop can be determined from the carrier density at x = +d:

$$V_{N+} = \frac{kT}{q} \ln\left[\frac{n(+d)}{N_D}\right] \quad \textbf{[8.42]}$$

Since the carrier concentration in the drift region at this interface is now given by Eq. [8.38]:

$$V_{N+} = \frac{kT}{q} \ln\left[\frac{n_{ieN+}}{N_D}\sqrt{\frac{J_T}{J_{SN+}}}\right] \quad \textbf{[8.43]}$$

In the presence of recombination in the end region, the on-state voltage drop for the SSD rectifier can be computed by utilizing the three components of the on-state voltage drop. Using Eq. [8.20] with Eq. [8.21], Eq. [8.41], and Eq. [8.43] yields:

$$V_{ON} = \frac{kT}{q} \ln\left(\frac{p(x_N)n(+d)}{n_i^2}\right) + \left\{\frac{2J_T d}{q(\mu_n + \mu_p)[n(+d) - p(x_N)]} - \frac{kT}{2q}\right\} \ln\left[\frac{n(+d)}{p(x_N)}\right] \quad \textbf{[8.44]}$$

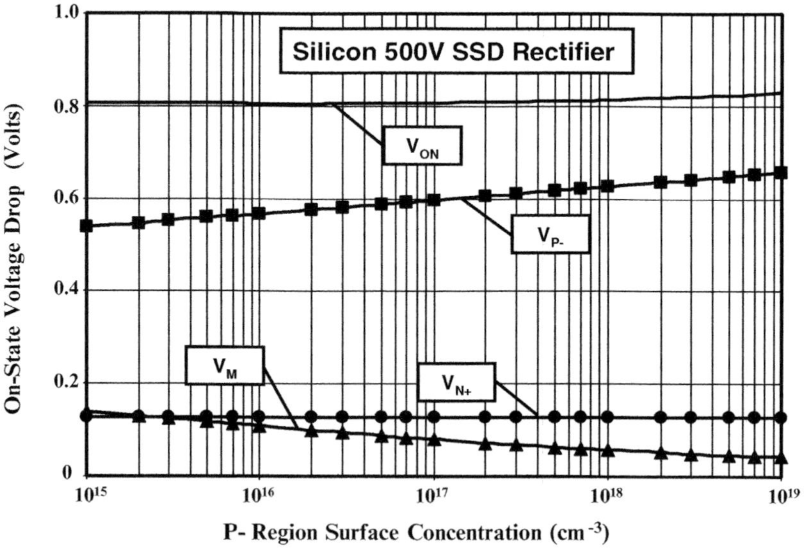


Fig. 8.9 Voltage Drops in the 500V Silicon SSD Rectifier when the $N^+$ End-Region Recombination is Dominant.

The on-state voltage drop for the SSD rectifier with $N^+$ end-region recombination dominant is a function of the doping concentration of the P- region via the hole concentration $p(x_N)$ at the P-/N junction. As an example, consider the case of a silicon SSD rectifier with a drift region doping concentration of 3.8 x $10^{14}$ $cm^{-3}$ and thickness of 70 microns capable of supporting 500 volts in the reverse blocking mode. The variation of the on-state voltage drop with surface concentration of the P- region for this case as predicted by the analytical model is provided in Fig. 8.9. It can be observed that the on-state voltage drop increases slightly as the doping concentration of the P- region is increased. In order to get a better understanding of this behavior the three components of the on-state voltage drop are also shown in the figure. The voltage drop at the $N/N^+$ interface is independent of doping concentration of the P- region. However, the voltage drop at the P-/N junction increases with increasing doping concentration of the P- region due to the larger hole concentration in the drift region at the junction. Concomitantly, the voltage drop in the drift (middle) region goes down with increasing doping concentration of the P- region because of the larger hole concentration within the drift region. The change in these two voltage drops is nearly equal resulting in the total on-state voltage drop increasing only slightly with increasing doping concentration of the P- region. The total on-state voltage drop at an on-state current density of 100 A/$cm^2$ with $N^+$ end-region recombination dominant is only slightly larger than the on-state voltage drop computed in the previous section. However, the rate of change in on-state voltage drop with increasing current density is quite different.

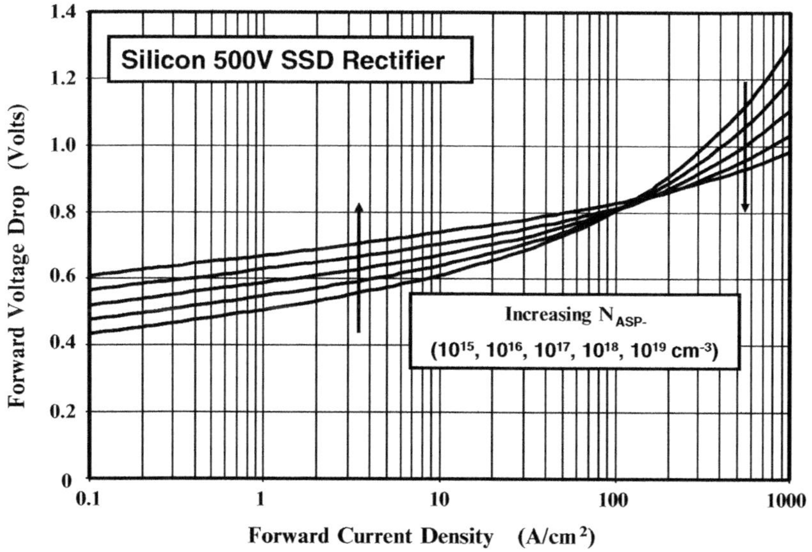


Fig. 8.10 On-State Characteristic for the Silicon SSD Rectifier after accounting for $N^+$ End Region Recombination.

The on-state characteristic for the 500 V silicon SSD rectifier obtained by using the above analytical model with $N^+$ end-region recombination is shown in Fig. 8.10 for the case of various surface doping concentration for the lightly-doped

P- region. The effective doping concentration ($N_{AEP-}$) for the P- region was assumed to be half the surface concentration during this analysis. In comparison with the on-state characteristics obtained without $N^+$ end-region recombination (see Fig. 8.7) a significant increase in the on-state voltage drop is observed at high on-state current density levels. It is worth pointing out that the analytical model with $N^+$ end-region recombination predicts a smaller on-state voltage drop at low on-state current densities when compared with the model without end-region recombination when the surface concentration of the P- region is small.

### Simulation Example

In order to gain further insight into the physics of operation for the SSD rectifier, the results of two-dimensional numerical simulations are provided in this section for a structure designed for supporting 500 volts. For this case, a drift region with doping concentration of 3.8 x $10^{14}$ $cm^{-3}$ was used with a thickness of 70 microns. The $P^+$ and $N^+$ end-regions had a surface concentration of 1 x $10^{19}$ $cm^{-3}$ and a depth of about 1 micron. In all cases, it was assumed that $\tau_{p0} = \tau_{n0}$ = 10 microseconds. The influence of band-gap narrowing, auger recombination, and carrier-carrier scattering was included during the numerical simulations. The depth of the P- region was held constant at 0.5 microns while varying its surface concentration.

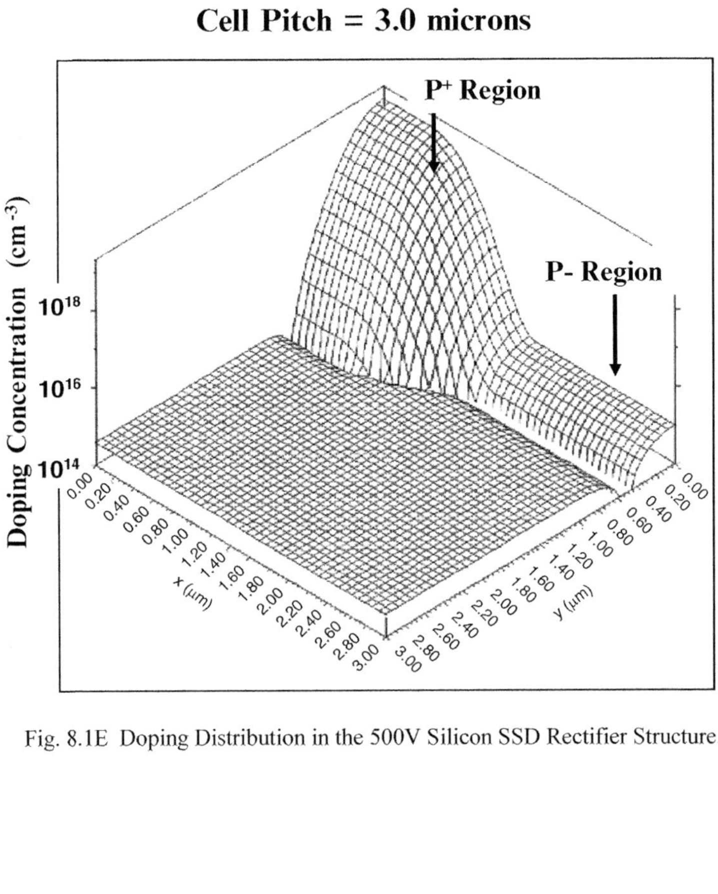


Fig. 8.1E Doping Distribution in the 500V Silicon SSD Rectifier Structure.

A three-dimensional view of the doping concentration in the upper portion of the silicon SSD device structure with cell pitch of 3 microns is provided in Fig. 8.1E. The $P^+$ region is located on the upper left-hand-side. It is formed by using a 1 micron deep diffusion from a 0.5 micron wide window in the cell. The lightly-doped P- region is formed using a Gaussian diffusion performed across the entire cell cross-section with a surface concentration of 1 x $10^{15}$ $cm^{-3}$. This approach simplifies the fabrication of the SSD rectifier structure because the $P^+$ and P- regions do not have be aligned during device processing. The $P^+$ region and the P- region have an equal area in the silicon MPS rectifier structure used for the simulations.

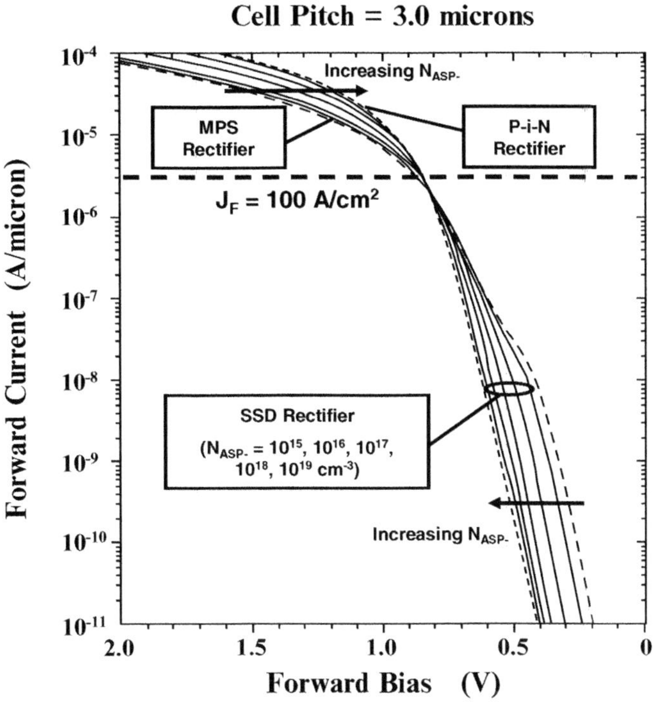


Fig. 8.2E On-State Characteristics of the 500V Silicon SSD Rectifiers.

The on-state characteristics for the silicon SSD rectifier structures obtained from the numerical simulations are shown in Fig. 8.2E by the solid lines for the case of various surface doping concentrations for the P- region. Several distinct regimes of operation are apparent in the shape of the characteristics for the case of low surface concentrations for the P- region. In order to relate these to the behavior of the MPS and P-i-N rectifiers, the on-state characteristics for these devices are shown in the figure by the dashed and dotted lines, respectively. A work function of 5.0 eV was used for the Schottky contact in the MPS rectifier structure. It can be observed that the on-state characteristic for the SSD rectifier structure with the surface concentration for the P- region of 1 x $10^{15}$ $cm^{-3}$ resembles that for the MPS rectifier. At lower current densities, its on-state voltage drop is substantially smaller than that for the P-i-N rectifier. At current densities above 100 $A/cm^2$, the on-state voltage drop for this SSD structure increases rapidly with increasing current density in a similar manner to that observed for the MPS rectifier. Consequently, the SSD rectifier structure with a low surface concentration for the P- region behaves like the MPS rectifier. In contrast, when the surface concentration for the P- region is made large (1 x $10^{19}$ $cm^{-3}$), the on-state characteristics become similar to that observed for the P-i-N rectifier. It is therefore convenient to tailor the characteristics of the SSD rectifier by adjusting the surface concentration of the P- region. The on-state voltage drop at a current density of 100 A/$cm^2$ obtained using the numerical simulation ranges from 0.84 to 0.86 volts for all the SSD rectifier structures. These values are slightly lower than that for the P-i-N rectifier. However, the stored charge in the SSD rectifier structure is smaller than that for the P-i-N rectifier resulting in superior turn-off characteristics.

The on-state *i-v* characteristics predicted by the analytical model (see Fig. 8.10) with the inclusion of recombination in the $N^+$ end-region are in remarkably good agreement with those observed with the simulations. However, the on-state characteristics predicted by the analytical model without end-region recombination (see Fig. 8.7) do not display the features observed in the on-state *i-v* characteristics obtained from the simulations. It is therefore important to account for recombination in the $N^+$ end-region for accurately describing the on-state characteristics of the SSD rectifier structure. The on-state voltage drop at an on-state current density of 100 A/$cm^2$ obtained using the analytical model with inclusion of recombination in the $N^+$ end-region is independent of the surface concentration of the P- region. This behavior is also in close agreement with the results of the numerical simulations providing further validation of the analytical model for the on-state voltage drop.

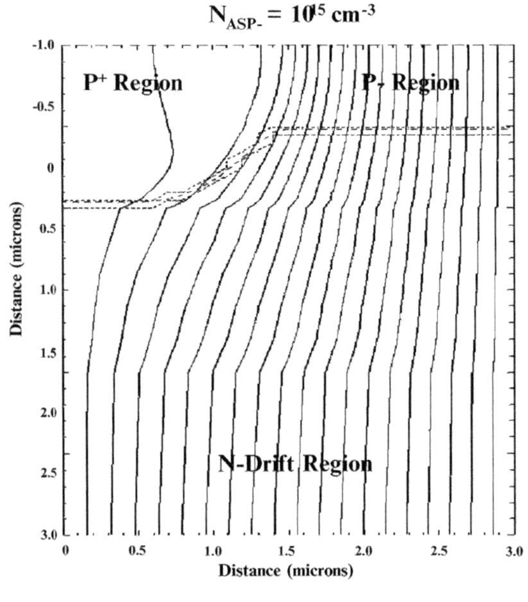


Fig. 8.3E Current Distribution within the 500V Silicon SSD Rectifier.

The current distribution within the silicon SSD rectifier structure with a P- region with a surface concentration of 1 x $10^{15}$ $cm^{-3}$ is shown in Fig. 8.3E at an on-state current density of 100 A/$cm^2$. It can be observed that most of the current flow occurs via the lightly-doped P- region. This is due to the smaller potential drop across this junction. This behavior allows modeling the SSD structure as a one-dimensional device when developing the analytical model with all the current flowing via the P- region. This is consistent with the using the cross-section along line A-A in Fig. 8.2 when formulating the analytical models for the carrier distribution and the on-state voltage drop.

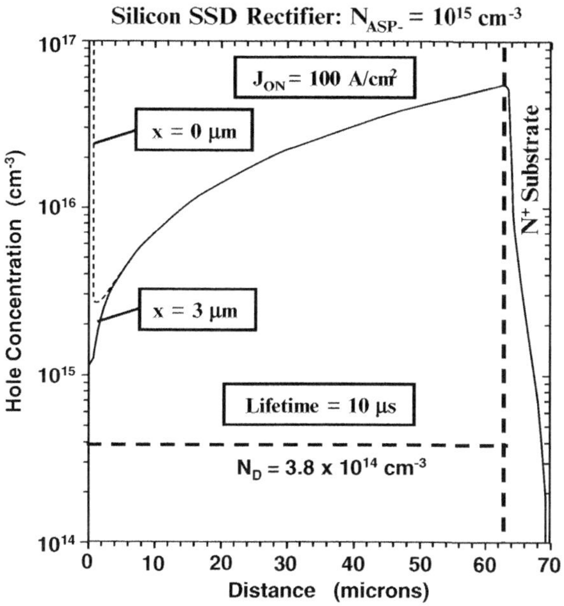


Fig. 8.4E Carrier Distribution within the 500V Silicon SSD Rectifier.

The carrier distribution within the silicon SSD rectifier at an on-state current density of 100 A/$cm^2$ is shown in Fig. 8.4E for the case of a P- region with surface concentration of 1 x $10^{15}$ $cm^{-3}$. Here, the hole concentration is shown with a solid line at the P- region (x = 3 microns) and at the $P^+$ region (x = 0 microns) by the dotted line. It can be seen that high-level injection conditions prevail in the drift region because the injected carrier concentration is much greater than the background doping concentration except in close proximity to the anode contact. The hole concentration profile obtained from the simulations is similar to that illustrated in Fig. 8.2 at the P- region. It can be observed from the figure that the hole distribution below the $P^+$ region is very similar to that below the P- region. This justifies utilization of a one-dimensional model for deriving the carrier distribution in the SSD rectifier structure. Due to the high-level injection conditions, the hole and electron concentrations are equal in magnitude throughout the drift region. From this figure, significant injection of holes can also be observed in the $N^+$ substrate region. It is therefore appropriate to include end region recombination when analyzing the current flow in the SSD rectifier.

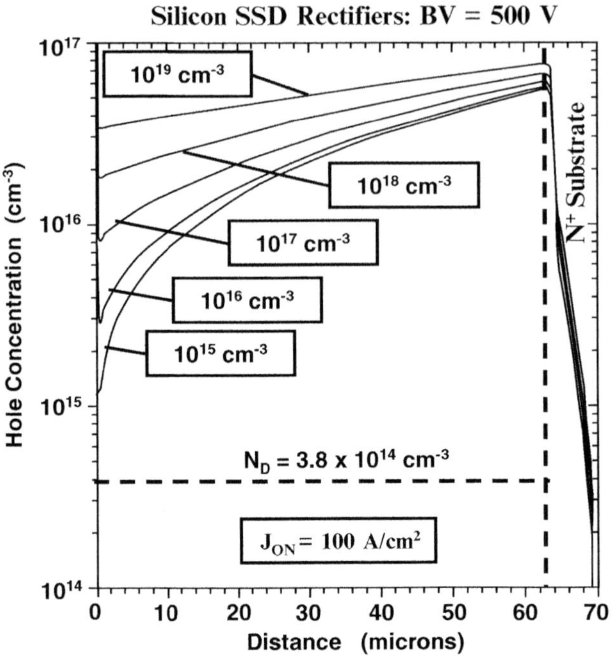


Fig. 8.5E Carrier Distribution within the Silicon SSD Rectifiers.

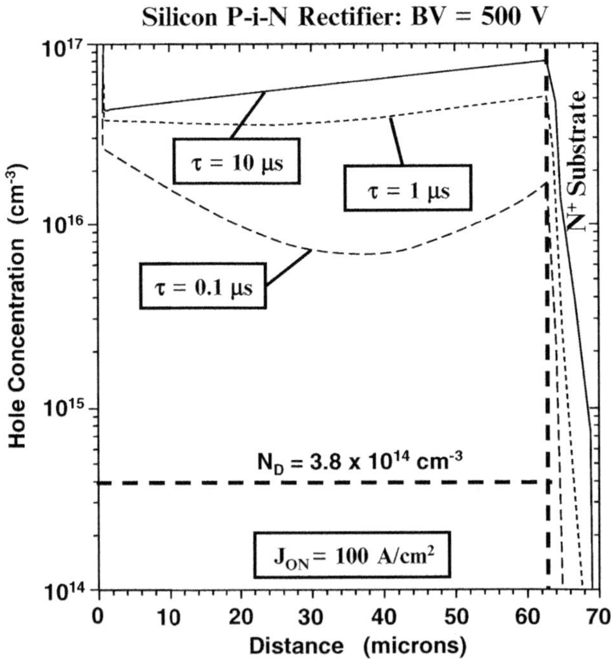


Fig. 8.6E Carrier Distribution within the Silicon P-i-N Rectifier.

The injection carrier concentration in the drift region increases when the surface concentration of the P- region is increased in the silicon SSD rectifier structure. This is illustrated in Fig. 8.5E where the hole concentration profile obtained by using numerical simulations is shown for five cases of the surface concentration of the P- region. The hole concentration at the P- region increases when the surface concentration of the P- region is increased. This behavior is described quite well by the analytical model with and without recombination in the $N^+$ end-region (see Fig. 8.4 ad Fig. 8.8). The hole concentration at the interface between the drift region and the $N^+$ substrate is observed to increase slightly in Fig. 8.5E. This behavior is intermediate to the predictions of the analytical models eveloped with and without recombination in the $N^+$ end-region (see Fig. 8.4 and Fig. 8.8).

The carrier distribution in the P-i-N rectifier obtained by numerical simulations is provided in Fig. 8.6E. A comparison of the hole concentration profiles for the SSD rectifier with a high surface concentration of 1 x $10^{19}$ $cm^{-3}$ for the P- region and that for the P-i-N rectifier with the same lifetime of 10 microseconds in the drift region can be performed using these figures. The carrier profiles are nearly identical demonstrating that the SSD rectifier behaves like a P-i-N rectifier when the surface concentration of the P- region is made very large.

## 8.2 Reverse Blocking

At first glance, the reverse leakage current in the SSD rectifier may be considered to like that for the P-i-N rectifier because the P-N junction extends across the entire cell cross-section. However, the P- region in the SSD structure must be lightly doped to achieve the desired reduction of carrier concentration at the anode side of the structure as demonstrated in the previous section. From this point of view, it is optimal to reduce the surface concentration of the P- region to only 1 x $10^{15}$ $cm^{-3}$. However, with such a low surface concentration for the P- region, the doping concentrations on the two sides of the junction become comparable. This promotes the extension of the depletion region on both sides of the junction. When the depletion region in the P- region extends through its width and reaches the ohmic contact at the anode metal surface, the leakage current for the SSD rectifier can become very large due to the large (infinite in the ideal case) surface recombination velocity at ohmic contacts. This will degrade the reverse blocking characteristics of the SSD rectifier. The surface concentration of the P- region must be sufficiently large to prevent the depletion of the entire P- region at the maximum reverse blocking voltage.

The depletion of the P- region can be analytically modeled by solving for the electric field distribution while taking into account the doping concentration and thickness of the P- region. A simple analytical model is developed here based up on assuming a uniform effective doping concentration for the P- region which is equal to half the surface doping concentration. For development of this model, the electric field profile for the portion of the SSD rectifier under the P- region is contrasted with the electric field profile for the portion of the SSD rectifier under the $P^+$ region in Fig. 8.11. The width of these p-type regions has been exaggerated to illustrate the shape of the electric field profile. In practice, the width of the $P^+$ and P- regions are small when compared to the depletion layer extension in the N drift region.

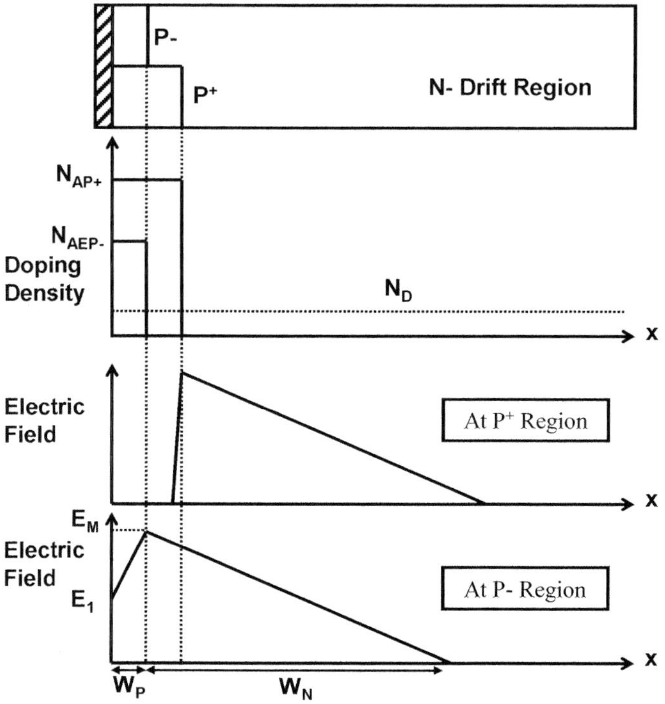


Fig. 8.11 Electric Field Profiles in the Silicon SSD Rectifier.

The electric field profile shown in Fig. 8.11 under the $P^+$ region is the familiar profile for an abrupt P-N junction[3]. The electric field profile at the P-region is different due to the low doping concentration of the P- region. An analytical computation of the electric field profile can be performed by determination of the maximum electric field ($E_M$) corresponding to an applied reverse bias ($V_R$). Under the assumption of a uniform effective doping concentration ($N_{AEP-}$) in the P- region, the electric field ($E_1$) at the surface (x = 0 in Fig. 8.11) is given by:

$$E_1 = E_M - \frac{qN_{AEP-}W_P}{\varepsilon_S} \quad [8.45]$$

where $W_P$ is the width of the P- region. The electric field within the P- region is then given by:

$$E(x) = E_1 + \frac{qN_{AEP-}}{\varepsilon_S}x \quad [8.46]$$

The electric field within the N drift region is related to the maximum electric field by:

$$E(x) = E_M - \frac{qN_D}{\varepsilon_S}(x - W_P) \qquad [8.47]$$

which is valid for x greater than $W_P$. The depletion layer width in the N drift region is also related to the maximum electric field:

$$W_N = \frac{E_M \varepsilon_S}{qN_D} \qquad [8.48]$$

Using the above expressions, the reverse bias supported by the SSD rectifier can be determined:

$$V_R = \frac{1}{2}(E_1 + E_M)W_P + \frac{1}{2}E_M W_N \qquad [8.49]$$

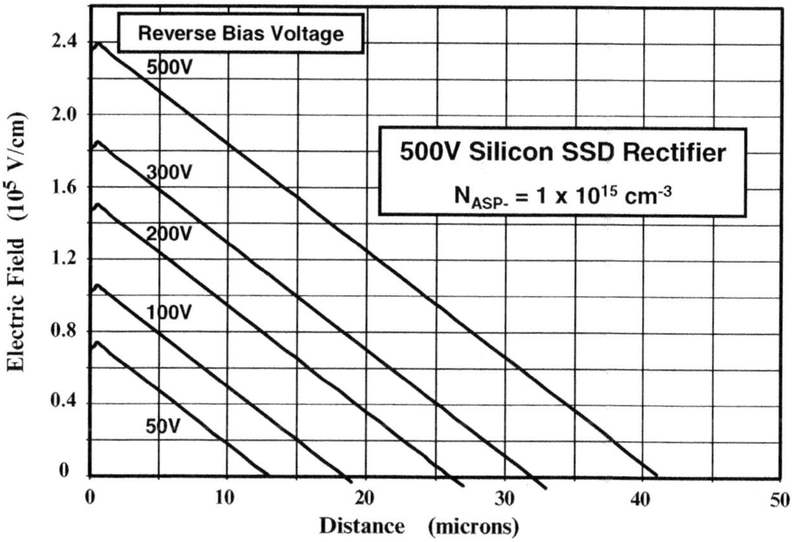


Fig. 8.12 Electric Field Profiles in the 500 V Silicon SSD Rectifier.

The electric field profile computed by using the above equations are shown in Fig. 8.12 at various reverse bias voltages for a 500V SSD rectifier structure with drift region doping concentration of 3.8 x $10^{14}$ cm$^{-3}$. In this case, a surface doping concentration ($N_{ASP-}$) of 1 x $10^{15}$ cm$^{-3}$ was used for the P- region with an effective doping concentration ($N_{AEP-}$) assumed to be one-half of this value in the analytical model. According to the analytical model, the maximum electric field occurs below the surface at the depth of the P- region. In the case of this low doping concentration for the P- region, the electric field punches through the P- region even at low reverse bias voltages. A high electric field ($E_1$) of 2.35 x $10^5$ V/cm is predicted by the model at the anode contact for a reverse bias of 500 volts. The development of the electric field within the P- region can be observed more clearly in Fig. 8.13 where the scale for the x-axis has been expanded.

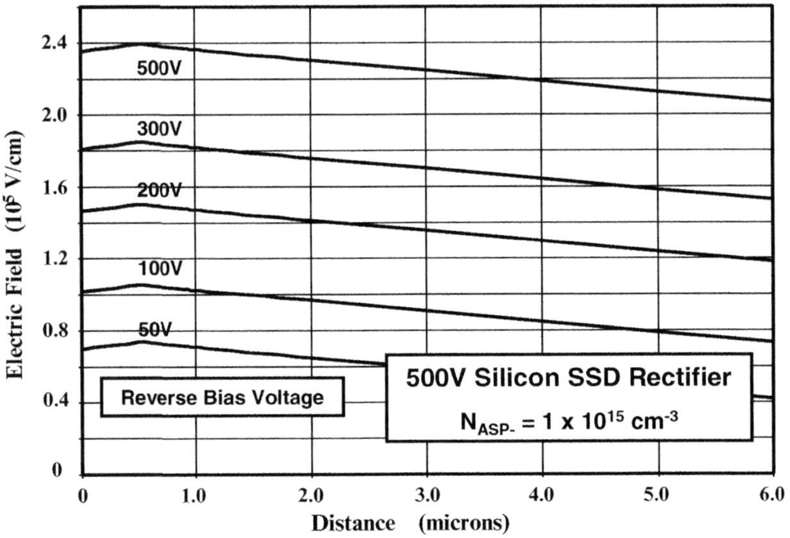


Fig. 8.13 Electric Field Profiles in the 500 V Silicon SSD Rectifier.

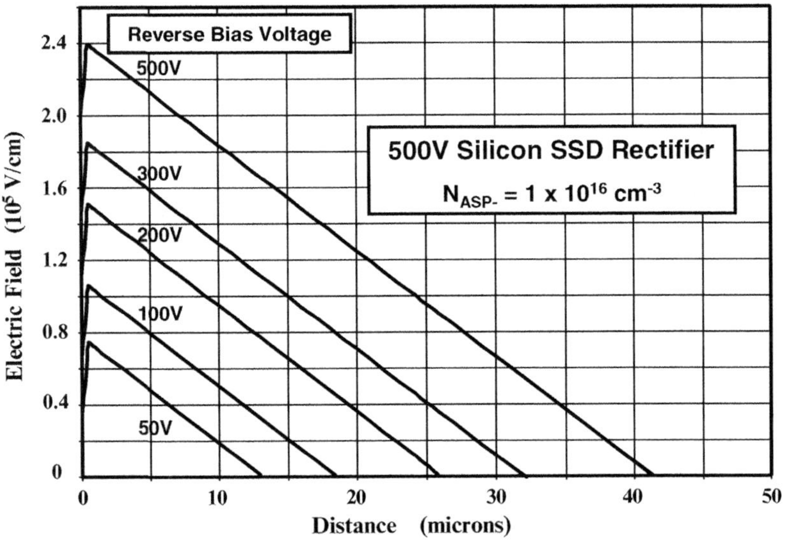


Fig. 8.14 Electric Field Profiles in the 500 V Silicon SSD Rectifier.

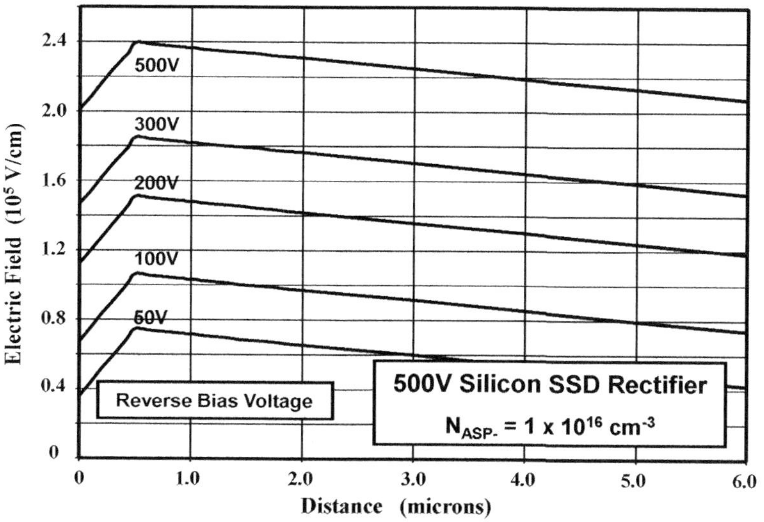


Fig. 8.15 Electric Field Profiles in the 500 V Silicon SSD Rectifier.

When the surface doping concentration of the P- region is increased to $1 \times 10^{16}$ cm$^{-3}$, the electric field within the P- region is reduced as shown in Fig. 8.14 and Fig. 8.15. In the case of this doping concentration for the P- region, the electric field still punches through the P- region even at low reverse bias voltages. A high electric field ($E_1$) of $2.01 \times 10^5$ V/cm is predicted by the model at the anode contact for a reverse bias of 500 volts. Thus, the surface doping concentration of $1 \times 10^{16}$ cm$^{-3}$ is insufficient to provide good reverse blocking capability in the silicon SSD rectifier.

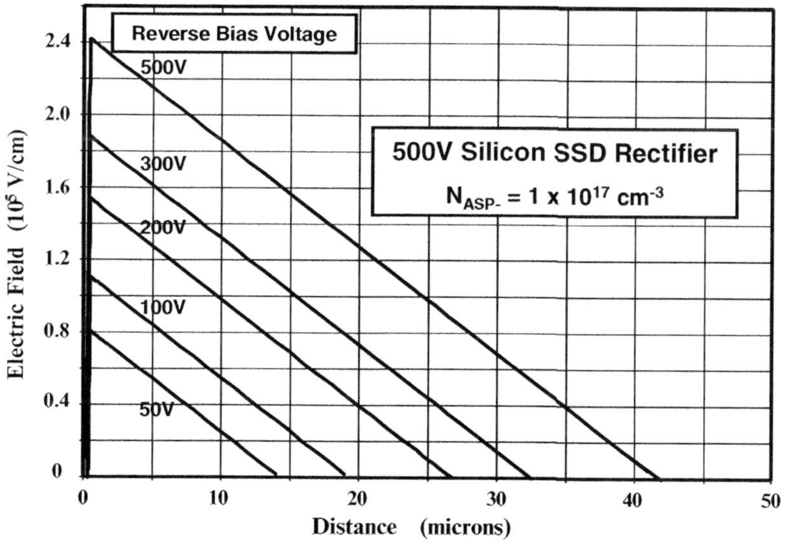


Fig. 8.16 Electric Field Profiles in the 500 V Silicon SSD Rectifier.

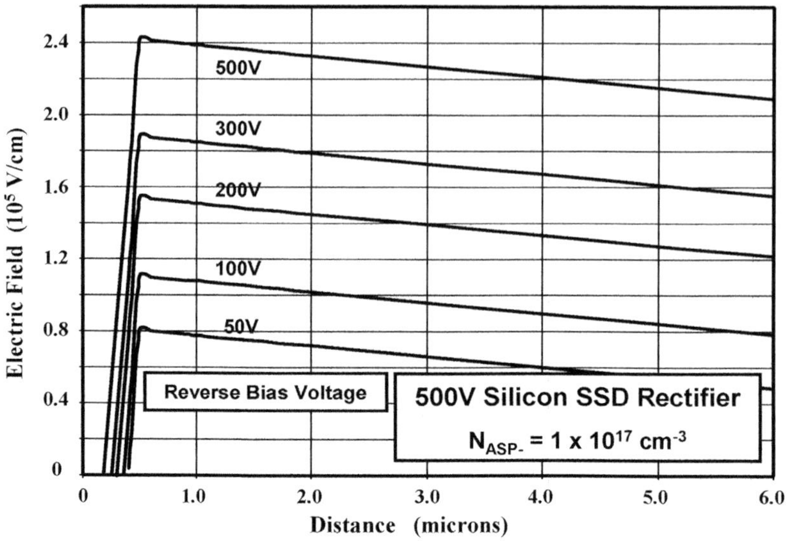


Fig. 8.17 Electric Field Profiles in the 500 V Silicon SSD Rectifier.

In order to suppress the depletion of the P- region under reverse blocking operation, it is necessary to increase the surface doping concentration of the P- region to 1 x $10^{17}$ cm$^{-3}$ as shown in Fig. 8.16 and Fig. 8.17. In the case of this doping concentration for the P- region, the electric field is truncated within the P- region even at high reverse bias voltages. The analysis of the electric field during reverse blocking operation of the SSD rectifier structure demonstrates that the ability to suppress the build up of free carriers in the vicinity of the anode is curtailed by the depletion of the P- region. It can therefore be concluded that the SSD structure cannot approach the performance of the MPS rectifier structure.

## Simulation Example

In order to validate the above model for the electric field distribution in the SSD rectifier in the reverse blocking mode, the results of two-dimensional numerical simulations on a 500 V silicon SSD rectifier structure are described here. The structure had a drift region with a doping concentration of 3.8 x $10^{14}$ cm$^{-3}$ and a thickness of 65 microns. The cell pitch for all structures was 3 microns with a $P^+$ region depth of 1 micron and an ion-implant window (dimension s in Fig. 3.5) of 0.5 microns. The P- region has a width ($W_P$) of 0.5 microns.

The electric field profile at the middle of the P- region in the silicon SSD rectifier structure with surface concentration of 1 x $10^{15}$ cm$^{-3}$ for the P- region is shown in Fig. 8.7E and Fig. 8.8E. At small reverse bias voltages, the electric field profile resembles that predicted by the model with the maximum electric field located at a depth of 0.5 microns. However, at larger reverse bias voltages, the maxima for the electric field profiles moves to a greater depth (about 3 microns). This phenomenon is due to the shielding of the P- region by depletion layer expansion at the $P^+$ region as observed in the MPS rectifier structure.

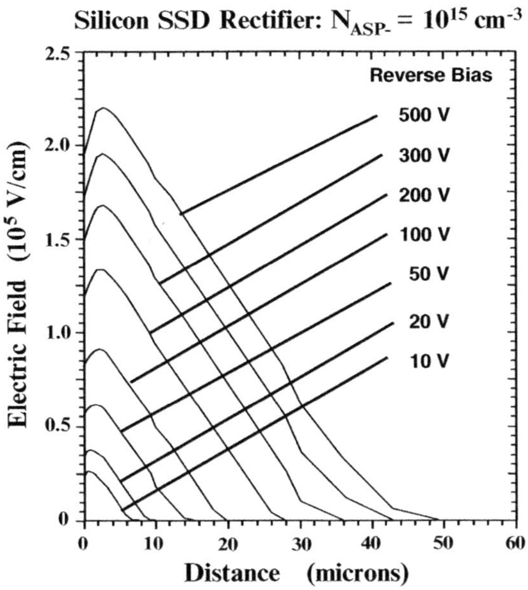


Fig. 8.7E Growth of the Electric Field at the Middle of the P- Region in a 500 V Silicon SSD Rectifier.

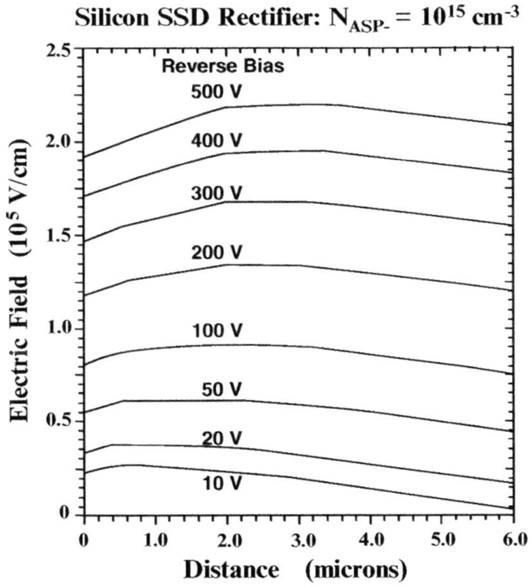


Fig. 8.8E Growth of the Electric Field at the Middle of the P- Region in a 500 V Silicon SSD Rectifier.

From the above figures, it is apparent that the electric field at the anode contact is large even for small reverse bias voltages. The electric field at the anode contact is observed to be 5.5 x $10^4$ V/cm at a reverse bias of 50 volts. This value is smaller than that predicted by the simple analytical model (see Fig. 8.13). Similarly, at a reverse bias of 500 volts, the electric field at the anode contact is observed to be 1.9 x $10^5$ V/cm which is significantly smaller than the value predicted by the simple analytical model. This difference is due to the shielding of the P- region by the depletion region from the $P^+$ region in the SSD structure.

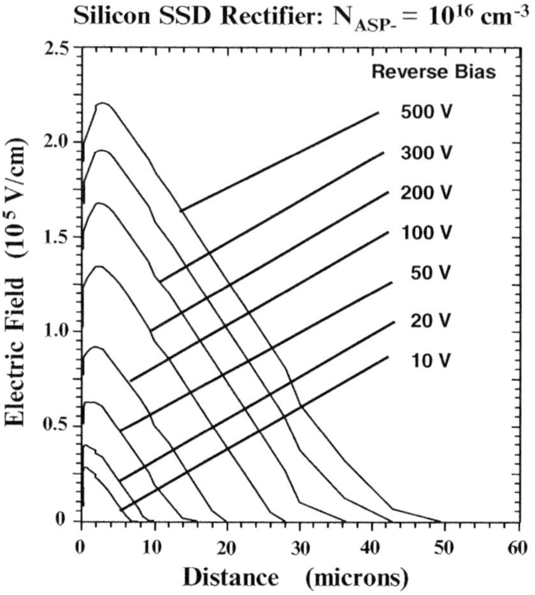


Fig. 8.9E Growth of the Electric Field at the Middle of the P- Region in a 500 V Silicon SSD Rectifier.

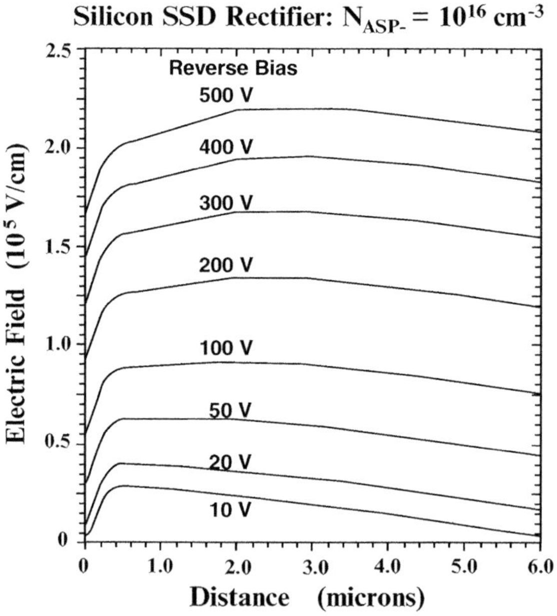


Fig. 8.10E Growth of the Electric Field at the Middle of the P- Region in a 500 V Silicon SSD Rectifier.

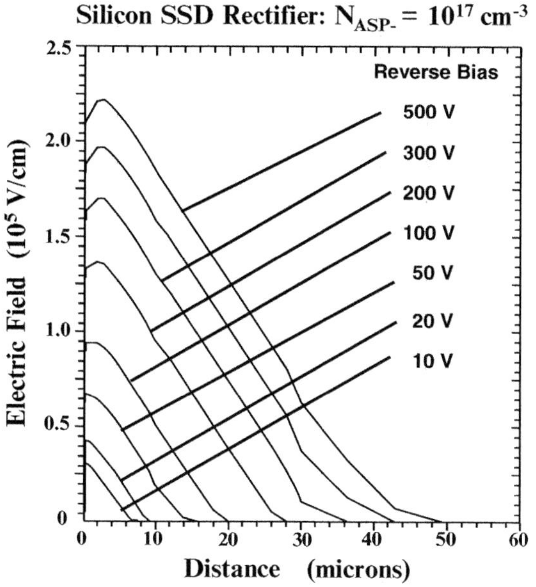


Fig. 8.11E Growth of the Electric Field at the Middle of the P- Region in a 500 V Silicon SSD Rectifier.

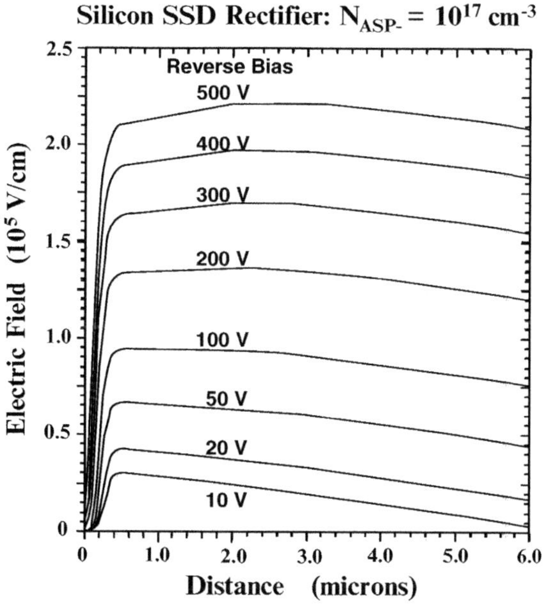


Fig. 8.12E Growth of the Electric Field at the Middle of the P- Region in a 500 V Silicon SSD Rectifier.

Numerical simulations of the SSD rectifier structure containing a P- region with a surface concentration of $1 \times 10^{16}$ cm$^{-3}$ show a reduction of the electric field at the anode contact as shown in Fig. 8.9E and Fig. 8.10E. In this case, the electric field ($E_1$) at the anode contact is found to be 1.67 x $10^5$ V/cm at a reverse bias of 500 volts. This value is substantially smaller than that predicted by the analytical model due to shielding from the $P^+$ region as indicated by the movement of the maximum electric field to 2 microns below the surface.

When the surface concentration of the P- region in the SSD rectifier structure is increased to 1 x $10^{17}$ $cm^{-3}$, the numerical simulations indicate that the P- region is no longer completely depleted at the maximum reverse bias of 500 volts. This can be observed in Fig. 8.11E and Fig. 8.12E. The results of the numerical simulations indicate that the electric field ($E_1$) is barely reduced to zero with this surface doping concentration for the P- region. From this observation, it can be concluded that the surface doping concentration of the P- region in the silicon SSD rectifier structure should be at least 1 x $10^{17}$ $cm^{-3}$. With this relatively high surface concentration for the P- region, the hole carrier concentration at the P-/N junction in the on-state is found to be quite large as shown in Fig. 8.5E. This results in a significantly inferior reverse recovery characteristics for the SSD rectifier structure when compared with the MPS rectifier structure as demonstrated in the next section of this chapter.

## 8.3 Switching Performance

Power rectifiers operate for part of the time in the on-state when the bias applied to the anode is positive and for the rest of the time in the blocking state when the bias applied to the anode is negative. During each operating cycle, the diode must be rapidly switched between these states to minimize power losses. Much greater power losses are incurred when the diode switches from the on-state to the reverse blocking state than when it is turned on. The presence of a large concentration of free carriers in the drift region during on-state current flow is responsible for the low on-state voltage drop of high voltage P-i-N, MPS and SSD rectifiers. The stored charge within the drift region of these power rectifiers produced by the on-state current flow must be removed before it is able to support high voltages. This produces a large reverse current for a short time duration. This phenomenon is referred to as *reverse recovery*.

### 8.3.1 Stored Charge

It is instructive to compare the stored charge in the SSD rectifier structure with that in the P-i-N and MPS rectifier structures because this provides a relative measure of the energy loss that will occur during the turn-off transient. In the case of the P-i-N rectifier, the drift region has almost a uniform (average) carrier concentration given by Eq. [6.23]. Using this equation for the case of the 500 V silicon P-i-N rectifier example in the previous sections, the average carrier concentration is found to be 9 x $10^{17}$ $cm^{-3}$ at an on-state current density of 100 A/$cm^2$ if the lifetime in the drift region is 10 microseconds. The total stored charge in the drift region based upon a uniform concentration of carriers is then found to be 1.01 mC/$cm^2$ by multiplying the carrier concentration and the thickness of the drift region. However, with a high lifetime in the drift region, the end-region recombination becomes dominant and the carrier concentration in the drift region is greatly reduced. From simulations, it is found that the average carrier concentration in the drift region is about 7 x $10^{16}$ $cm^{-3}$ leading to a total stored charge of 0.078 $mC/cm^2$.

In contrast, the carrier distribution in the MPS rectifier has a triangular shape with a maximum value at the interface between the drift region and $N^+$ substrate. The maximum carrier concentration in the 500V silicon MPS can be obtained by using either Eq. [7.21] or Eq. [7.31]. For a lifetime of 10 microseconds in the drift region, the maximum carrier concentration is found to be 6 x $10^{16}$ $cm^{-3}$ at an on-state current density of 100 $A/cm^2$. The total stored charge in the drift region of the silicon MPS rectifier is then found to be 0.034 $mC/cm^2$. Consequently, the stored charge in the silicon MPS rectifier is about 2-times smaller than for the P-i-N rectifier in spite of a lower on-state voltage drop.

The carrier distribution in the SSD rectifier structure depends up on the surface concentration of the P- region. As the surface concentration of the P- region is increased, the hole concentration at the P-/N junction increases as previously shown in Fig. 8.8. This carrier profile was derived under the assumption that the recombination in the drift region can be neglected. Under these assumptions, the carrier distribution has a linear shape as described Eq. [8.39] which can be re-written using the hole concentrations at the two extremes of the drift region:

$$p(x) = \frac{p(+d)}{2}\left(\frac{x}{d}+1\right) + \frac{p(x_N)}{2}\left(1-\frac{x}{d}\right) \quad \textbf{[8.50]}$$

The total stored charge in the drift region of the SSD rectifier in the on-state is given by integration of this hole concentration profile from x = –d to x= +d:

$$Q_S = q\left[\frac{p(-d)+p(+d)}{2}\right] \quad \textbf{[8.51]}$$

Using the hole concentration at x = –d given by Eq. [8.16] and the hole concentration at x = +d given by Eq. [8.38], the total stored charge in the drift region of the SSD rectifier is found to be:

$$Q_S = \frac{q\sqrt{J_T}}{2}\left[\sqrt{\frac{W_B N_{AEP-}}{qD_n}\left(1-\frac{D_p}{D_n}\right)} + \frac{n_{ieN+}}{\sqrt{J_{SN+}}}\right] \quad \textbf{[8.52]}$$

Based up on this expression, the stored charge in the SSD rectifier structure increases as the square-root of the on-state current density when recombination in the $N^+$ end-region is dominant.

The increase in the stored charge within the 500 volt silicon SSD rectifier structure computed by using the above analytical expression is shown in Fig. 8.18 with increasing surface concentration of the P- region. It can be observed that the stored charge doubles when the surface concentration of the P- region is increased from 1 x $10^{15}$ $cm^{-3}$ to 1 x $10^{19}$ $cm^{-3}$. The stored charge at the minimum surface concentration of 1 x $10^{17}$ $cm^{-3}$ for the P- region is only 8.6 percent larger than at the lowest P- region surface concentration. Although this appears favorable, the actual reverse recovery process is degraded by the significant hole concentration at the P-/N junction in the on-state as shown below.

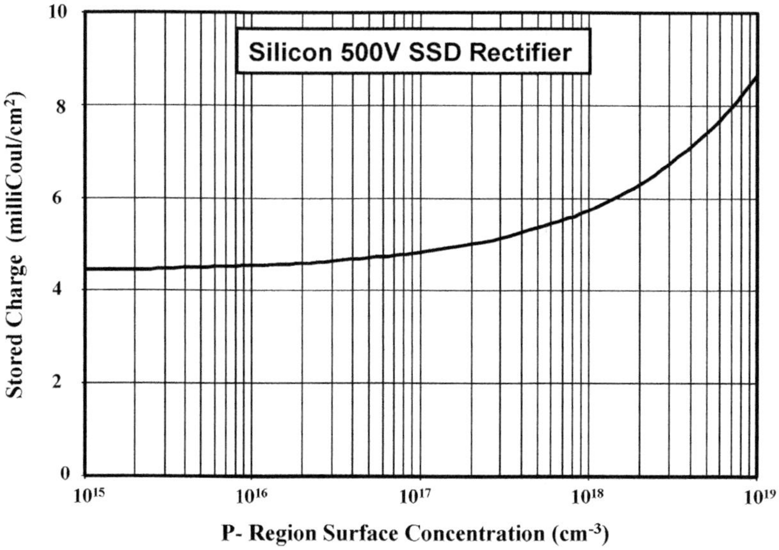


Fig. 8.18 Stored Charge in the 500 V Silicon SSD Rectifier.

### 8.3.2 Reverse Recovery

When the surface concentration for the P- region in the SSD rectifier structure is small, the on-state carrier distribution in the drift region resembles that for the MPS rectifier structure. The analytical theory for the reverse recovery process developed in chapter 7 can therefore be utilized to analyze the performance of the SSD rectifier structures with low surface concentration of the P- region. When the surface concentration for the P- region in the SSD rectifier structure is large, the on-state carrier distribution in the drift region resembles that for the P-i-N rectifier structure. The analytical theory for the reverse recovery process developed in chapter 6 can therefore be utilized to analyze the performance of the SSD rectifier structures with high surface concentration of the P- region. Consequently, an analytical formulation for the SSD rectifier structure is not developed here.

**Simulation Example**

The results of numerical simulations on the 500 V silicon SSD rectifier structure are provided here to demonstrate that it behaves like the MPS rectifier structure when the surface concentration of the P- region is low and like the P-i-N rectifier when the surface concentration of the P- region is large. The device structure had a drift region thickness of 65 microns with a doping concentration of $3.8 \times 10^{14}$ cm$^{-3}$. The surface concentration of the P- region was varied from $1 \times 10^{15}$ cm$^{-3}$ to $1 \times 10^{19}$ cm$^{-3}$ for the SSD rectifier structures. In all case, the reverse recovery process was performed with the cathode current ramped down from an on-state current density of 100 A/cm$^2$ at a rate of $3 \times 10^8$ A/cm$^2$-s. The reverse blocking voltage (supply voltage) was assumed to be 300 volts in all cases.

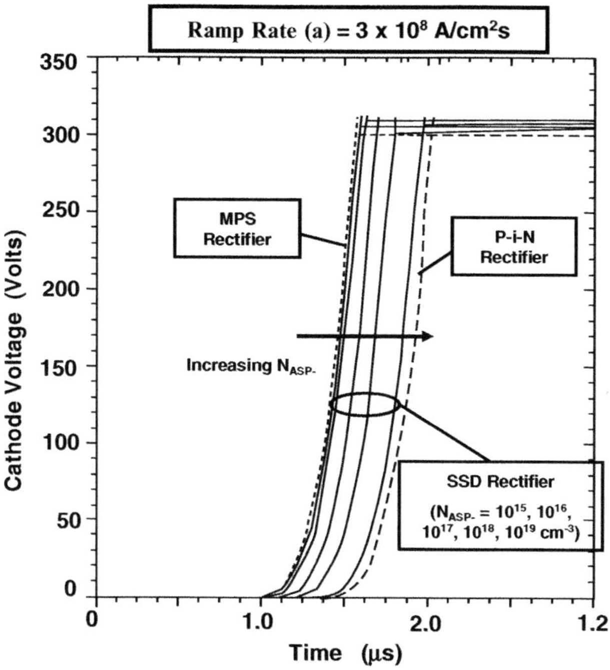


Fig. 8.13E Voltage Waveforms for a 500V Silicon SSD Rectifier during the Reverse Recovery Transient.

The voltage and current waveforms obtained with the aid of the numerical simulations of the reverse recovery for the silicon SSD rectifiers with various surface concentrations for the P- region are shown in Fig. 8.13E and 8.14E, respectively. In addition, the waveforms for the turn-off transient for the 500V silicon MPS rectifier (with work function of 5.0 for the Schottky contact) and the 500V P-i-N rectifier have been included in the plots for comparison with the SSD rectifier structures. The waveforms for SSD rectifiers fall in-between those for the MPS and P-i-N rectifiers. The waveforms for the SSD rectifier with the P- region with surface concentration of $1 \times 10^{15}$ cm$^{-3}$ are very close to those observed for the MPS rectifier. When the surface concentration for the P- region in the SSD rectifier structure is increased to $1 \times 10^{19}$ cm$^{-3}$, its characteristics become slightly superior to those of the P-i-N rectifier. Due to the problem with punch-through of the depletion layer in the P- region, its surface doping concentration must be at least $1 \times 10^{17}$ cm$^{-3}$. In this case, the performance of the SSD rectifier structure becomes significantly inferior to that of the MPS rectifier.

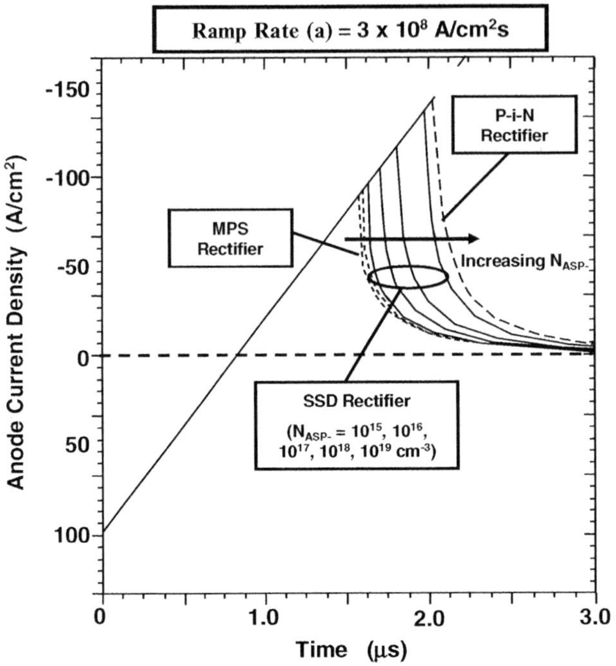


Fig. 8.14E Current Waveforms for a 500V Silicon SSD Rectifier during the Reverse Recovery Transient.

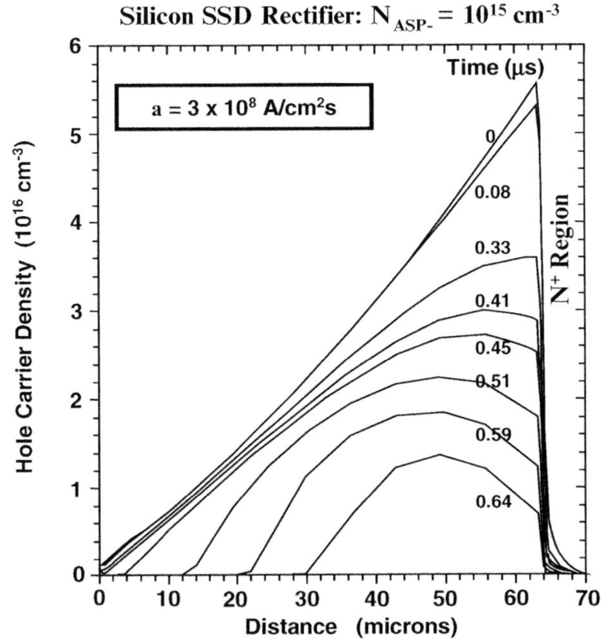


Fig. 8.15E Carrier Distribution in a 500V Silicon SSD Rectifier during the Reverse Recovery Transient.

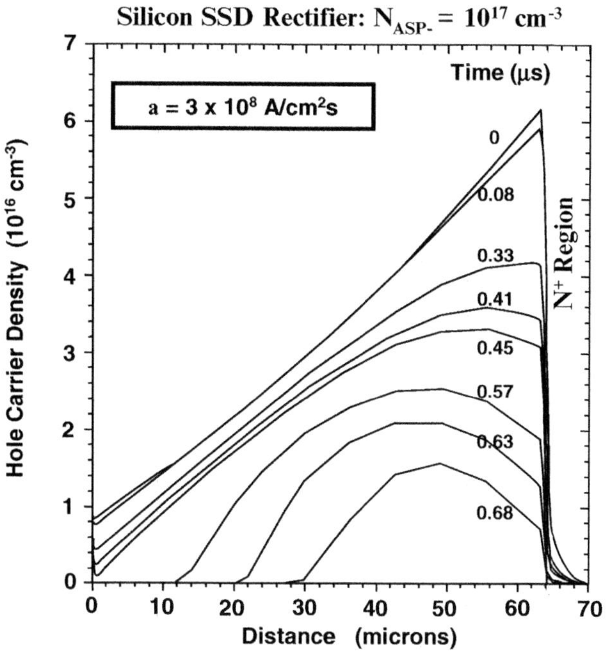


Fig. 8.16E Carrier Distribution in a 500V Silicon SSD Rectifier during the Reverse Recovery Transient.

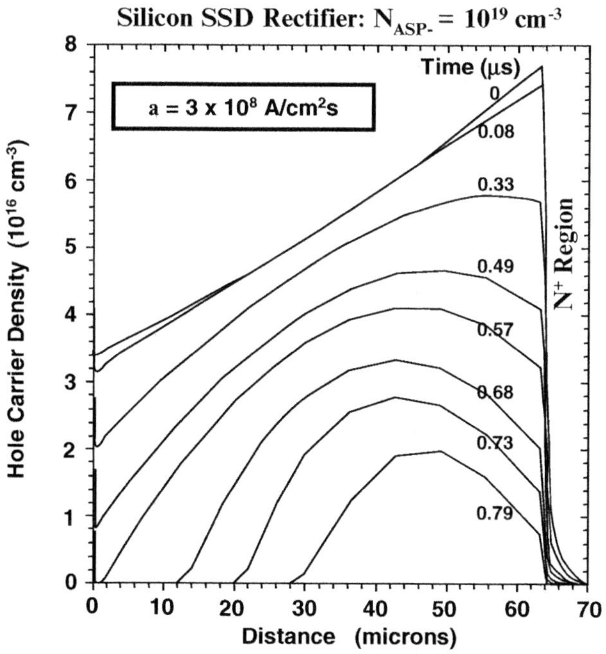


Fig. 8.17E Carrier Distribution in a 500V Silicon SSD Rectifier during the Reverse Recovery Transient.

A good understanding of the reverse recovery characteristics of the SSD rectifier can be obtained by examination of the free carrier profile in the drift region during the turn-off process. This information is provided in Fig. 8.15E, Fig. 8.16E, and Fig. 8.17E for the case of SSD rectifiers with P- regions having surface concentration of 1 x $10^{15}$ $cm^{-3}$, 1 x $10^{17}$ $cm^{-3}$ and 1 x $10^{19}$ $cm^{-3}$, respectively. From Fig. 8.15E, it can be seen than the free carrier distribution in the SSD rectifier is very similar to that observed for the MPS rectifier (see Fig. 7.39E) throughout the reverse recovery process. From Fig. 8.17E, it can be observed that the free carrier distribution in the SSD rectifier is more like that observed for the P-i-N rectifier (see Fig. 6.8E) throughout the reverse recovery process. For the optimal surface doping concentration of 1 x $10^{17}$ $cm^{-3}$ for the P- region in the SSD rectifier structure, the carrier distribution takes a form that falls in between these two cases. The presence of a significant hole concentration at the P-/N drift junction for the case of SSD rectifiers with P- regions having surface concentration of 1 x $10^{17}$ $cm^{-3}$ and 1 x $10^{19}$ $cm^{-3}$ prolongs the time at which these structures are able to support the reverse bias voltage resulting in larger peak reverse recovery current.

## 8.4 SSD Rectifier Trade-Off Curves

In the previous section, it was demonstrated that the peak reverse recovery current and the turn-off time can be reduced by utilizing the SSD rectifier structure. This enables reduction of the power losses during the switching transient. The peak reverse recovery current, reverse recovery turn-off time, and the reverse recovery charge for the SSD rectifier increase when the surface doping concentration of the P- region is increased. The reduction of the surface concentration for the P- region is limited to 1 x $10^{17}$ $cm^{-3}$ by the reach-through of the depletion region in the P- region during reverse blocking.

| Surface Concentration ($N_{ASP-}$) | $V_{ON}$ (100 A/$cm^2$) | $t_{rr}$ (μs) | $J_{PR}$ (A/$cm^2$) | $Q_{RR}$ (μC/$cm^2$) |
|---|---|---|---|---|
| 1 x $10^{17}$ $cm^{-3}$ | 0.843 | 0.557 | 103 | 28.7 |
| 1 x $10^{18}$ $cm^{-3}$ | 0.841 | 0.60 | 117 | 35.1 |
| 1 x $10^{19}$ $cm^{-3}$ | 0.845 | 0.66 | 138 | 45.5 |
| P-i-N | 0.870 | 0.697 | 140 | 48.8 |
| MPS | 0.847 | 0.50 | 86.7 | 21.7 |

Fig. 8.19 On-state Voltage Drop and Reverse Recovery Parameters obtained by Numerical Simulations of the 500V Silicon SSD Rectifiers.

In order to compare the performance of the silicon SSD rectifier with that of the silicon P-i-N and MPS rectifiers, the on-state voltage drop and reverse recovery characteristics were obtained for the 500 volt structures using numerical simulations. The data acquired from the simulations are tabulated in Fig. 8.19. For the same lifetime in the drift region, the peak reverse recovery current, the reverse recovery time, and the reverse recovery charge are substantially smaller for the silicon SSD rectifier when compared with the silicon P-i-N rectifier. However, even superior performance can be derived by utilizing the MPS rectifier structure. The best SSD rectifier structure has 30 percent larger reverse recovery charge than the MPS rectifier indicating that a significant reduction in reverse recovery power loss is possible with the MPS rectifier structure when compared to the SSD rectifier structure.

## 8.5 Silicon Carbide SSD Rectifiers

In principle, it is possible to extend the SSD rectifier concept to silicon carbide structures. However, the practical implementation of the concept is not feasible because the maximum electric field in the silicon carbide devices is much greater (~ 10-times) than that in silicon devices. The larger maximum electric fields in silicon carbide produce greater depletion of the P- region in the SSD rectifier structure. The concentration of the P- region has to be increased to much greater levels in the silicon carbide SSD structure when compared to the silicon structure. Due to the low diffusion rates for dopants in silicon carbide, the P- region can be assumed to have a uniform doping concentration. This is favorable for reducing the penetration of the depletion region within the P- region in silicon carbide SSD rectifier structures.

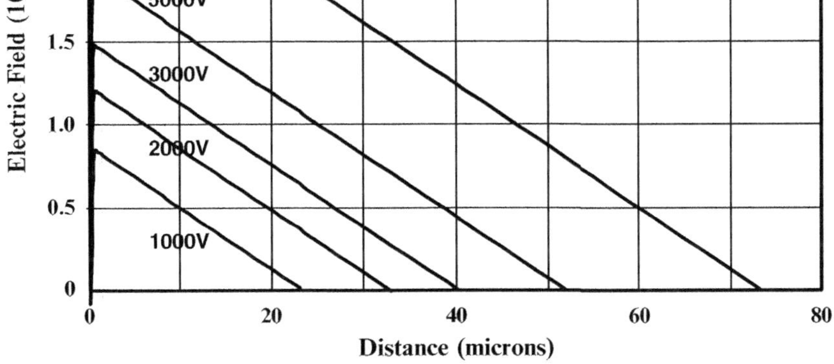


Fig. 8.20 Electric Field Profiles in the 10 kV 4H-SiC SSD Rectifier.

As an example, consider the 10 kV silicon carbide SSD rectifier structure with a drift region having a doping concentration of 2 x $10^{15}$ $cm^{-3}$ and thickness of 80 microns. The electric field profiles for the SSD rectifier structure with a doping concentration of 1 x $10^{17}$ $cm^{-3}$ for the P- region are shown in Fig. 8.20 and Fig. 8.21. In contrast to the silicon SSD rectifier structure with a surface concentration of 1 x $10^{17}$ $cm^{-3}$ for the P- region (see Fig. 8.16 and Fig. 8.17), there is strong reach-through of the depletion layer within the P- region for the silicon carbide SSD rectifier structure due to the much larger maximum electric field at the junction.

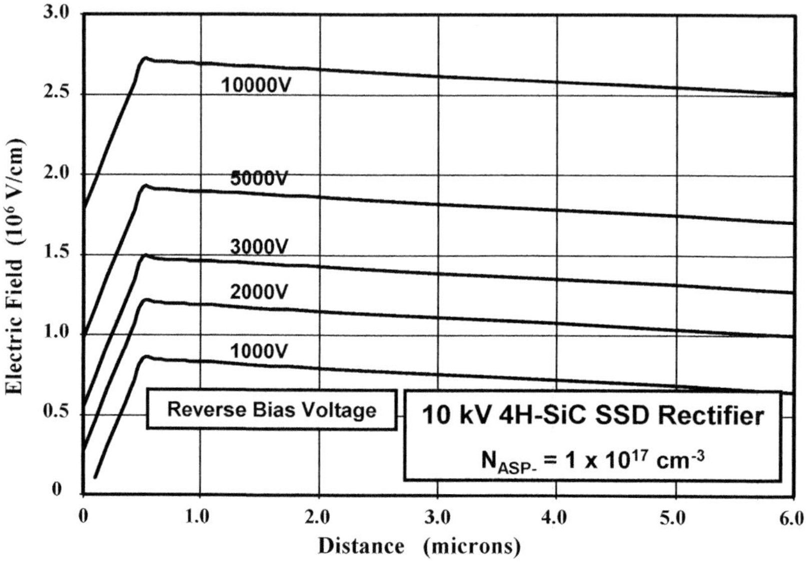


Fig. 8.21 Electric Field Profiles in the 10 kV 4H-SiC SSD Rectifier.

The doping concentration of the P- region in the silicon carbide SSD rectifier structure has to be increased relative to the silicon structure to suppress the reach-through of the depletion region in the P- region. In order to demonstrate this, the electric field profiles for the silicon carbide SSD rectifier structure with a doping concentration of 3 x $10^{17}$ $cm^{-3}$ for the P- region are shown in Fig. 8.22 and Fig. 8.23. It can be observed from Fig. 8.23 that the depletion region barely extends through the P- region at a reverse bias of 10,000 volts for the doping concentration of 3 x $10^{17}$ $cm^{-3}$ for the P- region. This indicates that a typical doping concentration of 5 x $10^{17}$ $cm^{-3}$ would be required for the P- region in practical silicon carbide SSD rectifier structures. This relatively high doping concentration for the P- region would negate most of the advantages of the SSD rectifier structure in the case of silicon carbide. It can therefore be concluded that the MPS rectifier structure is more suitable for improving the performance of high voltage silicon carbide power rectifiers.

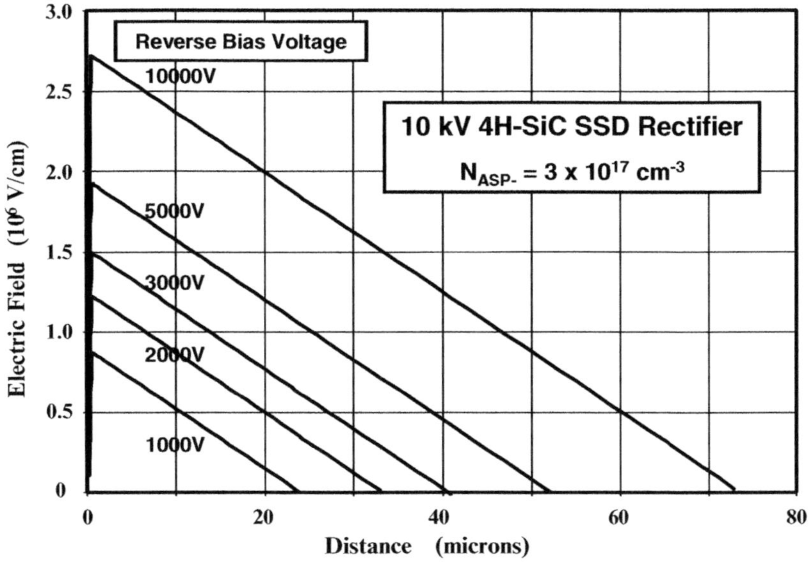


Fig. 8.22 Electric Field Profiles in the 10 kV 4H-SiC SSD Rectifier.

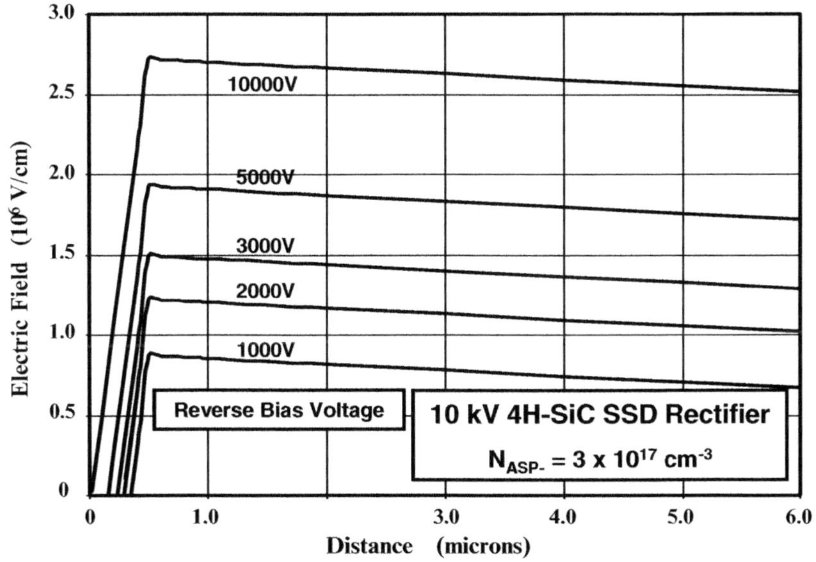


Fig. 8.23 Electric Field Profiles in the 10 kV 4H-SiC SSD Rectifier.

## 8.6 Summary

The physics of operation of the SSD rectifier structure has been analyzed in this chapter. Analytical expressions have been derived for the on-state carrier distribution and the on-state voltage drop. At on-state current levels, the injected minority carrier density in the drift region exceeds the relatively low doping concentration required to achieve high breakdown voltages. This high level injection in the drift region modulates its conductivity producing a reduction in the on-state voltage drop. The carrier concentration in the drift region of the SSD rectifier structures takes a form in between that for the P-i-N and MPS rectifiers.

As in the case of the P-i-N rectifier, the switching of the SSD rectifier from the on-state to the reverse blocking state is accompanied by a significant current flow in the reverse direction. However, the peak reverse recovery current and reverse recovery time are smaller than those observed in the P-i-N rectifier. When the doping concentration of the P- region of the SSD rectifier structure is made very small, its turn-off characteristics are like those of the MPS rectifier. However, such low doping levels in the P- region produce punch-through of the depletion layer to the anode contact which can result in a high leakage current during the reverse blocking mode. When the doping concentration of the P- region is made large, the SSD rectifier turn-off behavior approaches that for the P-i-N rectifier with large turn-off power losses. Consequently, it can be concluded that the MPS rectifier structures offers a better solution for obtaining a low on-state voltage drop and low reverse recovery switching losses.

## References

[1] B.J. Baliga, "Analysis of the High-Voltage Merged P-i-N/Schottky (MPS) Rectifier", IEEE Electron Device Letters, Vol. EDL-8, pp. 407-409, 1987.

[2] Y. Shimizu, et al, "High-Speed, Low-Loss P-N Diode having a Channel Structure", IEEE Transactions on Electron Devices, Vol. ED-31, pp. 1314-1319, 1984.

[3] B.J. Baliga, "Fundamentals of Power Semiconductor Devices", Springer Science, New York, 2008.

# Chapter 9

# Synopsis

Power devices are required for systems that operate over a broad spectrum of power levels and frequencies as discussed in the textbook[1]. The wide range of operating frequency for these applications was illustrated in Fig. 1.1. Another classification for the applications that is useful is based up on the operating voltage level as shown in Fig. 9.1. At lower voltages (< 100 voltages), a large number of power rectifiers are needed for computer power supplies and in automotive electronics. Rectifiers with larger voltage ratings are required for many other applications, such as motor drives, shown in the figure.

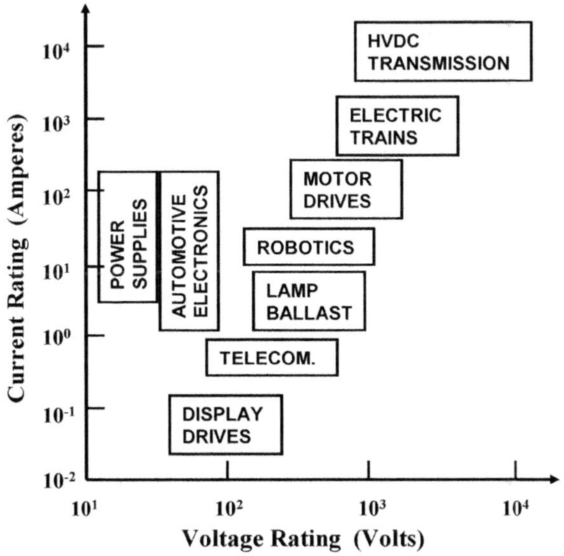


Fig. 9.1 System ratings for Power Devices.

B.J. Baliga, *Advanced Power Rectifier Concepts*, DOI: 10.1007/978-0-387-75589-2_9,

A variety of advanced power rectifier structures were discussed in previous chapters of this book for serving the applications shown in Fig. 9.1. The choice of the optimum device suitable for each application depends upon the device voltage rating and the circuit switching frequency. In the case of applications with low operating voltages (< 100 volts), silicon Schottky rectifiers have been commonly used due to their low on-state voltage drop. A typical example is the DC-to-DC buck converter circuit shown in Fig. 1.12 used as Voltage Regulator Modules (VRMs). However, the reduction of power losses in this application is limited by the large leakage current for silicon Schottky rectifiers. The performance of these devices can be improved up on by using the Junction Barrier controlled Schottky (JBS) rectifier concept discussed in chapter 3.

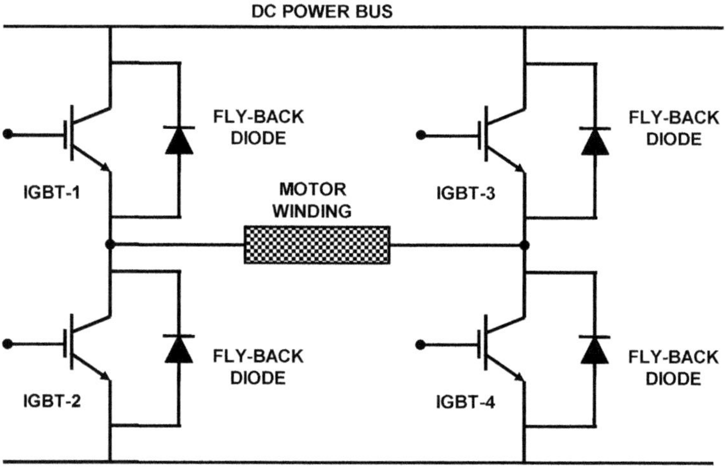


Fig. 9.2 Typical H-Bridge Topology for Motor Control.

Power rectifiers are also commonly used in motor control circuits. The most prevalent applications for commercial and industrial systems utilize silicon IGBTs as the power switch and silicon P-i-N rectifiers as the fly-back diodes with an H-bridge configuration as shown in Fig. 9.2. The operating voltages for these applications typically range from 300 volts to 6000 volts. At these voltage levels, the silicon MPS rectifier, discussed in chapter 7 of this book, provides an alternative to the P-i-N rectifier for reducing the power loss not only in the rectifiers but also in the IGBTs. In addition, an even greater improvement in performance can be obtained by utilizing the silicon carbide JBS rectifier that was discussed in chapter 3.

In this concluding chapter, an example of the performance enhancement that can be obtained by replacing the silicon Schottky rectifier with the silicon JBS rectifier in VRMs is provided. This is followed by an example of a typical motor control application to quantify the benefits of replacing the silicon P-i-N rectifier with the silicon MPS rectifier and the silicon carbide JBS rectifier.

## 9.1 DC-to-DC Buck Converter Application

The basic DC-to-DC buck converter circuit is shown in Fig. 1.12. In this circuit, current flow through the output inductor is provided for a part of the cycle through the power MOSFET. During the rest of the cycle, the current in the inductor circulates through the rectifier. The current and voltage waveforms for the rectifier in this circuit are illustrated in Fig. 9.3.

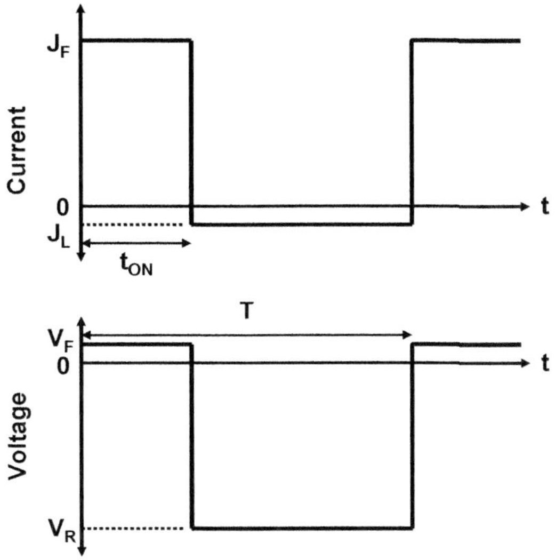


Fig. 9.3 Typical Switching Waveforms for the Rectifier.

As shown in Fig. 9.3, typical Schottky rectifiers exhibit a voltage drop ($V_F$) during current conduction in the forward direction. This results in power dissipation per unit area in the on-state given by:

$$P_L(on) = \delta . J_F . V_F \qquad [9.1]$$

where $J_F$ is the on-state current density. In this expression, δ is referred to as the duty cycle given by:

$$\delta = t_{ON} / T \qquad [9.2]$$

where $t_{ON}$ is the on-state duration and T is the time period (the reciprocal of the operating frequency). The on-state power dissipation decreases with increasing temperature because the on-state voltage drop decreases with increasing temperature for Schottky rectifiers.

The power dissipation per unit area in the off-state is given by:

$$P_L(off) = (1 - \delta) . J_L . V_R \qquad [9.3]$$

where $J_L$ is the leakage current density exhibited by the device in its off-state due to supporting a reverse bias ($V_R$). The power dissipation in the off-state increases with temperature due to a rapid increase in the leakage current for Schottky rectifiers.

The total power dissipation incurred in the diode is obtained by combining these terms:

$$P_L(total) = P_L(on) + P_L(off) \tag{9.4}$$

As the temperature of the diode is increased from room temperature, the on-state power dissipation decreases resulting in a reduction of the total power dissipation because the leakage current is small. However, the leakage current increases rapidly at high temperatures resulting in an increase in the power dissipation with temperature. Consequently, the power dissipation in the Schottky rectifier goes through a minimum with increasing temperature as illustrated in Fig. 9.4 for the case of a device with breakdown voltage of 50 volts. A duty cycle of 10 percent was used in this example as representative of the buck converter operating with a ratio of the DC input voltage to output DC voltage of 10. The Schottky barrier height was adjusted to obtain a minimum power dissipation at 400 °K. A reverse bias voltage of 50 volts and an on-state current density of 100 A/cm$^2$ were assumed for this example. The impact of Schottky barrier lowering and pre-breakdown avalanche multiplication were taken into account during the analysis.

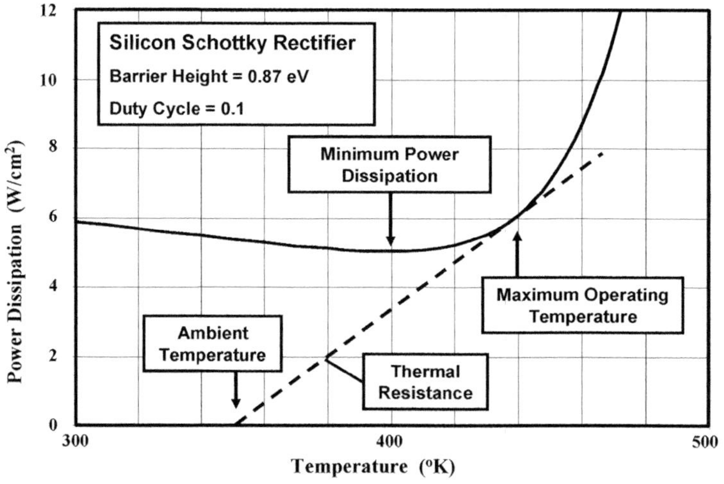


Fig. 9.4 Typical Power Dissipation for a Silicon Schottky Rectifier.

The maximum stable operating temperature for the Schottky rectifier is limited by the thermal impedance of the package and heat sink. If a tangent is drawn from the ambient temperature to the power dissipation curve as shown in Fig. 9.4, the maximum stable operating temperature of 440 °K is obtained as shown in the figure. Although stable operation is theoretically predicted below this temperature point, it is prudent to keep the maximum operating temperature below the point of minimum power dissipation indicated in the figure. The lowest power dissipation is 5 W/cm$^2$ for the Schottky rectifier in this example.

The replacement of the silicon Schottky rectifier with the JBS rectifier enables a reduction of the power dissipation because the leakage current is suppressed in the JBS rectifier structure. Judicious choice of the structural dimensions of the JBS rectifier structure is required to achieve an improved performance. Consider the case of a JBS rectifier with the same breakdown voltage of 50 volts as the Schottky rectifier and operating in the same DC-to-DC converter with the same duty cycle of 10 percent (a ratio of the DC input voltage to output DC voltage of 10). Assuming the same reverse bias voltage of 50 volts and an on-state current density of 100 A/cm$^2$ and after including the impact of less Schottky barrier lowering due to the shielding by the junction, the calculated power dissipation is shown in Fig. 9.5 as a function of temperature. Three device structures were considered in this analysis with different cell pitch while keeping the width of the diffusion window (s) at 0.25 microns and the depth of P$^+$ region at 0.5 microns. The Schottky barrier height in the JBS rectifier structure was reduced to 0.77 eV to achieve a minimum power dissipation at the same temperature (400 $^o$K) as observed for the Schottky rectifier. It can be concluded that the lowest power dissipation is 4.5 W/cm$^2$ for the JBS rectifier with a cell pitch of 1.25 to 1.5 microns in this example. This demonstrates that a reduction in power losses by about 10 percent is feasible using this technology. Note that when the cell pitch is reduced to 1.00 microns in the JBS rectifier structure, the minimum power dissipation exceeds that observed for the Schottky rectifier.

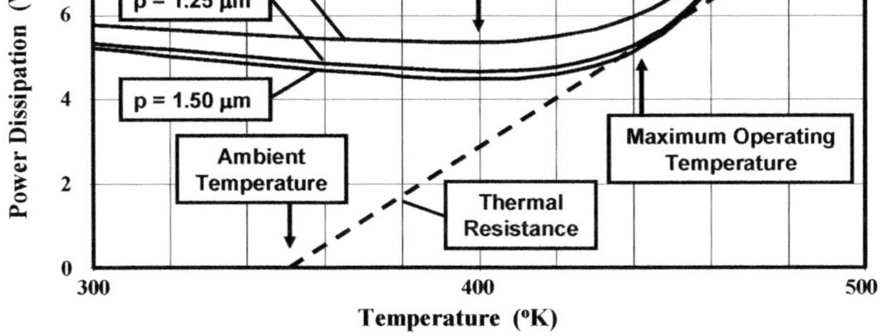


Fig. 9.5 Power Dissipation for Silicon JBS Rectifiers.

## 9.2 Typical Motor Control Application

The control of motors using PWM circuits is typically performed using the H-bridge configuration shown in Fig. 9.2. In this figure, the circuit has been implemented using four IGBT devices as the switches and four P-i-N rectifiers as the fly-back diodes. This is the commonly used topology for medium and high power motor drives where the DC bus voltage exceeds 200 volts. The direction of the current flow in the motor winding can be controlled with the H-bridge configuration. If IGBT-1 and IGBT-4 are turned-on while maintaining IGBT-2 and IGBT-3 in their blocking mode, the current in the motor will flow from the left-side to the right-side in the figure. The direction of the current flow can be reversed if IGBT-3 and IGBT-2 are turned-on while maintaining IGBT-1 and IGBT-4 in their blocking mode. Alternately, the magnitude of the current flow can be increased or decreased by turning on the IGBT devices in pairs. This method allows synthesis of a sinusoidal waveform across the motor windings with a variable frequency that is dictated by the PWM circuit[2].

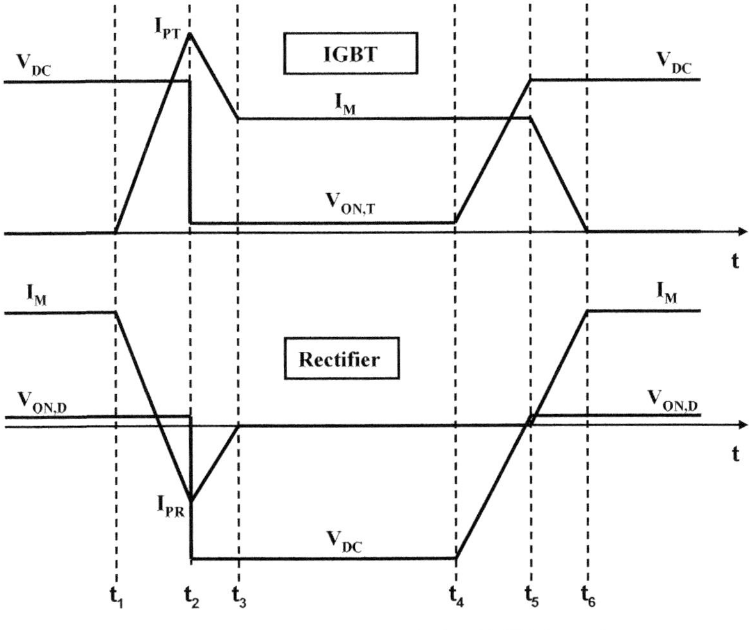


Fig. 9.6 Typical Waveforms during PMW Operation.

The typical waveforms for the current and voltage across the power transistor and the fly-back diode are illustrated in Fig. 9.6 during just one cycle of the PWM operation. These waveforms have been linearized for simplification of the analysis[3]. The cycle begins at time $t_1$ when the transistor is turned-on by its gate drive voltage. Prior to this time, the transistor is supporting the DC supply voltage and the fly-back diode is assumed to be carrying the motor current. As the transistor turns-on, the motor current is transferred from the diode to the transistor during the time interval from $t_1$ to $t_2$. In the case of high DC bus voltages, where P-i-N rectifiers are utilized, the fly-back diode will not be able to support voltage until the stored charge in its drift region is removed as discussed in chapter 6. In order to achieve this, the P-i-N rectifier must undergo its reverse recovery process. During reverse recovery, substantial reverse current flows through the rectifier with a peak value $I_{PR}$ reached at time $t_2$. The large reverse recovery current produces significant power dissipation in the diode. Moreover, the current in the IGBT at time $t_2$ is the sum of the motor winding current $I_M$ and the peak reverse recovery current $I_{PR}$. This produces substantial power dissipation in the transistor during the turn-on transient. The power dissipation in both the transistor and the diode are therefore governed by the reverse recovery characteristics of the power rectifier.

The power transistor is turned-off at time $t_4$ allowing the motor current to transfer from the transistor to the diode. In the case of an inductive load, such as motor windings, the voltage across the transistor increases before the current is reduced, as illustrated in Fig. 9.6 during the time interval from $t_4$ to $t_5$. Subsequently, the current in the transistor reduces to zero during the time interval from $t_5$ to $t_6$. The turn-off durations are governed by the physics of the transistor structure as discussed in previous chapters. Consequently, the power dissipation in both the transistor and the diode during the turn-off event are determined by the transistor switching characteristics.

In addition to the power losses associated with the two basic switching events within each cycle, power loss is incurred within the diode and the transistor during their respective on-state operation due to a finite on-state voltage drop. It is common practice to trade off a larger on-state voltage drop to obtain a smaller switching loss in the bipolar power devices. Consequently, the on-state power loss cannot be neglected especially if the operating frequency is low. The leakage current for the devices is usually sufficiently small so that the power loss in the blocking mode can be neglected.

The total power loss incurred in the power transistor can be obtained by summing four components:

$$P_{L,T}(total) = P_{L,T}(on) + P_{L,T}(off) + P_{L,T}(turnon) + P_{L,T}(turnoff) \qquad \textbf{[9.5]}$$

The power loss incurred in the transistor during the on-state duration from time $t_3$ to $t_4$ is given by:

$$P_{L,T}(on) = \frac{(t_4 - t_3)}{T}.I_M.V_{ON,T} \qquad \textbf{[9.6]}$$

The power loss incurred in the transistor during the off-state duration beyond time $t_6$ until the next turn-on event is given by:

$$P_{L,T}(off) = \frac{(T - t_6)}{T}.I_{L,T}.V_{DC} \qquad \textbf{[9.7]}$$

The leakage current ($I_{L,T}$) for the transistors is usually very small allowing this term to be neglected during the power dissipation analysis.

The power loss incurred in the transistor during the turn-on event from time $t_1$ to $t_3$ can be obtained by analysis of the segments between the time intervals $t_1$ to $t_2$ and $t_2$ to $t_3$. The power loss incurred during the first segment is given by:

$$P_{L,T-1}(turnon) = \frac{1}{2}\frac{(t_2 - t_1)}{T}.I_{PT}.V_{DC} \qquad [9.8]$$

where the peak transistor current is dependent on the peak reverse recovery current of the P-i-N rectifier:

$$I_{PT} = I_M + I_{PR} \qquad [9.9]$$

In the power loss analysis, it will be assumed that the time duration ($t_2 - t_1$) is determined by the reverse recovery behavior of the P-i-N rectifier and is independent of the operating frequency. The power loss incurred during the second segment is given by:

$$P_{L,T-2}(turnon) = \frac{1}{2}\frac{(t_3 - t_2)}{T}.\left(\frac{I_{PT} + I_M}{2}\right).V_{DC} \qquad [9.10]$$

In the power loss analysis, it will be assumed the time duration ($t_3 - t_2$) is also determined by the reverse recovery behavior of the P-i-N rectifier and is independent of the operating frequency.

The power loss incurred in the transistor during the turn-off event from time $t_4$ to $t_6$ can be obtained by analysis of the segments between the time intervals $t_4$ to $t_5$ and $t_5$ to $t_6$. The power loss incurred during the first segment is given by:

$$P_{L,T-1}(turnoff) = \frac{1}{2}\frac{(t_5 - t_4)}{T}.I_M.V_{DC} \qquad [9.11]$$

The time interval ($t_5 - t_4$) is determined by the time taken for the transistor voltage to rise to the DC power supply voltage. The power loss incurred during the second segment is given by:

$$P_{L,T-2}(turnoff) = \frac{1}{2}\frac{(t_6 - t_5)}{T}.I_M.V_{DC} \qquad [9.12]$$

The time interval ($t_6 - t_5$) is determined by the time taken for the transistor current to decay to zero.

In a similar manner, the total power loss incurred in the power rectifier can be obtained by summing four components:

$$P_{L,R}(total) = P_{L,R}(on) + P_{L,R}(off) + P_{L,R}(turnon) + P_{L,R}(turnoff) \qquad [9.13]$$

The power loss incurred in the power rectifier during the on-state duration from time $t_6$ to the end of the period is given by:

$$P_{L,R}(on) = \frac{(T - t_6)}{T}.I_M.V_{ON,R} \quad \textbf{[9.14]}$$

In writing this expression, it is assumed that the cycle begins at time $t_1$. The power loss incurred in the power rectifier during the off-state time duration ($t_4 - t_3$) is given by:

$$P_{L,R}(off) = \frac{(t_4 - t_3)}{T}.I_{L,R}.V_{DC} \quad \textbf{[9.15]}$$

The leakage current ($I_{L,R}$) for the power rectifier will be assumed to be very small allowing this term to be neglected during the power dissipation analysis.

The power loss incurred in the power rectifier during the turn-on event from time $t_1$ to $t_3$ can be obtained by analysis of the segments between the time intervals $t_1$ to $t_2$ and $t_2$ to $t_3$. The power loss incurred during the first segment is much smaller than during the second segment due to the small on-state voltage drop for the power rectifiers. The power loss incurred during the second segment is given by:

$$P_{L,R-2}(turnon) = \frac{1}{2}\frac{(t_3 - t_2)}{T}.I_{PR}.V_{DC} \quad \textbf{[9.16]}$$

The power loss incurred in the power rectifier during the turn-off event from time $t_4$ to $t_6$ can be obtained by analysis of the segments between the time intervals $t_4$ to $t_5$ and $t_5$ to $t_6$. The power loss incurred during the first segment is negligible due to the low leakage current for the power rectifier. The power loss incurred during the second segment is given by:

$$P_{L,R-2}(turnoff) = \frac{1}{2}\frac{(t_6 - t_5)}{T}.I_M.V_{ON,D} \quad \textbf{[9.17]}$$

This power loss is also small due to the low on-state voltage drop of power rectifiers.

In this section, the above power loss analysis is applied to a motor control application using a medium DC-bus voltage with a duty cycle of 50 percent. The DC-bus voltage ($V_{DC}$) will be assumed to be 400 volts as pertains to the power source in a hybrid electric-car. In this case, the device blocking voltage rating is typically 600 volts. The current being delivered to the motor winding ($I_M$) will be assumed to be 20 amperes. Due to the larger blocking voltage required for this application, it is commonly implemented using silicon bipolar devices, namely the IGBT as the power switch and the P-i-N rectifier as the fly-back diode.

| Characteristics | Silicon IGBT |
|---|---|
| On-State Voltage Drop (V) | 1.8 |
| Turn-Off Time ($t_5 - t_4$) (μs) | 0.1 |
| Turn-Off Time ($t_6 - t_5$) (μs) | 0.2 |

Fig. 9.7 Characteristics of IGBTs with 600-V Blocking Voltage Rating.

In the textbook[1], it was demonstrated that the most suitable device for this motor control application is the silicon IGBT. The characteristics of the IGBT that are pertinent to the analysis of the power loss are provided in Fig. 9.7 when the device is operated at an on-state current density of 100 A/cm$^2$.

| Characteristics | Silicon P-i-N | Silicon MPS | 4H-SiC JBS |
|---|---|---|---|
| On-State Voltage Drop (V) | 0.85 | 0.87 | 1.0 |
| Turn-On Time ($t_2 - t_1$) (μs) | 0.35 | 0.25 | 0.01 |
| Turn-On Time ($t_3 - t_2$) (μs) | 0.35 | 0.25 | 0.01 |
| Peak Reverse Recovery Current (A) | 28 | 17 | 2.0 |

Fig. 9.8 Characteristics of Rectifiers with 600-V Blocking Voltage Rating.

It was also demonstrated in the textbook[1] that the performance of motor control systems can be greatly improved by replacing the silicon P-i-N rectifier with a silicon carbide Schottky rectifier. However, the leakage current for the silicon carbide Schottky rectifier increases by 6 orders of magnitude with increasing reverse bias voltage. This problem can be overcome by using the JBS rectifier concept as shown in chapter 3 of this book. The silicon carbide JBS rectifier is therefore an excellent candidate for motor control applications. However, the cost of silicon carbide rectifiers is at present much greater than that for silicon devices. Consequently, a more attractive approach to reducing power losses in the motor control applications is by replacing the silicon P-i-N rectifier with a silicon MPS rectifier that was discussed in chapter 7.

The characteristics for the power rectifiers that are pertinent to the analysis of the power loss are provided in Fig. 9.8. All the devices are assumed to be operated at an on-state current density of 100 A/cm$^2$. The on-state voltage drop, reverse recovery time, and peak reverse recovery current for the silicon P-i-N rectifier and the MPS rectifier are taken from the data provided earlier in chapter 7 (see Fig. 7.35 and Fig. 7.36) for the case of a lifetime of 10 microseconds in the drift region. In the case of the 4H-SiC JBS rectifier structure, the on-state voltage drop is limited by the barrier height for the metal-semiconductor contact due to the low specific on-resistance for the drift region (see Fig. 3.10 for the case of a cell pitch of more than 1.2 microns). The reverse recovery current for the JBS rectifier is assumed to be 10 percent of the on-state current because of the displacement current[4,5].

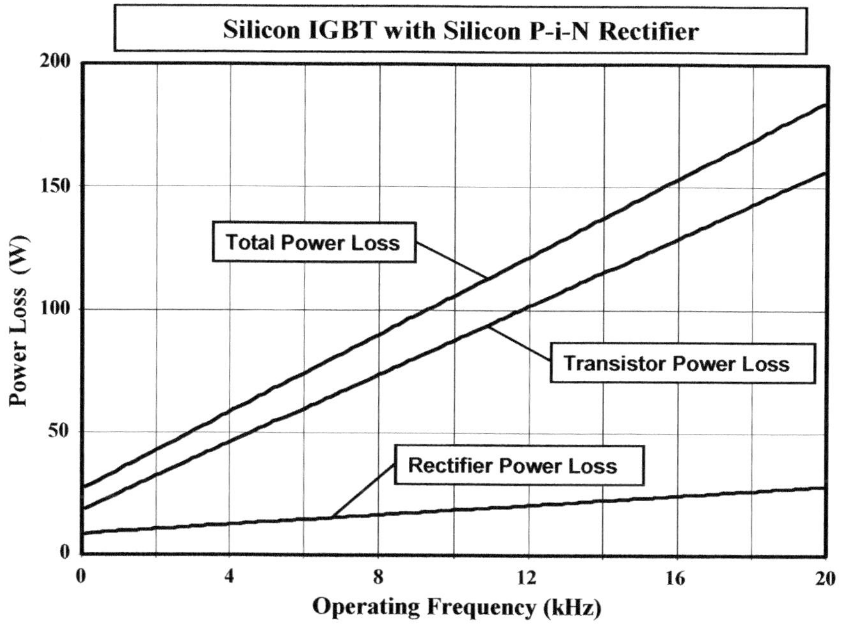


Fig. 9.9 Power Losses during Motor Control with 400-V DC-Bus: Silicon IGBT with Silicon P-i-N Rectifier.

The power losses in the case of the silicon IGBT as the power switch with a silicon P-i-N rectifier as the diode is provided in Fig. 9.9 for frequencies ranging to 20 kHz. The power losses in the transistor are dominant in this case. The power loss in the transistor is larger than in the rectifier at low operating frequencies due to its larger on-state voltage drop. The power loss in the transistor increases rapidly with increasing frequency and becomes the dominant power loss mechanism in this case. The total power loss for this combination of a silicon IGBT as the switch and the silicon P-i-N rectifier as the diode is 185 watts at 20 kHz when 8000 watts of power is delivered to the load.

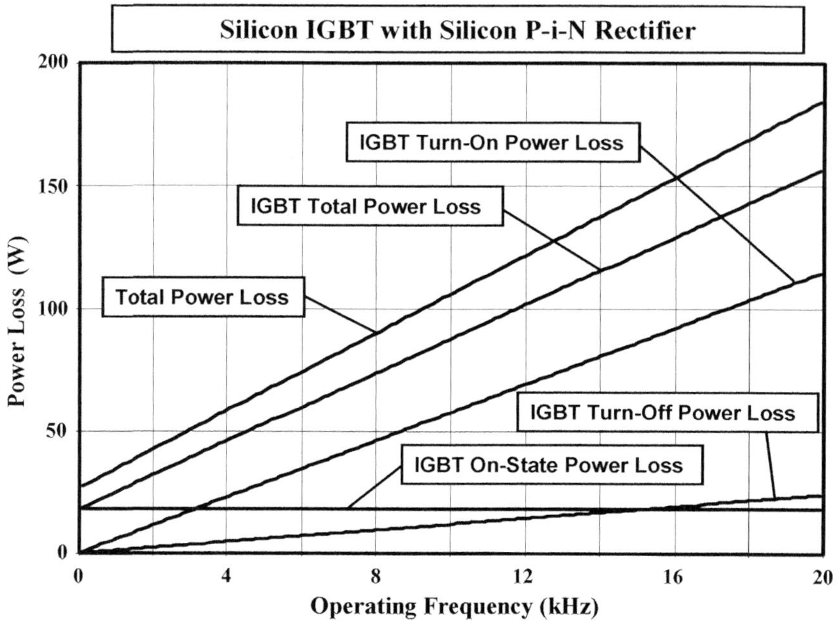


Fig. 9.10 IGBT Power Losses during Motor Control with 400-V DC-Bus: Silicon IGBT with Silicon P-i-N Rectifier.

The components of the power loss within the silicon IGBT are provided in Fig. 9.10 as a function of the operating frequency. From this graph, it is clear that the on-state power loss is dominant up to about 3 kHz. At higher frequencies, the turn-on power loss becomes dominant because it increases rapidly with frequency. This graph demonstrates that the turn-on losses in the IGBT are dominant at high frequencies for this combination of a silicon IGBT as the switch and the silicon P-i-N rectifier as the diode. The high turn-on power losses in the IGBT are produced by the large peak reverse recovery current of the P-i-N rectifier. The large peak reverse recovery current also produces a high current density in the IGBT structure making it prone to latch-up induced failure during operation in the motor control circuit. Based up on these observations, it is clear that improvements in the performance of the power rectifier are required to reduce power losses in the motor control application.

The power losses in the motor control circuit can be reduced by replacing the silicon P-i-N rectifier with a silicon MPS rectifier. The power losses in the case of the silicon IGBT as the power switch with a silicon MPS rectifier as the diode is provided in Fig. 9.11 for frequencies ranging to 20 kHz. The power losses in the transistor are still dominant in this case. The power loss in the transistor is larger than in the rectifier at low operating frequencies due to its larger on-state voltage drop. The power loss in the transistor increases rapidly with increasing frequency and becomes the dominant power loss mechanism in this case. The total power loss for this combination of a silicon IGBT as the switch and the silicon MPS rectifier as the diode is however reduced to 125 watts at 20 kHz when 8000 watts of power is delivered to the load.

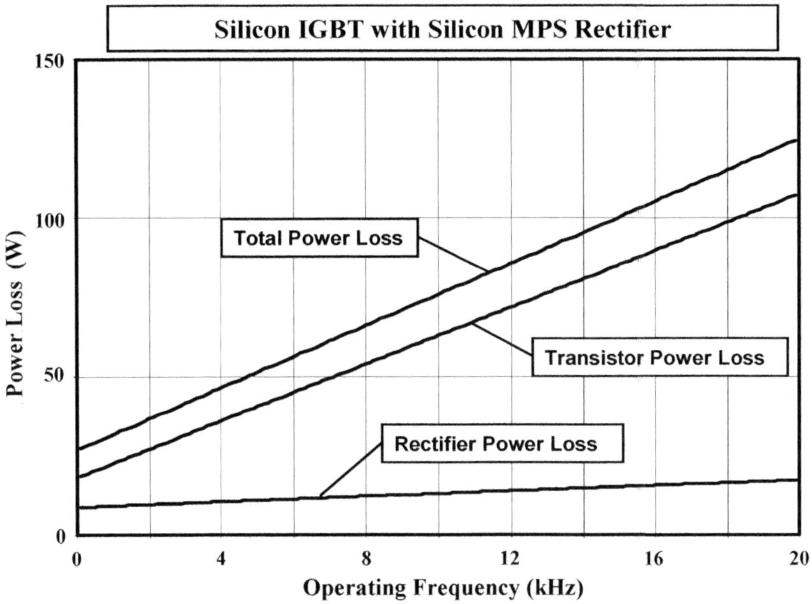


Fig. 9.11 Power Losses during Motor Control with 400-V DC-Bus: Silicon IGBT with Silicon MPS Rectifier.

The components of the power loss within the IGBT are provided in Fig. 9.12 as a function of the operating frequency for the above combination of a silicon IGBT as the switch and the silicon MPS rectifier as the diode. From this graph, it is clear that the on-state power loss is dominant up to about 5 kHz. At higher frequencies, the turn-on power loss becomes dominant because it increases rapidly with frequency. This graph demonstrates that the turn-on losses in the IGBT are still dominant at high frequencies even for this combination of an IGBT as the switch and the MPS rectifier as the diode. The high turn-on power losses in the IGBT are produced by the peak reverse recovery current of the MPS rectifier even though it is smaller than that for the P-i-N rectifier. The reduced peak reverse recovery current of the MPS rectifier makes the IGBT less prone to latch-up induced failure during operation in the motor control circuit. Based up on these observations, it is clear that further improvements in the performance of the power rectifier are required to reduce power losses in the motor control application.

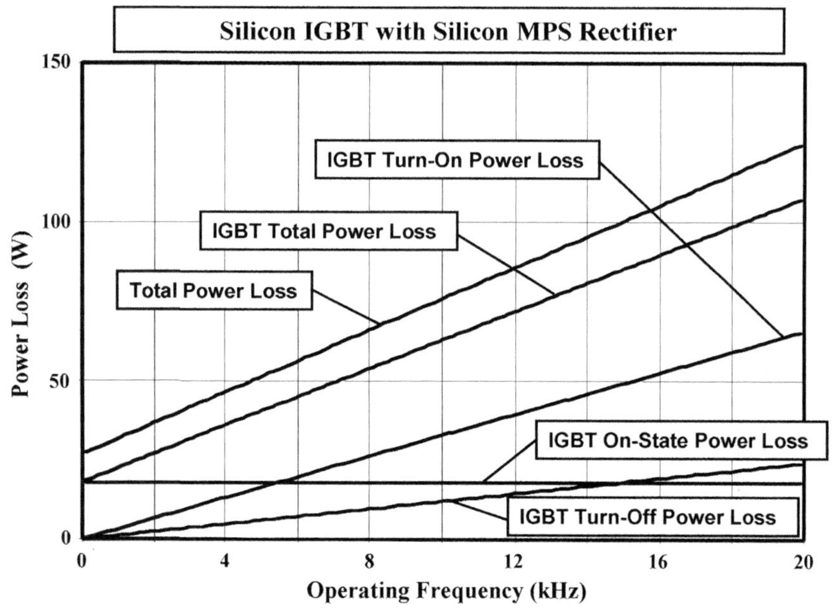


Fig. 9.12 IGBT Power Losses during Motor Control with 400-V DC-Bus: Silicon IGBT with Silicon MPS Rectifier.

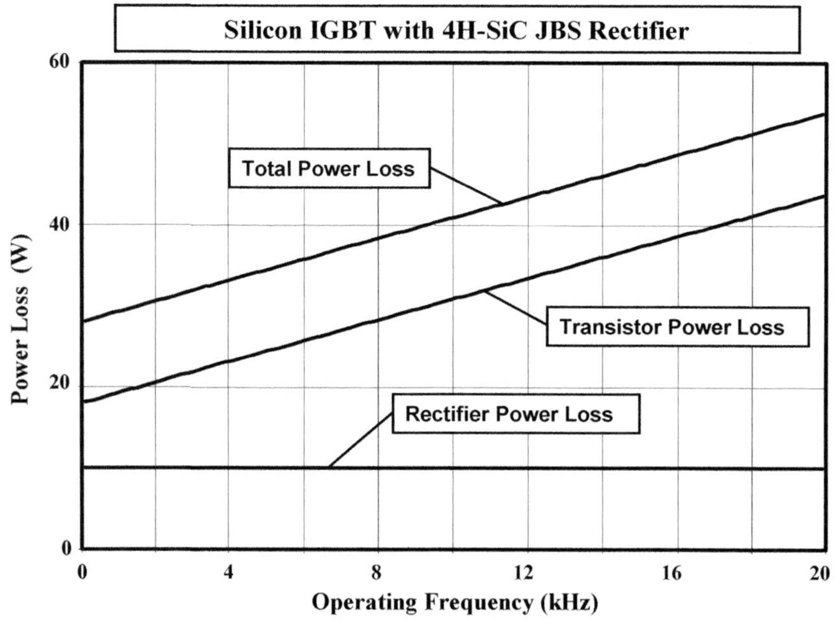


Fig. 9.13 Power Losses during Motor Control with 400-V DC-Bus: Silicon IGBT with Silicon Carbide JBS Rectifier.

An even greater improvement in performance of the motor control system was proposed by replacing the silicon P-i-N rectifier with a silicon carbide Schottky rectifier[3]. Due to the high leakage of the silicon carbide Schottky rectifier, it is only appropriate to consider the silicon carbide JBS rectifier for the motor control applications. The power losses in the case of the silicon IGBT as the power switch with a silicon carbide JBS rectifier as the diode is provided in Fig. 9.13 for frequencies ranging to 20 kHz. The power losses in the transistor are still dominant in this case. The power loss in the transistor is larger than in the rectifier at low operating frequencies due to its larger on-state voltage drop. It can be observed that the power loss in the transistor increases less rapidly with increasing frequency than for the case of the silicon power rectifiers. The total power loss for this combination of a silicon IGBT as the switch and the silicon carbide JBS rectifier as the diode is therefore reduced to 54 watts at 20 kHz when 8000 watts of power is delivered to the load.

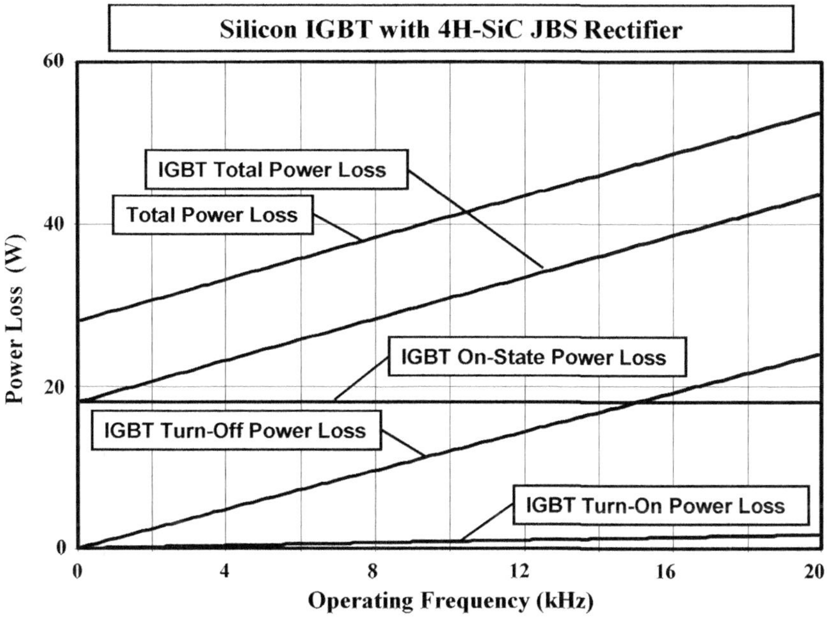


Fig. 9.14 IGBT Power Losses during Motor Control with 400-V DC-Bus: Silicon IGBT with Silicon Carbide JBS Rectifier.

The components of the power loss within the IGBT are provided in Fig. 9.14 as a function of the operating frequency for the above combination of a silicon IGBT as the switch and the silicon carbide JBS rectifier as the diode. From this graph, it is clear that the on-state power loss is dominant up to about 15 kHz. More significantly, the turn-on power loss in the IGBT is now much smaller than the turn-off power loss. This graph demonstrates that the turn-off losses in the IGBT are dominant at high frequencies for this combination of a silicon IGBT as the switch and the silicon carbide JBS rectifier as the diode. This indicates that further improvements in the performance of the rectifier will not benefit the motor control application. However, replacement of the silicon IGBT with the silicon carbide power MOSFET will produce a significant further reduction of the power losses as shown in the textbook[1].

## 9.3 Summary

This chapter provides a comparison of the benefits of utilizing various advanced power rectifier concepts that have been described in this book for the typical low voltage DC-to-DC converters for Voltage Regulator Modules (VRMs) and medium voltage motor control applications. It can be concluded that silicon JBS rectifiers can provide a reduction of losses in the VRM application. For the motor control application working from a medium DC-bus voltage, it is advantageous to utilize the silicon MPS rectifier as a low cost technology in the short time frame. In the future, for these applications, silicon carbide JBS rectifiers will greatly reduce power losses when used in conjunction with the silicon IGBT devices as power switches.

## References

[1] B.J. Baliga, "Fundamentals of Power Semiconductor Devices", Springer-Science, New York, 2008.

[2] B.K. Bose, "Power Electronics and Variable Frequency Drives", IEEE Press, 1997.

[3] B.J. Baliga, "Power Semiconductor Devices for Variable-Frequency Drives", Proceedings of the IEEE, Vol. 82, pp. 1112-1122, 1994.

[4] P. Brosselard, et al, "Bipolar Conduction Impact on Electrical Characteristics and Reliability of 1.2 and 3.5 kV 4H-SiC JBS Diodes", IEEE Transactions on Electron Devices, Vol. 55, pp. 1847-1856, 2008.

[5] B.A. Hull, et al, "Performance and Stability of Large-Area 4H-SiC 10-kV JBS Rectifiers", IEEE Transactions on Electron Devices, Vol. 55, pp. 1864-1870, 2008.

Professor Baliga is internationally recognized for his leadership in the area of power semiconductor devices. In addition to over 500 publications in international journals and conference digests, he has authored and edited 15 books ("*Power Transistors*", IEEE Press 1984; "*Epitaxial Silicon Technology*", Academic Press 1986; "*Modern Power Devices*", John Wiley 1987; "*High Voltage Integrated Circuits*", IEEE Press 1988; "*Solution Manual: Modern Power Devices*", John Wiley 1988; "*Proceedings of the 3rd Int. Symposium on Power Devices and ICs*", IEEE Press 1991; "*Modern Power Devices*", Krieger Publishing Co. 1992; "*Proceedings of the 5th Int. Symposium on Power Devices and ICs*", IEEE Press 1993; "*Power Semiconductor Devices*"; PWS Publishing Company 1995; "*Solution Manual: Power Semiconductor Devices*"; PWS Publishing Company 1996; "*Cryogenic Operation of Power Devices*", Kluwer Press 1998; "*Silicon RF Power MOSFETs*", World Scientific Publishing Company 2005; "*Silicon Carbide Power Devices*", World Scientific Publishing Company 2006; "*Fundamentals of Power Semiconductor Devices*", Springer Science, 2008; "*Solution Manual: Fundamentals of Power Semiconductor Devices*", Springer Science, 2008. In addition, he has contributed chapters to another twenty books. He holds 120 U.S. Patents in the solid-state area. In 1995, one of his inventions was selected for the *B.F. Goodrich Collegiate Inventors Award* presented at the *Inventors Hall of Fame*.

Professor Baliga obtained his Bachelor of Technology degree in 1969 from the Indian Institute of Technology, Madras, India. He was the recipient of the *Philips*

*India Medal* and the *Special Merit Medal (as Valedictorian)* at I.I.T, Madras. He obtained his Masters and Ph.D. degrees from Rensselaer Polytechnic Institute, Troy NY, in 1971 and 1974, respectively. His thesis work involved Gallium Arsenide diffusion mechanisms and pioneering work on the growth of InAs and GaInAs layers using Organometallic CVD techniques. At R.P.I., he was the recipient of the *IBM Fellowship* in 1972 and the *Allen B. Dumont Prize* in 1974.

From 1974 to 1988, Dr. Baliga performed research and directed a group of 40 scientists at the General Electric Research and Development Center in Schenectady, NY, in the area of Power Semiconductor Devices and High Voltage Integrated Circuits. During this time, he pioneered the concept of MOS-Bipolar functional integration to create a new family of discrete devices. He is the *inventor of the IGBT* which is now in production by many International Semiconductor companies. This invention is widely used around the globe for air-conditioning, home appliance (washing machines, refrigerators, mixers, etc) control, factory automation (robotics), medical systems (CAT scanners, uninterruptible power supplies), and electric street-cars/bullet-trains, as well as for the drive-train in electric and hybrid-electric cars under development for reducing urban pollution. The U.S. Department of Energy has released a report that the variable speed motor drives enabled by IGBTs produce an energy savings of 2 quadrillion btus per year (equivalent to 70 Giga-Watts of power). The widespread adoption of Compact Fluorescent Lamps (CFLs) in place of incandescent lamps is producing an additional power savings of 30 Giga-Watts. The cumulative impact of these energy savings on the environment is a reduction in Carbon Dioxide emissions from Coal-Fired power plants by over One Trillion pounds per year. Most recently, the IGBT has enabled fabrication of very compact, light-weight, and inexpensive defibrillators used to resuscitate cardiac arrest victims. When installed in fire-trucks, paramedic vans, and on-board airlines, it is projected by the American Medical Association (AMA) to save 100,000 lives per year in the US. For this work, *Scientific American Magazine* named him one of the eight heroes of the semiconductor revolution in their 1997 special issue commemorating the Solid-State Century.

Dr. Baliga is also the originator of the concept of merging Schottky and p-n junction physics to create a new family of power rectifiers that are commercially available from various companies. In 1979, he theoretically demonstrated that the performance of power MOSFETs could be enhanced by several orders of magnitude by replacing silicon with other materials such as gallium arsenide and silicon carbide. This is forming the basis of a new generation of power devices in the 21st Century.

In August 1988, Dr. Baliga joined the faculty of the Department of Electrical and Computer Engineering at North Carolina State University, Raleigh, North Carolina, as a Full Professor. At NCSU, in 1991 he established an international center called the *Power Semiconductor Research Center* (PSRC) for research in the area of power semiconductor devices and high voltage integrated circuits, and has served as its Founding Director. His research interests include the modeling of novel device concepts, device fabrication technology, and the investigation of the impact of new materials, such as GaAs and Silicon Carbide, on power devices. In 1997, in recognition of his contributions to NCSU, he was given the highest university faculty rank of *Distinguished University Professor of Electrical Engineering*.

In 2008, Professor Baliga was a key member of an NCSU team - partnered with four other universities - that was successful in being granted an Engineering Research Center from the National Science Foundation for the development of micro-grids that allow integration of renewable energy sources. Within this program, he is responsible for the fundamental sciences platform and the development of power devices from wide-band-gap semiconductors for utility applications.

Professor Baliga has received numerous awards in recognition for his contributions to semiconductor devices. These include two *IR 100 awards* (1983, 1984), the *Dushman* and *Coolidge Awards* at GE (1983), and being selected among the *100 Brightest Young Scientists in America* by Science Digest Magazine (1984). He was elected *Fellow of the IEEE* in 1983 at the age of 35 for his contributions to power semiconductor devices. In 1984, he was given the *Applied Sciences Award* by the world famous sitar maestro Ravi Shankar at the Third Convention of Asians in North America. He received the 1991 *IEEE William E. Newell Award*, the highest honor given by the Power Electronics Society, followed by the 1993 *IEEE Morris E. Liebman Award* for his contributions to the emerging *Smart Power Technology*. In 1992, he was the first recipient of the BSS Society's *Pride of India Award*. At the age of 45, he was elected as Foreign Affiliate to the prestigious *National Academy of Engineering*, and was one of only 4 citizens of India to have the honor at that time (converted to regular Member in 2000 after taking U.S. Citizenship). In 1998, the University of North Carolina system selected him for the *O. Max Gardner Award*, which recognizes the faculty member among the 16 constituent universities who has made the greatest contribution to the welfare of the human race. In December 1998, he received the *J.J. Ebers Award*, the highest recognition given by the IEEE Electron Devices Society for his technical contributions to the Solid-State area. In June 1999, he was honored at the Whitehall Palace in London with the *IEEE Lamme Medal*, one of the highest forms of recognition given by the IEEE Board of Governors, for his contributions to development of an apparatus/technology of benefit to society. In April 2000, he was honored by his Alma Mater as a *Distinguished Alumnus*. In November 2000, he received the *R.J. Reynolds Tobacco Company Award for Excellence in Teaching, Research, and Extension* for his contributions to the College of Engineering at North Carolina State University.

In 1999, Prof. Baliga founded a company, *Giant Semiconductor Corporation*, with seed investment from Centennial Venture Partners, to acquire an exclusive license for his patented technology from North Carolina State University with the goal of bringing his NCSU inventions to the marketplace. A company, *Micro-Ohm Corporation*, subsequently formed by him in 1999, has been successful in licensing the GD-TMBS power rectifier technology to several major semiconductor companies for world-wide distribution. These devices have application in power supplies, battery chargers, and automotive electronics. In June 2000, Prof. Baliga founded another company, *Silicon Wireless Corporation*, to commercialize a novel super-linear silicon RF transistor that he invented for application in cellular base-stations and grew it to

41 employees. This company (renamed *Silicon Semiconductor Corporation*) is located at Research Triangle Park, N.C. It received an investment of $10 Million from *Fairchild Semiconductor Corporation* in December 2000 to co-develop and market this technology. Based upon his additional inventions, this company has also produced a new generation of Power MOSFETs for delivering power to microprocessors in notebooks and servers. This technology was licensed by his company to Linear Technologies Corporation with transfer of the know-how and manufacturing process. Voltage Regulator Modules (VRMs) using his transistors are currently available in the market for powering microprocessor and graphics chips in laptops and servers.

# Index

GPSR Compliance

*The European Union's (EU) General Product Safety Regulation (GPSR) is a set of rules that requires consumer products to be safe and our obligations to ensure this.*

*If you have any concerns about our products, you can contact us on ProductSafety@springernature.com*

In case Publisher is established outside the EU, the EU authorized representative is:

Springer Nature Customer Service Center GmbH
Europaplatz 3
69115 Heidelberg, Germany

**Batch number: 10755284**

Printed by Printforce, the Netherlands